# A Contour-Oriented Approach to Shape Analysis

# A Contour-Oriented Approach to Shape Analysis

Peter J. van Otterloo
*Philips Research Laboratories, Eindhoven, The Netherlands*

**Prentice Hall**
New York London Toronto Sydney Tokyo Singapore

First published 1991 by
Prentice Hall International (UK) Ltd
66 Wood Lane End, Hemel Hempstead
Hertfordshire HP2 4RG
A division of
Simon & Schuster International Group

Typeset in 11pt Times
by Pentacor PLC

Printed in Great Britain by Billing & Sons Ltd, Worcester

Library of Congress Cataloging-in-Publication Data

van Otterloo, P. J., 1952–
A contour-oriented approach to shape analysis/by P. J. van Otterloo.
p. cm.
Includes bibliographical references and index.
ISBN 0–13–173840–2 (alk. paper)
1. Image processing—Digital techniques. I. Title.
TA1632.V3 1991
621.36′7—dc20 90–23087
CIP

British Library Cataloguing in Publication Data

van Otterloo, P. J. (Peter J.) *1952–*
A contour oriented approach to shape analysis.
1. Digital image processing
I. Title
621.38045

ISBN 0–13–173840–2

1 2 3 4 5 95 94 93 92 91

# Contents

# Summary

The shape of objects or of regions of interest plays a fundamental role in digital image analysis. Over the past few decades many approaches to the characterization of shape and shape similarity have appeared in the literature. An important class of shape analysis techniques is based on the representation of the outer boundaries of objects. This class of techniques is studied in detail in this book.

Various methods for representing the outer boundaries of two-dimensional objects have been proposed. The main goal of this book is to provide a unified theoretical basis for shape similarity analysis on the basis of parametric contour representations; unified in the sense that various contour representations, some of which have been considered previously in the literature, are presented and their relationships derived. Thereby the theoretical framework, into which these contour representations fit, is made clear.

Measures of dissimilarity, based on parametric contour representations, are defined. Apart from establishing relations between the contour representations themselves, we also attempt to establish relations between the measures of dissimilarity. Furthermore, the possibilities for contour representation normalization are evaluated and the trade-off between optimization and normalization, necessary to achieve the desired invariance properties in the proposed dissimilarity measures, is discussed.

We relate the concept of symmetry in plane figures with that of similarity under symmetry transformations. This enables us to define measures for the quantification of symmetry, or for the lack of symmetry (for which we use the term dissymmetry), in a plane figure on the basis of previously defined dissimilarity measures.

Since the Fourier coefficients of parametric contour representations have been given much attention in the literature, we formulate, throughout this thesis, the Fourier coefficient-based counterparts of the dissimilarity and dissymmetry measures. Where possible we relate both types of measures. Some considerations are given for the practical implementation of the proposed techniques. A short survey of the contents of the individual chapters is given below.

In Chapter 1 we give a short survey of approaches to shape analysis in the context of two-dimensional image analysis and outline the approach that is followed in this book.

Chapter 2 introduces the parametric contour representations that we consider and establishes their relations. It also defines the concept of similarity, which is subsequently reformulated as a relation that exists between the contour representations of a pair of similar objects. Finally it discusses two types of plane symmetry: mirror-symmetry and rotational symmetry. The conditions that a contour representation must satisfy, in order to represent a contour that has these types of symmetry, are formulated.

Chapter 3 follows the same lines as Chapter 2, but in terms of the Fourier coefficients of the contour representations. It also discusses the consequences of normalized arc length parametrization on Fourier series expansions of contour representations and gives bounds on truncation errors of finite Fourier series expansions.

In Chapter 4 the measures for dissimilarity and dissymmetry are defined and their theoretical properties evaluated. Relations between these measures are formulated. These relations will be helpful in solving the design problem, i.e. out of a multitude of possibilities, which contour representation and which measure, or combination of measures, should be selected in a given application. To this end also a number of experiments are performed and evaluated. Furthermore, the trade-off between optimization and normalization, necessary to achieve the desired invariance of the proposed measures for contour position, orientation and size and for the position of the parametric starting point, is discussed in this chapter.

In Chapter 5 a discussion of the results obtained in the previous chapters is presented, a number of open problems are indicated and some suggestions for further research are given.

Finally, this book contains three appendices. Appendix A deals with some mathematical concepts and properties relevant to the contents. In Appendix B a computationally efficient method for the computation of the moments of polygonal regions is described and evaluated. Appendix C discusses the estimation of contour representations through piecewise polynomial approximation.

# Glossary of special symbols and notations[1]

| Symbol | Description |
|---|---|
| $A$ | area of the region enclosed by a contour |
| **AC** | space of absolutely continuous functions |
| a.e. | almost everywhere |
| $a_p[n]$ | real part of $c_p[n]$ |
| $b_p[n]$ | imaginary part of $c_p[n]$ |
| **BV** | space of functions of bounded variation |
| CPCC$(\cdot, \cdot)$ | Cophenetic correlation coefficient |
| $C(\cdot)$ | truncation error bound for chains |
| $\mathbf{C}$, $\mathbf{C}^0$ | space of continuous functions |
| $\mathbf{C}^k$ | space of $k$ times continuously differentiable functions |
| $\mathbf{C}^\infty$ | space of infinitely-differentiable functions |
| **CBV** | space of continuous functions of bounded variation |
| $\boldsymbol{C}(p, q)$ | matrix of data independent coefficients $c(\gamma, \delta; p, q)$ |
| $\boldsymbol{C}_1$ | cophenetic matrix resulting from Single Linkage Clustering on $\boldsymbol{D}$ |
| $\boldsymbol{C}_2$ | cophenetic matrix resulting from Complete Linkage Clustering on $\boldsymbol{D}$ |
| $\boldsymbol{C}_3$ | cophenetic matrix resulting from Average Linkage Clustering (UPGMA) on $\boldsymbol{D}$ |

[1] This glossary lists the major special symbols and notations used in this book. Some symbols and notations, that are used only locally and explained there, have been left out of this list.

| | |
|---|---|
| $\mathbb{C}$ | the set of complex numbers |
| $c(\gamma, \delta; p, q)$ | data independent coefficient in the computation of the moment $m_{pq}$ of an $N$-sided polygon |
| $c_p[n]$ | $p$-th coefficient of the polynomial $\pi_n(t)$ |
| $\mathbf{c}_0$ | space of sequences that converge to zero |
| $D(\cdot, \cdot)$ | dissimilarity coefficient, generated by a pair of dissimilarity matrices |
| $\boldsymbol{D}$ | matrix of dissimilarity coefficients $D(\cdot, \cdot)$ |
| $\boldsymbol{D}_.$ | dissimilarity matrix |
| $\langle \boldsymbol{D}_. \rangle$ | dissimilarity matrix average |
| $\boldsymbol{D}(p, q)$ | matrix of data independent coefficients $d(\alpha, \beta; p, q)$ |
| $\mathscr{D}_\zeta$ | translation operator that displaces a contour over $\zeta$ |
| $d$ | metric |
| $d(\alpha, \beta; p, q)$ | data independent coefficient in the computation of the moment $m_{pq}$ of an $N$-sided polygon |
| $d^{(p)}(f_1, f_2)$ | dissimilarity measure of index $p$ based on the contour representation $f$ |
| $\tilde{d}^{(p)}(f_1, f_2)$ | mirror-dissimilarity measure of index $p$ based on the contour representation $f$ |
| $d^{(p)}[f_1, f_2]$ | discrete dissimilarity measure of index $p$ based on the discrete contour representation $f[\ ]$ |
| $\tilde{d}^{(p)}[f_1, f_2]$ | discrete mirror-dissimilarity measure of index $p$ based on the discrete contour representation $f[\ ]$ |
| $d^{(p)}(\hat{f}_1, \hat{f}_2)$ | dissimilarity measure of index $p$ based on the Fourier representation $\hat{f}$ |
| $\tilde{d}^{(p)}(\hat{f}_1, \hat{f}_2)$ | mirror-dissimilarity measure of index $p$ based on the Fourier representation $\hat{f}$ |
| $d^{(p)}[\hat{f}_1, \hat{f}_2]$ | discrete dissimilarity measure of index $p$ based on the truncated Fourier representation $\hat{f}[\ ]$ |

| | |
|---|---|
| $\tilde{d}^{(p)}[\hat{f}_1, \hat{f}_2]$ | discrete mirror-dissimilarity measure of index $p$ based on the truncated Fourier representation $\hat{f}[\ ]$ |
| $d_*^{(p)}(f_1, f_2)$ | normalized dissimilarity measure of index $p$ based on the contour representation $f$ |
| $\tilde{d}_*^{(p)}(f_1, f_2)$ | normalized mirror-dissimilarity measure of index $p$ based on the contour representation $f$ |
| $d_*^{(p)}[f_1, f_2]$ | normalized discrete dissimilarity measure of index $p$ based on the discrete contour representation $f[\ ]$ |
| $\tilde{d}_*^{(p)}[f_1, f_2]$ | normalized discrete mirror-dissimilarity measure of index $p$ based on the discrete contour representation $f[\ ]$ |
| $d_{PF}^{(2)}(\hat{z}_1, \hat{z}_2)$ | dissimilarity measure based on the Fourier representation $\hat{z}$, as defined by Persoon and Fu |
| $d^{(p)}(f; \boldsymbol{m})$ | measure of dissymmetry $\boldsymbol{m}$ of index $p$ based on the contour representation $f$ |
| $d^{(p)}[f; \boldsymbol{m}]$ | discrete measure of dissymmetry $\boldsymbol{m}$ of index $p$ based on the discrete contour-representation $f[\ ]$ |
| $d^{(p)}(\hat{f}; \boldsymbol{m})$ | measure of dissymmetry $\boldsymbol{m}$ of index $p$ based on the Fourier representation $\hat{f}$ |
| $d^{(p)}[\hat{f}; \boldsymbol{m}]$ | discrete measure of dissymmetry $\boldsymbol{m}$ of index $p$ based on the truncated Fourier representation $\hat{f}[\ ]$ |
| $d^{(p)}(f; \boldsymbol{n}, m)$ | $m$-th component of dissymmetry $\boldsymbol{n}$ of index $p$ based on the contour representation $f$ |
| $d^{(p)}(\hat{f}; \boldsymbol{n}, m)$ | $m$-th component of dissymmetry $\boldsymbol{n}$ of index $p$ based on the Fourier representation $\hat{f}$ |
| $d^{(p,q)}(f; \boldsymbol{n})$ | measure of dissymmetry $\boldsymbol{n}$ of index pair $(p, q)$ based on the contour representation $f$ |
| $d^{(p,q)}[f; \boldsymbol{n}]$ | discrete measure of dissymmetry $\boldsymbol{n}$ of index pair $(p, q)$ based on the discrete contour representation $f[\ ]$ |
| $d^{(p,q)}(\hat{f}; \boldsymbol{n})$ | measure of dissymmetry $\boldsymbol{n}$ of index pair $(p, q)$ based on the Fourier representation $\hat{f}$ |
| $d^{(p,q)}[\hat{f}; \boldsymbol{n}]$ | discrete measure of dissymmetry $\boldsymbol{n}$ of index pair $(p, q)$ based on the truncated Fourier representation $\hat{f}[\ ]$ |

| | |
|---|---|
| $E$ | Young's modulus |
| $E_n^{(p)}(f)$ | infimum of the $\mathbf{L}^p$-norm of the approximation error of a function $f$ over all trigonometric polynomials $\mathbf{T}_n$ of degree at most $n$ |
| $E_n(f)$ | $E_n^{(\infty)}(f)$ |
| ess sup | essential supremum |
| $e_n^{(p)}(f)$ | infimum of the $\mathbf{L}^p$-norm of the approximation error of a function $f$ over all trigonometric polynomials $\mathbf{t}_n$ of degree at most $n$, free of a constant term |
| $e_n(f)$ | $e_n^{(\infty)}(f)$ |
| $f$ | generic symbol for any of the contour representations $z$, $\dot{z}$, $\ddot{z}$, $\psi$ and $K$ (as indicated the context) |
| $\hat{f}$ | sequence of Fourier coefficients generated by $f$ |
| $f^*$ | translation- and scale-normalized version of contour representation $f$ |
| $f[\ ]$ | discrete version of contour representation $f$ |
| $\hat{f}[\ ]$ | truncated version of Fourier representation $\hat{f}$ |
| $\langle f \rangle$ | contour average of the function $f$ |
| $\lVert f \rVert_p$ | norm on $\mathbf{L}^p(2\pi)$ |
| $\lVert \hat{f} \rVert_p$ | norm on $\ell^p(\mathbb{Z})$ |
| GCD | greatest common divisor |
| $g$ | generic symbol for any of the real-valued contour representations $\psi$ and $K$ (as indicated in the context) |
| $H(t)$ | Heaviside unit step function |
| $H_p(\omega)$ | frequency response of $h_p[n]$ |
| $h_p[n]$ | coefficients of a polynomial FIR filter for the computation of the $p$-th coefficient $c_p[n]$ |
| $h_{..}(\tau; f)$ | cyclic convolution function based on contour representations of type $f$ |

| | |
|---|---|
| $h_{..}[\tau; f]$ | discrete cyclic convolution function based on discrete contour representations of type $f[\ ]$ |
| Im | imaginary part of a complex number |
| $I$ | moment of inertia |
| inf | infimum |
| $K$ | curvature function |
| $K_a[n]$ | approximated curvature function sample |
| $K_{max}$ | maximum curvature of a curve or contour |
| $\boldsymbol{k}$ | curvature vector function |
| LCM | least common multiple |
| $L$ | perimeter of a contour |
| $L_{ps}(\cdot)$ | truncation error bound for piecewise smooth contours |
| $L_r(\cdot)$ | truncation error bound for regular contours |
| $L_{wr}(\cdot)$ | truncation error bound for weakly regular contours |
| $\mathbf{L}^p$, $\mathbf{L}^\infty$ | Lebesgue spaces |
| lim | limit |
| $\ell^p$, $\ell^\infty$ | sequence spaces |
| $\mathcal{M}_x$ | mirror-reflection operator that performs the mirror-reflection of a contour about the $x$-axis |
| max | maximum |
| min | minimum |
| $m_{pq}$ | moment of order $(p + q)$ |
| $\boldsymbol{m}$ | (element of) mirror-symmetry |
| $\mathbb{N}$ | the set of natural numbers: $\{1, 2, 3, \ldots\}$ |
| $\boldsymbol{n}$ | (element of) $\boldsymbol{n}$-fold rotational symmetry |
| $\boldsymbol{n}\cdot\boldsymbol{m}$ | (elements of) $\boldsymbol{n}$-fold compositional symmetry |
| $\boldsymbol{n}(t)$ | unit normal function |

| | |
|---|---|
| $O$ | Landau order symbol 'big oh' |
| $o$ | Landau order symbol 'little oh' |
| $P$ | partition |
| $P_n$ | trigonometric polynomial of degree at most $n$ |
| $\boldsymbol{P}(t)$ | center of curvature vector at $\boldsymbol{z}(t)$ |
| $\mathcal{P}[a, b]$ | set of all partions of the interval $[a, b]$ |
| $p_n$ | trigonometric polynomial of degree at most $n$, free of a constant term |
| $\boldsymbol{p}(t)$ | unit tangent vector function |
| Re | real part of a complex number |
| $R(\xi)$ | polar representation of a contour |
| $R'(\xi')$ | signed polar representation of a contour |
| $R(\cdot, \cdot)$ | correlation coefficient of a pair of dissimilarity matrices |
| $\boldsymbol{R}$ | matrix of correlation coefficients $R(\cdot, \cdot)$ |
| $\mathbb{R}$ | the set of real numbers |
| $\mathbb{R}^+$ | the set of positive real numbers: $\{x: x \in \mathbb{R} \text{ and } x > 0\}$ |
| $\mathbb{R}^2$ | the Cartesian product of $\mathbb{R}$ and $\mathbb{R}$: $\mathbb{R}^2 = \mathbb{R} \times \mathbb{R} = \{(x, y): x \in \mathbb{R} \text{ and } y \in \mathbb{R}\}$ |
| $\mathcal{R}_\alpha$ | rotation operator that rotates a contour over an angle $\alpha$ in counterclockwise direction |
| $r$ | radial distance function |
| $r'$ | signed radial distance function |
| $S_n f$ | partial Fourier sum of degree $n$ |
| $S_N(\gamma, \delta; p, q)$ | data dependent coefficient in the computation of the moment $m_{pq}$ of an $N$-sided polygon |
| $\mathcal{S}_\beta$ | scaling operator that scales a contour by a factor $\beta$ |
| sgn | sign function |

| | |
|---|---|
| sup | supremum |
| $s$ | arc length parameter |
| $\dot{s}$ | d$s$/d$t$ |
| $T_N(\alpha, \beta; p, q)$ | data dependent coefficient in the computation of the moment $m_{pq}$ of an $N$-sided polygon |
| $\mathbf{T}_n$ | set of all trigonometric polynomials of degree at most $n$ |
| $\mathcal{T}_\tau$ | parametric shift operator that causes a forward shift of the parameter of a contour representation over $\tau$ |
| $t$ | normalized arc length parameter |
| $t^+$ | $\lim_{\delta \downarrow 0} (t + \delta)$ |
| $t^-$ | $\lim_{\delta \downarrow 0} (t - \delta)$ |
| $\mathbf{t}_n$ | set of all trigonometric polynomials of degree at most $n$, free of a constant term |
| $U$ | elastic energy or bending energy per unit length |
| $U_{tot}$ | total bending energy, necessary to deform one thin elastic beam into another |
| $u$ | parameter of an analytic form of a position function |
| $\boldsymbol{u}_x$ | unit $x$-vector |
| $\boldsymbol{u}_y$ | unit $y$-vector |
| $V(\cdot)$ | truncation error bound based on total variation |
| $Var\,(f)$ | total variation of $f$ |
| $x$ | $x$-component of the position function $z$ |
| $y$ | $y$-component of the position function $z$ |
| $\mathbb{Z}$ | the set of integers: $\{\ldots, -1, 0, 1, \ldots\}$ |
| $z$ | position function |
| $\dot{z}$ | tangent function |
| $\ddot{z}$ | acceleration function |

| | |
|---|---|
| $z_a[n]$ | approximated position function sample |
| $\dot{z}_a[n]$ | approximated tangent function sample |
| $\ddot{z}_a[n]$ | approximated acceleration function sample |
| $\boldsymbol{z}$ | position vector function |
| $\dot{\boldsymbol{z}}$ | tangent vector function |
| $\ddot{\boldsymbol{z}}$ | acceleration vector function |
| $\alpha$ | rotation angle |
| $\alpha^*$ | rotation normalization parameter |
| $\beta$ | scaling coefficient |
| $\beta^*$ | scale normalization parameter |
| $\boldsymbol{\Gamma}$ | class of simple closed curves |
| $\boldsymbol{\Gamma}_{pr}$ | class of piecewise regular simple closed curves |
| $\boldsymbol{\Gamma}_{ps}$ | class of piecewise smooth simple closed curves |
| $\boldsymbol{\Gamma}_{pwr}$ | class of piecewise weakly regular simple closed curves |
| $\boldsymbol{\Gamma}_r$ | class of regular simple closed curves |
| $\boldsymbol{\Gamma}_s$ | class of smooth simple closed curves |
| $\boldsymbol{\Gamma}_{wr}$ | class of weakly regular simple closed curves |
| $\gamma$ | contour |
| $\Delta$ | difference operator |
| $\Delta^2$ | second order difference operator |
| $\delta(t)$ | Dirac delta function |
| $\zeta$ | displacement of a contour in the plane; reference position with respect to a contour |
| $\zeta_c^*$ | translation normalization parameter based on the contour average |
| $\zeta_r^*$ | translation normalization parameter based on the regional average over the area enclosed by a contour |

| | |
|---|---|
| $\eta(k)$ | argument of the Fourier coefficient $\hat{z}(k)$ |
| $\theta$ | tangent angle function |
| $\dot{\theta}$ | $d\theta/dt$ |
| $\theta_a[n]$ | approximated tangent angle function sample |
| $\mathbf{\Lambda}$ | space of functions that satisfy a uniform Lipschitz condition |
| $\lambda$ | Lipschitz constant |
| $\lambda_n$ | $n$-th Lebesgue constant |
| $\lambda(\psi)$ | normalization parameter for $\psi$ |
| $\mu_{pq}$ | central moment of order $(p + q)$ |
| $\xi$ | angle of revolution |
| $\xi'$ | accumulated angle of revolution |
| $\pi_n(t)$ | polynomial centered at $z[n]$ |
| $\varrho(t)$ | radius of curvature |
| $\varrho_{..}(\tau; f)$ | cyclic correlation function based on contour representations of type $f$ |
| $\varrho_{..}[\tau; f]$ | discrete cyclic correlation function based on discrete contour representations of type $f[\ ]$ |
| $\tau$ | forward shift of a contour representation parameter |
| $\tau^*$ | starting point normalization parameter |
| $\varphi$ | cumulative angular function |
| $\varphi_a[n]$ | approximated cumulative angular function sample |
| $\chi_R(x, y)$ | characteristic function of a region $R \subset \mathbb{R}^2$ |
| $\psi$ | periodic cumulative angular function |
| $\psi_a[n]$ | approximated periodic cumulative angular function sample |
| $\omega$ | normalized frequency parameter |
| $\lvert\cdot\rvert$ | magnitude |

| | |
|---|---|
| $\|\cdot\|$ | norm |
| $\nabla.$ | gradient operator |
| $*$ | (cyclic) convolution |
| $\forall$ | for all |
| $\in$ | is an element of |
| $\cup, \bigcup$ | union |
| $\cap, \bigcap$ | intersection |
| $\subset$ | is a proper subset of |
| $\subseteq$ | is a subset of |

# Acknowledgements

On the long awaited completion of this book I would like to thank the institutions and all the people that have contributed in one way or another.

First of all, I wish to thank my former colleagues and students of the Information Theory Group of the Department of Electrical Engineering of the Delft University of Technology for their support and for the stimulating environment during the earliest years of research.

Discussions with Dr. Ian T. Young, especially in the summer of 1980 during my stay at the Lawrence Livermore National Laboratory, Livermore, CA, USA, have had a profound influence on the contents of this book. The hospitality of the Biomedical Sciences Division of the Lawrence Livermore National Laboratory as well as the funding support of the National Cancer Institute, Grant Number CA-28833, and the US Department of Energy contract with the Lawrence Livermore National Laboratory, Number W-7405-ENG-48, are gratefully acknowledged.

I am particularly grateful to the management of Philips' Research Laboratories for providing me with the environment and the facilities to complete this book.

Dr. Adrian A. Gerbrands, then of the Institute of Cultural Anthropology of the University of Leyden, stimulated my interest for quantifying symmetry in objects. The contents of Chapter 3 have benefited a great deal from my cooperation with Michel Dekking. The clustering experiments in Chapter 4 were performed with the help of Hein Haas, who also created the clustering software. The mathematical expertise of Guido Janssen on many mathematical issues has been indispensible.

Several people were of great help by reading, discussing and criticizing various parts of this book. I wish to thank Ad van den Enden, Hein Haas, Bas Michielsen, Ying-Lie O, Kiran Varshneya-Rohra and Ludo Tolhuizen in this respect.

Jouke Lakerveld took care of a smooth transfer of my software from the computing facilities of the Information Theory Group of the Delft University of Technology to those of the Digital Signal Processing Group of Philips' Research Laboratories. Ton Barten performed helpful literature searches.

The conversion of the handwritten manuscript into print was a tremendous task. The enthusiastic support of Theo Schoenmakers, who organized it all, has been crucial in achieving the present result. Without his continuous efforts this book would have never achieved its present appearance. Dick van den Abeele and Hein Haas spent many hours of their spare time on the proofreading of typeset versions of the manuscript and made many helpful suggestions. The editorial help of Elma Kleikamp-Rethans has been invaluable. I also wish to thank Henny Alblas, who took care of the drawings, Rik van den Wildenberg, who took the photographs and Carin Groenen-Lamers, who helped to produce the index.

Finally I wish to express my gratitude to my wife Tineke and my children Nienke and Jos, for their continuous support and patience during the many years it took to complete this book. Without them this work might never have been finished.

# Chapter 1

# Introduction

## 1.1 Shape analysis: a classical problem in the analysis of image data

This book is devoted to digital shape analysis, and in particular to its quantitative aspects. Our visual system uses shape as an important feature to recognize and order the things in the world that surrounds us. Therefore it is not surprising that, in the attempts to equip machines with recognition capabilities, shape analysis has always been an important topic. This statement appears to be true not only when it concerns a rather simple task of printed character recognition, where shape analysis can be one of the final steps in the recognition process, but also in complex, wide-ranging computer vision tasks in an artificial intelligence context, where shape analysis in general provides only an intermediate result.

Our visual system is remarkably capable of associating and recognizing shapes. Probably in part as a result of the ease with which we recognize shapes, we have not developed a rich vocabulary for describing shape, let alone ways of quantifying shape or differences between shapes. The latter is also caused by the fact that our visual system is very bad at assessing population variance. Our descriptions of shape are usually of a qualitative nature. In fact, the development of quantitative methods for shape analysis and comparison, that do not yield results that are in conflict with our perception, is hampered by our own limited abilities to quantify shape and shape differences.

According to *The Shorter Oxford English Dictionary* [1975], shape stands for 'external form' or 'contour' or, more precisely, 'that quality of a material object or geometrical figure which depends on constant relations of position and proportionate position among all the points composing its outline or external surface.' In this book we will deal only with planar shapes, since images in digital image analysis usually portray two-dimensional projections of three-dimensional scenes. We will not study external surfaces that determine the shape of three-dimensional objects, despite of the growing interest in three-dimensional shape analysis and representations in areas such as computer vision (Ballard and Brown [1982], Horn [1986]) and computer graphics (Newman and Sproull [1979], Foley and van Dam [1982]).

In many applications, such as for example character recognition, computer analysis of microscopic slides and many industrial inspection tasks, two-dimensional shape analysis turns out to be adequate. Furthermore, the complexity of three-dimensional shape analysis often still does not allow its application in practice. As mentioned earlier, we usually have only two-dimensional information at our disposal. Sometimes, if the shape analysis problem concerns only a limited set of known objects, a 'dictionary' of two-dimensional perspective projections is used to match or interpolate the shapes to be analyzed (Richard and Hemàmi [1974], Wallace and Wintz [1980], Wallace and Mitchell [1980], Sarvarayudu [1982]).

In stereology, many problems are of an inherently three-dimensional nature, whereas the information available is two-dimensional, e.g. thin slices of material. Three-dimensional information is extrapolated from the results of two-dimensional image and shape analysis, using techniques from integral geometry and statistics (DeHoff and Rhines [1968], Weibel [1979], Weibel [1980], Serra [1982]).

Following the verbal formulation of the concept of shape just given, we will not consider the internal structure of an object, such as its brightness, colour, texture, etc., to be part of its shape. In studying the shape of an object, we will merely deal with its geometrical properties. Furthermore, we will assume that shape is invariant under the following transformations:

- translation
- scaling
- rotation.

These transformations, the *equiform transformations* or *similarity transformations*, form a group: the equiform group. As a result we may form equivalence classes of shapes that can be mapped onto one another by the similarity transformations. Equivalent shapes are called *similar*. Note that, if we were interested in the stronger property of congruence among shapes, we would have to discard the property of invariance under scaling.

The importance of shape as a tool for analysis, ordering and classification has led to the study of shape in many, diverse fields of science and has resulted in an abundant literature on this topic. This abundance causes a useful review, that would do justice to the many approaches and contributions, to be outside the scope of this book, even when such a review would be restricted to the context of pattern classification and digital image analysis. Therefore we will mention only some important or interesting texts, along with some historical remarks, that will provide access to the enormous amount of literature available. We will also list some major distinctions in approaches to shape analysis by computer and outline the approach that we will use.

An early study of shape in the context of biology and evolution is that of Thompson (Thompson [1942], Thompson [1961]). This important work remains a source of inspiration to this day. See for example Bookstein [1978], which also contains many references to the work of followers of Thompson.

Understandably, shape and especially shape perception constitute a popular topic in the psychology literature. Some early reports on the quantitative study of shape, in the context of psychology, can be found in Attneave [1954], Attneave and Arnoult [1956] and Hake [1957]. Visual perception of shape is dealt with in books by Zusne [1970], Cornsweet [1970] and Rock [1973]. Concerning the mathematical modelling of visual perception we mention Zeeman [1962], Moore [1971] and Moore, Seidl and Parker [1975]. A coding-type theory of visual perception is proposed in Leeuwenberg [1968].

Shape is used as a tool for the seriation of objects in archeology and in the history of art (cf. e.g. Clarke [1968], Plomp [1979]), and for grouping objects or designs in anthropology.

In particle analysis shape is used as a parameter to determine physical properties of particles (cf. e.g. Schwarcz and Shane [1969], Ehrlich and Weinberg [1970], Beddow [1980], Beddow and Meloy [1980], Beddow [1984a], Beddow [1984b]).

In the context of providing machines with recognition capabilities, Minsky and Papert's book [1969] is a classic text, emphasizing concepts from topology and computational geometry. Texts of a more general nature, that pay considerable attention to shape analysis and representation by computer, are Duda and Hart [1973], Gonzalez and Wintz [1977], Ballard and Brown [1982] and Levine [1985]. Pavlidis [1977a] is almost entirely devoted to shape analysis, while Pavlidis [1982] deals for the larger part with shape representation. The latter book emphasizes the interrelations between image processing, pattern recognition and computer graphics. We note that shape is a common issue in the processing, recognition and generation of pictorial data. We also note the importance of the computer graphics literature as a source of information about the representation and processing of shape information. In this context we already mentioned Newman and Sproull [1979] and Foley and van Dam [1982]. Also the work of Hou [1983], which deals with digital document processing, is a valuable reference in this respect.

With the practice of stereology in mind, Serra [1982] presents an in-depth study of the effects of digitization and operators and the size of these operators upon the topological and geometrical properties of objects in the images under study. In this respect we also mention Ahuja and Schachter [1983].

In many applications, such as cell analysis, chromosome analysis and particle analysis, the shape of the objects of interest changes with the resolution at which they are observed and, consequently, can only be defined by convention. A detailed exposé of such phenomena and their mathematical modelling is given by Mandelbrot (Mandelbrot [1977], Mandelbrot [1982a]). The mathematical models proposed by Mandelbrot have been applied, for example, in computer graphics (e.g. Carpenter [1980], Fournier, Fussel and Carpenter [1982], Mandelbrot [1982b], Kajiya [1983], Pentland [1983]). The consequences of the dependence of shape on resolution for digital shape analysis largely remain to be studied.

Pavlidis has published two survey papers on shape analysis by computer: Pavlidis [1978] reviews digital shape analysis in general, while Pavlidis [1980a] is devoted to contour-oriented approaches to digital shape analysis. The latter paper is commented upon in Wallace [1981]. A supplementary review of digital shape analysis literature can be found in Sarvarayudu [1982]. Though completeness may not be expected from these surveys, they provide for a distinction between a number of shape analysis techniques and for access to the literature. A most useful entry into the literature on digital shape analysis is provided by the extensive bibliographies on picture processing by Rosenfeld, published annually since 1972 in the journal *Computer Graphics and Image Processing*. In 1980 this journal had its name changed to *Computer Vision, Graphics and Image Processing*.

We now present some criteria to distinguish methods of digital shape analysis that will enable us to classify the approach that will be used in this book.

One criterion to distinguish shape analysis techniques is the subdivision into *information-preserving* and *information-nonpreserving* techniques, depending upon whether a shape can be reconstructed with a controllable level of precision from the representation, used in its analysis, or not.

A second criterion to distinguish shape analysis techniques is the subdivision into *region-oriented* and *contour-oriented* techniques, also referred to as *internal* and *external* techniques, respectively (Pavlidis [1980a]).

A third criterion is the discrimination between techniques that map the pictorial data, containing the shape information, into a set of numbers and those that map these data into another picture. This distinction between *scalar transform techniques* and *space domain techniques* in Pavlidis [1978] is rather subtle and with some shape analysis techniques it can hardly be made. Virtually all shape analysis techniques, at some stage in the analysis, transform the pictorial information into a set of numbers or symbols to represent the shape information.

The distinction between shape analysis techniques that are concerned with *local shape properties* and those that perform *global shape analysis* constitutes a fourth discriminatory criterion.

Further we mention as a fifth criterion for discrimination, the distinction between techniques that use a *deterministic approach* to shape analysis and those that use *statistical techniques*, based on a *stochastic model*.

We note that tools from many different mathematical disciplines have been used in various approaches to digital shape analysis: elements from set theory, algebra, topology, mathematical analysis, differential geometry, integral geometry, probability theory, graph theory, formal languages and automata theory, etc. can all be found in the literature on shape analysis.

## 1.2 Scope of this book and an outline of its contents

Our starting point for digital shape analysis will be the segmented image, in which the individual connected components have been identified and labelled (cf. Gonzalez and Wintz [1977], Pavlidis [1982], Danielsson [1982]). These components will be the subject of shape analysis. As we do not consider the internal structure of these components in the original picture to be part of their shape, this is a reasonable starting point. In this book we will concentrate for the major part on information-preserving, contour-oriented, global, deterministic techniques. The techniques that we will study can also clearly be classified as scalar transform techniques, as defined by Pavlidis [1978].

A number of reasons why contour-oriented shape analysis techniques are popular can be given. Shape information is contained in the contours of objects (Attneave [1954], Attneave and Arnoult [1956]). This point of view is confirmed by observations that edge detection constitutes an important aspect of shape recognition by the human visual system (Zusne [1970], Shapley and Tolhurst [1973], Marr [1976], Marr and Hildreth [1980], Marr [1982]) and by psychovisual experiments studying eye movements (Zusne [1970], Noton and Stark [1971a], Noton and Stark [1971b]).

Information-preserving contour representations allow for a reconstruction of the segmented image. If additional processing has been performed on the contours in the segmented image or on their representations, then a good approximation of the segmented image can still be obtained. By means of contour representation we obtain in general a considerable data reduction, compared with the segmented picture, without loss of information. Furthermore, shape analysis techniques, that are based on parametric contour representations, are intimately related with well-founded mathematical disciplines such as mathematical analysis and differential geometry. At the application level, methods from numerical analysis and digital signal processing can be readily applied.

Region-oriented shape analysis techniques, on the other hand, are often very time-consuming. With some region-based techniques, such as template matching, invariance for orientation and for scaling is hard to accomplish.

Shape analysis techniques that give rise to graph-like structures as a means of representation (e.g. Fischler and Elschlager [1973], Pavlidis [1977a], Shapiro [1980], Shapiro and Haralick [1981], Bunke and Allermann [1983]), usually lead to computationally intensive matching problems, such as (sub)graph isomorphism problems. See for example Read and Corneil [1977] and McGregor [1979] for a discussion of (sub)graph isomorphism problems. Noise and distortion usually are problems that are difficult to deal with in structural shape analysis techniques. The same holds for syntactic shape analysis techniques, which are discussed, for example, in Fu [1974], Fu [1977] and You and Fu [1979]. Though a number of parsing techniques have been developed for syntactic shape analysis, both in structural and in syntactic shape analysis, the inference problem remains largely unsolved.

With contour-oriented shape analysis techniques, noise in contour representations can be reduced, using techniques from numerical analysis or digital signal processing or by means of techniques that have been developed especially for contour-oriented digital shape analysis (cf. Bowie and Young [1977a], van Otterloo [1978]).

The methods that we will describe for shape analysis provide measures for the geometrical aspects of symmetry in objects and similarity between objects. We will not deal with any field of application in particular. In image analysis applications, where the geometrical aspects of similarity between objects is of importance, the methods described here can prove to be useful. This does not mean that we claim that the methods described in this book can be usefully utilized in every application: different applications will require different approaches, thereby ruling out the thought of a uniquely optimal approach. The results of our symmetry analysis methods can be used as properties of the shapes under study. The results of our similarity analysis methods can be applied in shape-clustering algorithms, in order to determine shape classes. They can also serve as intermediate results in structural shape analysis methods and in complex computer vision tasks, or they can be used directly for shape classification.

Global shape analysis techniques sometimes pose problems in the case of heavily distorted shapes. For example, due to bad signal-to-noise ratio conditions in the original picture and/or due to imperfections resulting from the segmentation procedure, parts of objects in the original picture may have been assigned to other connected components in the segmented image. Overlapping objects constitute another problem that is usually not yet taken care of at the segmentation stage of the analysis. In such cases it may be desirable to incorporate a feedback from the shape analysis stage to the segmentation stage. A number of methods have been developed for the detection and handling of overlapping objects (cf. e.g. Arcelli and Levialdi [1973], Eccles, McQueen and Rosen [1977], Sychra et al. [1978], Dessimoz et al. [1979], Lester et al. [1978], Bengtsson et al. [1981], Kailay, Sadananda and Das [1981], Bhanu and Faugeras [1981], Turney, Mudge and Volz [1984]). In such cases, where the segmented picture contains deficient shapes, the possibility to perform similarity analysis on partial shapes, or, equivalently, on contour segments that represent incomplete shapes, may be required. Similarity measurement methods based on coefficients generated by global transformations are obviously of an inherently global nature: local shape properties cannot be taken into account with these methods. Similarity measurement methods based on the Fourier coefficients or on the Walsh coefficients of parametric contour representations, which are among the most popular of all contour-oriented shape analysis techniques, belong to this class. However, the parametric contour representations themselves can be linked directly, and thus locally, with the contours in the space domain. Therefore, similarity measurement methods based on parametric contour representations themselves can almost immediately be used for the analysis of similarity between shape segments. In this book we will formulate shape similarity measurement only for closed contours, but the generalization of such a measurement to shape segments is simple and straightforward.

Region-oriented shape analysis techniques distinguish themselves from contour-oriented techniques in that the former techniques can deal directly with topologically nonsimple components, i.e. components with holes. In such cases, contour-oriented techniques will have to deal with the outer boundaries of the components and of the holes separately. At a higher hierarchical level in the analysis the results of the analysis of the individual boundaries of the components must be linked with information about the relative positions, sizes and orientations of these boundaries. At these higher levels in image analysis labelled graphs are useful to represent the information extracted from the image, despite the computational complexity when it comes to matching graph structures (Read and Corneil [1977]).

The main goal of this book is to provide a unified theoretical basis for shape similarity analysis on the basis of parametric contour representations; unified in the sense that various contour representations, some of which have been considered previously in the literature, will be presented and their relationships derived. Thereby the theoretical framework, into which these contour representations fit, will be made clear.

Apart from establishing relations between the contour representations themselves, we will also attempt to establish relations between the measures of dissimilarity that are based on these contour representations. Further, the trade-off between optimization and normalization, necessary in order to achieve the desired invariance properties in the proposed dissimilarity measures, will be evaluated.

We will relate the concept of symmetry in plane figures with that of similarity under symmetry transformations. This enables us to define measures for the quantification of symmetry, or for the lack of symmetry (for which we will use the term dissymmetry), in a plane figure on the basis of previously defined dissimilarity measures.

Since the Fourier coefficients of parametric contour representations have been given so much attention in the literature, we will, throughout this book, formulate the Fourier coefficient-based counterparts of the dissimilarity and dissymmetry measures. Where possible we will relate both types of measures. Some considerations will be given for the practical implementation of the proposed techniques. A short survey of the contents of the individual chapters is given below.

Chapter 2 introduces the parametric contour representations that we will consider and establishes their relations. It also defines the concept of similarity, which is subsequently reformulated as a relation that exists between the contour representations of a pair of similar objects. Finally it discusses two types of plane symmetry: mirror-symmetry and rotational symmetry. The conditions that a contour representation must satisfy, in order to represent a contour that has these types of symmetry, are formulated.

Chapter 3 follows the same lines as Chapter 2, but in terms of the Fourier coefficients of the contour representations. It also discusses the consequences of normalized arc length parametrization on Fourier series expansions of contour representations and gives bounds on truncation errors of finite Fourier series expansions.

In Chapter 4 the measures for dissimilarity and dissymmetry are defined and their theoretical properties evaluated. Relations between these measures are formulated. These relations will be helpful in solving the design problem, i.e. out of a multitude of possibilities, which contour representation and which measure, or combination of measures, should be selected in a given application. To this end also a number of experiments are performed and evaluated. Furthermore, the trade-off between optimization and normalization, necessary in order to achieve the desired invariance for translation, rotation, scaling and parametric starting point in the proposed measures, is evaluated in this chapter.

In Chapter 5 a discussion of the results, obtained in the previous chapters, is presented, a number of open problems are indicated and some suggestions for further research are given.

Finally, this book contains three appendices. Appendix A deals with some mathematical concepts and properties relevant to the contents. In Appendix B a computationally efficient method for the computation of the moments of polygonal regions is described and evaluated. Appendix C discusses the estimation of contour representations through piecewise polynomial approximation.

## References

Ahuja, N. and B.J. Schachter [1983]
*Pattern Models*, New York: John Wiley and Sons, Inc.

Arcelli, C. and S. Levialdi [1973]
'On Blob Reconstruction', Comp. Graph. and Im. Proc. **2**: 22-38.

Attneave, F. [1954]
'Some Informational Aspects of Visual Perception', Psychol. Rev. **61**: 183-193.

Attneave, F. and M.D. Arnoult [1956]
'The Quantitative Study of Shape and Pattern Perception', Psychol. Bull. **53**: 452-471.
Reprinted in: *Pattern Recognition*, L. Uhr (Ed.): 123-141, New York: John Wiley and Sons, Inc., 1966.

Ballard, D.H. and C.M. Brown [1982]
*Computer Vision*, Englewood Cliffs, NJ: Prentice-Hall, Inc.

Beddow, J.K. [1980]
'Particle Morphological Analysis'. In: *Advanced Particulate Morphology*, J.K. Beddow and T.P. Meloy (Eds.): 1-84, Boca Raton, FL: C.R.C. Press.

Beddow, J.K. (Ed.) [1984a]
*Particle Characterization in Technology, Volume 1: Applications and Microanalysis*, Boca Raton, FL: C.R.C. Press.

Beddow, J.K. (Ed.) [1984b]
*Particle Characterization in Technology, Volume 2: Morphological Analysis*, Boca Raton, FL: C.R.C. Press.

Beddow, J.K. and T.P. Meloy (Eds.) [1980]
*Advanced Particulate Morphology*, Boca Raton, FL: C.R.C. Press.

Bengtsson, E., P. Eriksson, J. Holmquist, T. Jarkrans, B. Nordin and B. Stenkvist [1981]
'Segmentation of Cervical Cells: Detection of Overlapping Cell Nuclei', Comp. Graph. and Im. Proc. **16**: 382-394.

Bhanu, B. and O.D. Faugeras [1981]
'Recognition of Occluded Two Dimensional Objects', Proc. of the Second Scandinavian Conference on Image Analysis, Helsinki, Finland: 72–77.

Bookstein, F.L. [1978]
*The Measurement of Biological Shape and Shape Change*, Lecture Notes in Biomathematics, **24**, New York: Springer-Verlag.

Bowie, J.E. and I.T. Young [1977a]
'An Analysis Technique for Biological Shape-II', Acta Cytol. **21**: 455-464.

Bunke, H. and G. Allermann [1983]
'Inexact Graph Matching for Structural Pattern Recognition', Patt. Recogn. Lett. **1**: 245-253.

Carpenter, L.C. [1980]
'Computer Rendering of Fractal Curves and Surfaces'. In: Siggraph '80 Conference Proceedings Supplement: 9-15, New York: Association for Computing Machinery (ACM).

Clarke, D.L. [1968]
*Analytical Archaeology*, London: Methuen and Co.

Cornsweet, T.N. [1970]
*Visual Perception*, New York: Academic Press.

Danielsson, P.-E. [1982]
'An Improved Algorithm for Binary and Nonbinary Images', IBM Journ. of Res. and Dev. **26**: 698-707.

DeHoff, R.T. and F.N. Rhines [1968]
*Quantitative Microscopy*, New York: McGraw-Hill Book Co., Inc.

Dessimoz, J.-D., M. Kunt, J.M. Zurcher and G.H. Granlund [1979]
'Recognition and Handling of Overlapping Industrial Parts', Proc. 9th Intern. Symp. on Industr. Robots, Washington, DC: 357-366.

Duda, R.O. and P.E. Hart [1973]
*Pattern Classification and Scene Analysis*, New York: John Wiley and Sons, Inc.

Eccles, M.J., M.P.C. McQueen and D. Rosen [1977]
'Analysis of the Digitized Boundaries of Planar Objects', Patt. Recogn. **9**: 31-41.

Ehrlich, R. and B. Weinberg [1970]
'An Exact Method for Characterization of Grain Shape', Journ. of Sedim. Petrol. **40**: 205-212.

Fischler, M.A. and R.A. Elschlager [1973]
'The Representation and Matching of Pictorial Structures', IEEE Trans. Comp. **C-22**: 67-92.

Foley, J.D. and A. van Dam [1982]
*Fundamentals of Interactive Computer Graphics*, Reading, MA: Addison-Wesley.

Fournier, A., D. Fussel and L.C. Carpenter [1982]
'Computer Rendering of Stochastic Models', Comm. ACM **25**: 371-384.

Fu, K.S. [1974]
*Syntactic Methods in Pattern Recognition*, New York: Springer-Verlag.

Fu, K.S. (Ed.) [1977]
*Syntactic Pattern Recognition Applications*, New York: Springer-Verlag.

Gonzalez, R.C. and P.A. Wintz [1977]
*Digital Image Processing*, Reading, MA: Addison-Wesley.

Hake, H.W. [1957]
'Contributions of Psychology to the Study of Pattern Vision', WADC Techn. Report 57-621, Astia Doc. No. AD 142035: 60-79, 99-118. Wright Air Development Center, U.S. Air Force Wright-Patterson Air Force Base, Dayton, OH.
Reprinted as: 'Form Discrimination and the Invariance of Form'. In: *Pattern Recognition*, L. Uhr (Ed.): 142-173, New York: John Wiley and Sons, Inc., 1966.

Horn, B.K.P. [1986]
*Robot Vision*, Cambridge, MA: The MIT Press.

Hou, H.S. [1983]
*Digital Document Processing*, New York: John Wiley and Sons, Inc.

Kailay, B.C., R. Sadananda and J.R. Das [1981]
'An Algorithm for Segmenting Juxtaposed Objects', Patt. Recogn. **13**: 347-351.

Kajiya, J.T. [1983]
'New Procedure for Ray Tracing Procedurally Defined Objects', ACM Trans. Graph. **2**: 161-181.

Leeuwenberg, E.L.J. [1968]
*Structural Information of Visual Patterns* (Ph.D. Thesis, University of Nijmegen, Nijmegen, The Netherlands), The Hague, The Netherlands: Mouton and Co.

Levine, M.D. [1985]
*Vision in Man and Machine*, New York: McGraw-Hill Book Co., Inc.

Lester, J.M., H.A. Williams, B.A. Weintraub and J.F. Brenner [1978]
'Two Graph Searching Techniques for Boundary Finding in White Blood Cell Images', Comput. in Biol. and Med. **8**: 293-308.

McGregor, J.J. [1979]
'Relational Consistency Algorithms and Their Application in Finding Subgraph and Graph Isomorphisms', Inform. Sci. **19**: 229-250.

Mandelbrot, B.B. [1977]
*Fractals: Form, Chance and Dimension*, San Francisco, CA: W.H. Freeman and Co.

Mandelbrot, B.B. [1982a]
*The Fractal Geometry of Nature*, San Francisco, CA: W.H. Freeman and Co.

Mandelbrot, B.B. [1982b]
'Comment on Computer Rendering of Fractal Stochastic Models', Comm. ACM **25**: 581-584.

Marr, D. [1976]
'Early Processing of Visual Information', Philos. Trans. Roy. Soc. London **B 275**: 483-524.

Marr, D. [1982]
*Vision*, San Francisco, CA: W.H. Freeman and Co.

Marr, D. and E. Hildreth [1980]
'Theory of Edge Detection', Trans. Roy. Soc. London **B 207**: 187-217.

Minsky, M.L. and S. Papert [1969]
*Perceptrons, an Introduction to Computational Geometry*, Cambridge, MA: MIT Press.

Moore, D.J.H. [1971]
'A Theory of Form', Intern. Journ. of Man-Mach. Stud. **3**: 31-59.

Moore, D.J.H., R.A. Seidl and D.J. Parker [1975]
'A Configurational Theory of Visual Perception', Intern. Journ. of Man-Mach. Stud. **7**: 449-509.

Newman, W.M. and R.F. Sproull [1979]
*Principles of Interactive Computer Graphics*, 2nd Edition, New York: McGraw-Hill Book Co., Inc.

Noton, D. and L. Stark [1971a]
'Eye Movements and Visual Perception', Scient. Amer. **224**, No. 6: 34-43.

Noton, D. and L. Stark [1971b]
'Scanpaths in Saccadic Eye Movements While Viewing and Recognizing Patterns', Vision Res. **11**: 929-942.

Pavlidis, T. [1977a]
*Structural Pattern Recognition*, New York: Springer-Verlag.

Pavlidis, T. [1978]
'A Review of Algorithms for Shape Analysis', Comp. Graph. and Im. Proc. **7**: 243-258.

Pavlidis, T. [1980a]
'Algorithms for Shape Analysis of Contours and Waveforms', IEEE Trans. Patt. Anal. and Mach. Intell. **PAMI-2**: 301-312.

Pavlidis, T. [1982]
*Algorithms for Graphics and Image Processing*, Berlin: Springer-Verlag.

Pentland, A. [1983]
'Fractal-Based Description of Natural Scenes', Proc. IEEE Comp. Soc. Conf. on Comp. Vision and Patt. Recogn., Washington, DC: 201-209.

Plomp, R. [1979]
*Spring-Driven Dutch Pendulum Clocks, 1657-1710*, Schiedam, The Netherlands: Interbook International, B.V.

Read, R.C. and D.G. Corneil [1977]
'The Graph Isomorphism Disease', Journ. of Graph Theory **1**: 339-363.

Richard, Jr., C.W. and H. Hemami [1974]
'Identification of Three-Dimensional Objects Using Fourier Descriptors of the Boundary Curve', IEEE Trans. Syst., Man and Cybern. **SMC-4**: 371-378.

Rock, I. [1973]
*Orientation and Form*, New York: Academic Press.

Sarvarayudu, G.P.R. [1982]
*Shape Analysis Using Walsh Functions* (Ph.D. Thesis, Dept. of Electron. and Electr. Commun. Engin., Indian Institute of Technology, Kharagpur, India).

Schwarcz, H.P. and K.C. Shane [1969]
'Measurement of Particle Shape by Fourier Analysis', Sedimentology **13**: 213-231.

Serra, J. [1982]
*Image Analysis and Mathematical Morphology*, New York: Academic Press.

Shapiro, L.G. [1980]
'A Structural Model of Shape', IEEE Trans. Patt. Anal. and Mach. Intell. **PAMI-2**: 111-126.

Shapiro, L.G. and R.M. Haralick [1981]
'Structural Descriptions and Inexact Matching', IEEE Trans. Patt. Anal. and Mach. Intell. **PAMI-3**: 504-519.

Shapley, R.M. and D.I. Tolhurst [1973]
'Edge Detectors in Human Vision', Journ. of Physiol. **229**: 165-183.

*The Shorter Oxford English Dictionary on Historical Principles*, 2 Volumes [1975]
Third Edition, C.T. Onions (Ed.), Oxford: Oxford University Press.

Sychra, J.J., P.H. Bartels, M. Bibbo, J. Taylor and G.L. Wied [1978]
'Computer Recognition of Binucleation with Overlapping Epithelial Cells', Acta Cytol. **22**: 22-28.

Thompson, D'Arcy W. [1942]
*On Growth and Form*, 2 Volumes, Cambridge, England: Cambridge University Press. [Original Edition: 1917].

Thompson, D'Arcy W. [1961]
*On Growth and Form, Abridged Edition*, Cambridge, England: Cambridge University Press.

Turney, J.L., T.N. Mudge and R.A. Volz [1984]
'Experiments in Occluded Parts Recognition'. In: *Intelligent Robots: Third International Conference on Robot Vision and Sensory Controls RoViSeC3*, D.P. Casasent and E.L. Hall (Eds.), Proc. SPIE **449**: 719-725.

van Otterloo, P. J. [1978]
*A Feasibility Study of Automated Information Extraction from Anthropological Pictorial Data* (Thesis, Dept. of Elec. Engin., Delft University of Technology, Delft, The Netherlands).

Wallace, T.P. [1981]
'Comments on Algorithms for Shape Analysis of Contours and Waveforms, IEEE Trans. Patt. Anal. and Mach. Intell. **PAMI-3**: 593.

Wallace, T.P. and O.R. Mitchell [1980]
'Analysis of Three-Dimensional Movements Using Fourier Descriptors', IEEE Trans. Patt. Anal. and Mach. Intell. **PAMI-2**: 583-588.

Wallace, T.P. and P.A. Wintz [1980]
'An Efficient Three-Dimensional Aircraft Recognition Algorithm Using Normalized Fourier Descriptors', Comp. Graph. and Im. Proc. **13**: 99-126.

Weibel, E.R. [1979]
*Stereological Methods, Volume 1: Practical Methods for Biological Morphometry*, New York: Academic Press.

Weibel, E.R. [1980]
*Stereological Methods, Volume 2: Theoretical Foundations*, New York: Academic Press.

You, K.C. and K.S. Fu [1979]
'A Syntactic Approach to Shape Recognition Using Attributed Grammars', IEEE Trans. Syst., Man and Cybern. **SMC-9**: 334-345.

Zeeman, E.C. [1962]
'The Topology of the Brain and Visual Perception'. In: *Topology of 3-Manifolds and Related Topics*, Proc. 1961 Topology Institute, M.K. Fort, Jr. (Ed.): 240-256, Englewood Cliffs, NJ: Prentice-Hall, Inc.

Zusne, L. [1970]
*Visual Perception of Form*, New York: Academic Press.

# Chapter 2

# Parametric contour representation, similarity and symmetry

## 2.1 Parametric contour representation

In Section 1.2 we gave a short survey of two-dimensional shape analysis techniques and provided a number of references to the literature on this subject. Following the arguments we put forth in that section, we restrict our attention to techniques that make use of the shape information contained in the outer boundaries of objects/regions in two-dimensional images.

The class of curves, that represent object contours or outlines, is formed by the class of simple closed curves in the plane. It is the purpose of this section and Section 2.2 to present mathematical tools for the representation of such curves. We are particularly interested in representations that are *information-preserving*, i.e. representations that allow for an exact reconstruction of a shape. Special attention will be given to the mathematical relations that exist between these representations.

For digital shape analysis it is important to have a proper model for the smoothness of the contours to be analyzed. Therefore, in this section and in Chapter 3 we will define a number of smoothness classes of contours.

We consider the shape of an object/region to be invariant for translation, rotation and scaling. Operators, that produce these operations, will be introduced in Section 2.2. The effects of these operators upon various contour representations will also be described.

In Section 2.3 we will define the concepts of geometric similarity and geometric mirror-similarity. Subsequently these concepts will be formulated as relations between pairs of contour representations.

The concepts of geometric mirror-symmetry, $\boldsymbol{n}$-fold geometric rotational symmetry and $\boldsymbol{n}$-fold geometric compositional symmetry will be defined in Section 2.4. These concepts will then be formulated as special properties of contour representations.

In Section 2.5 we will review the results of this chapter.

The most direct and flexible description of a curve $\gamma$ in the plane is a parametric representation

$$x = x(t), \quad y = y(t), \quad t \in [a, b], \tag{2.1.1}$$

where $x(t)$ and $y(t)$ are real-valued continuous functions of the real parameter $t$. Since both $x(t)$ and $y(t)$ belong to *the set of real numbers* $\mathbb{R}$, a point on the curve is represented as an element of $\mathbb{R} \times \mathbb{R}$ or, equivalently, of $\mathbb{R}^2$. The values of the parameter $t$ serve to distinguish different points on the curve $\gamma$, i.e. for each value of $t \in [a, b]$ there exists one and only one point $(x(t), y(t))$ on $\gamma$. In this way $\gamma$ is defined as the image of a continuous mapping of the interval $[a, b]$ onto the curve in the plane. The points on the curve are ordered according to increasing values of $t$.

In view of the parametric representation of a curve, as defined in Eq. 2.1.1, a curve can be represented in an illustrative way by a vector function in $\mathbb{R}^2$

$$\boldsymbol{z}(t) = x(t)\boldsymbol{u}_x + y(t)\boldsymbol{u}_y, \quad t \in [a, b], \tag{2.1.2}$$

where

$$\boldsymbol{u}_x = (1, 0), \tag{2.1.3a}$$

$$\boldsymbol{u}_y = (0, 1) \tag{2.1.3b}$$

is a pair of orthogonal unit vectors that spans $\mathbb{R}^2$. We will call $\boldsymbol{z} = \boldsymbol{z}(t)$ the *position vector function* of a curve $\gamma$. With this representation $\gamma$ is the locus of the endpoints of the vectors $\boldsymbol{z}(t)$ as the parameter $t$ traces out the interval $[a, b]$.

In many situations it is convenient to identify each point $(x(t), y(t)) \in \mathbb{R}^2$ with $z(t) = x(t) + \mathrm{i}y(t) \in \mathbb{C}$, *the set of complex numbers* (Yaglom [1968]). In that case a curve $\gamma$ is represented in the complex plane by the equation

$$z = z(t) = x(t) + \mathrm{i}y(t), \quad t \in [a, b]. \tag{2.1.4}$$

We call $z = z(t)$ the *position function* of a curve $\gamma$.

It will depend upon the particular situation whether we represent $\gamma$ by a real-valued two-dimensional vector function or by a complex-valued function. A vector formulation gives good insight into the nature of some other contour representations, that will be defined in the next section, while the formulation of some of these contour representations as complex-valued functions provides more analytical convenience. Scalar contour representations are of course formulated as real-valued functions.

The continuity of the mapping from the $t$-interval onto $\gamma$ guarantees that points that are 'close' in the $t$-interval are also 'close' on $\gamma$. As we remarked before, the points on $\gamma$ are ordered because the values of the parameter $t$ are ordered in the interval $[a, b]$. This enables us to define a *sense* on $\gamma$, thus making $\gamma$ an oriented curve. It is customary to define the positive sense on the curve in the direction of increasing $t$.

The parametrization of a curve $\gamma$ can be accomplished in many ways. Any monotonic continuous function $\tau = \chi(t)$, $t \in [a, b]$, defines a parameter $\tau$ such that $x$ and $y$ are continuous functions of $\tau$ and that different values of $\tau$ correspond to different points on $\gamma$. If $\tau$ is a monotonic increasing function of $t$, then $\tau \in [\chi(a), \chi(b)]$ and the sense on $\gamma$ will be preserved. If, however, $\tau$ is a monotonic decreasing function of $t$, then $\tau \in [\chi(b), \chi(a)]$ and the sense on $\gamma$ will be reversed. The change in parameter is reversible *iff* (if and only if) $\chi(t)$ is either strictly increasing or strictly decreasing (Ahlfors [1953]).

A curve $\gamma$ is said to be a *simple curve* or a *Jordan curve* if it does not intersect itself

$$z(t_1) = z(t_2) \quad \text{iff} \quad t_1 = t_2, \qquad t_1, t_2 \in [a, b]. \tag{2.1.5}$$

A curve $\gamma$ is closed if the initial point $a$ and the terminal point $b$ of the parameter interval are mapped onto the same point of $\gamma$

$$z(a) = z(b). \tag{2.1.6}$$

**Definition 2.1.** *Simple closed curve.*
A curve $\gamma$ is a simple closed curve iff there exists a continuous mapping $z$ of the parameter interval $[a, b]$ onto $\gamma$ that satisfies both Eq. 2.1.5 and Eq. 2.1.6. □

The class of simple closed curves will be indicated by $\boldsymbol{\Gamma}$. Usually the sense or positive orientation of a curve $\gamma \in \boldsymbol{\Gamma}$ is chosen to be counterclockwise, i.e. if we travel along the curve in the direction of increasing $t$, then the *interior of* $\gamma$, i.e. the region bounded by $\gamma$, will be on the left of $\gamma$.

Within $\boldsymbol{\Gamma}$ we can discriminate several subclasses of simple closed curves. First we define the class $\boldsymbol{\Gamma}_s$ of *smooth simple closed curves.*

**Definition 2.2.** *Smooth simple closed curve.*
A curve $\gamma$ is a smooth simple closed curve iff:

- $\gamma \in \boldsymbol{\Gamma}$, (2.1.7a)
- $\dot{z}$ exists and is absolutely continuous everywhere in the interval $[a, b]$, i.e. $\dot{z} \in \mathbf{AC}[a, b]$ (cf. Appendix A), (2.1.7b)
- $\dot{z} \neq 0$ everywhere in the interval $[a, b]$, (2.1.7c)
- $\dot{z}(a) = \dot{z}(b)$. (2.1.7d)

□

**Remark.**
Instead of the requirement in Eq. 2.1.7b, mere continuity of $\dot{z}$ in the interval $[a, b]$ is sufficient in the usual definition of a smooth simple closed curve. We require $\dot{z}$ of a smooth simple closed curve to be absolutely continuous to ensure that the curve can be reconstructed from the second derivative of its position function $z$, as we will show in the next section.

□

In the above

$$\dot{z} = \dot{z}(t) = \frac{\mathrm{d}z(t)}{\mathrm{d}t} \tag{2.1.8}$$

denotes differentiation of the curve representation $z$ with respect to its parameter $t$. If no confusion can arise we will in general delete the parameter. By Eq. 2.1.7c it is guaranteed that the mapping from the $t$-interval into the plane is locally topological, i.e. the mapping sets up a point-to-point correspondence that is continuous in both directions. For simple closed curves that satisfy the condition in Eq. 2.1.7c the mapping is even globally topological: the curve is the topological equivalent of the interval $[a, b]$ (Stoker [1969]). An even more restricted class of curves is the class $\boldsymbol{\Gamma}_r$ of *regular simple closed curves.*

**Definition 2.3.** *Regular simple closed curve.*
A curve $\gamma$ is a regular simple closed curve iff:

- $\gamma \in \boldsymbol{\Gamma}_s$, (2.1.9a)

- $\ddot{z}$ exists and is continuous everywhere in the interval $[a, b]$, (2.1.9b)

- $\ddot{z}(a) = \ddot{z}(b)$. (2.1.9c)

□

In Definition 2.3 $\ddot{z} = \ddot{z}(t)$ stands for the second derivative of $z$ with respect to its parameter $t$. The boundary conditions, as expressed by Eqs. 2.1.7d and 2.1.9c, are enforced in order to ensure that the properties of $\gamma$ will not depend upon a particular choice on the curve of an image $z(a)$ of the initial point of the parameter interval, $a$. In mechanical engineering and in computer graphics, where the modelling of curves by splines is rather popular, regular curves are frequently encountered since cubic splines belong to this class (De Boor [1978]).

The class $\boldsymbol{\Gamma}_{ps}$ of *piecewise smooth simple closed curves* consists of curves that satisfy Eqs. 2.1.7a, c and that satisfy Eq. 2.1.7b, except for a finite number of points in $[a, b]$. In an analogous way the class $\boldsymbol{\Gamma}_{pr}$ of *piecewise regular closed curves* is defined. The class of *simple closed polygons* is an important subset of $\boldsymbol{\Gamma}_{pr}$, since it provides for a mathematically tractable approximation to curves encountered in practice. We remark that our notion of a piecewise smooth simple closed curve is more restrictive than the usual concept of a *simple closed contour* in the mathematical literature (cf. e.g. Ahlfors [1953], Churchill, Brown and Verhey [1974]). We will use the term contour to indicate the outer boundary of a two-dimensional object and assume that its mathematical properties correspond to those of piecewise smooth simple closed curves. The results we derive, however, are often valid for a wider class of simple closed curves.

From the definitions in the foregoing the following inclusion relations follow immediately:

$$\boldsymbol{\Gamma}_{\mathrm{r}} \subset \boldsymbol{\Gamma}_{\mathrm{s}} \subset \boldsymbol{\Gamma}_{\mathrm{ps}} \subset \boldsymbol{\Gamma} \tag{2.1.10}$$

and

$$\boldsymbol{\Gamma}_{\mathrm{r}} \subset \boldsymbol{\Gamma}_{\mathrm{pr}} \subset \boldsymbol{\Gamma}_{\mathrm{ps}} \subset \boldsymbol{\Gamma}. \tag{2.1.11}$$

The *arc length* of a curve $\gamma$, parametrized by $t$ on the interval $[a, b]$, is defined as the integral

$$s(a, b) = \int_a^b |\dot{z}| \mathrm{d}t = \int_a^b [\dot{x}^2 + \dot{y}^2]^{1/2} \mathrm{d}t. \tag{2.1.12}$$

The class of curves for which this integral exists forms the class of *rectifiable curves* (Courant and John [1965]). The class $\boldsymbol{\Gamma}_{\mathrm{ps}}$ is a subset of the class of rectifiable curves: the arc length or *perimeter* of a curve $\gamma \in \boldsymbol{\Gamma}_{\mathrm{ps}}$ is the sum of the arc lengths of the smooth arcs of $\gamma$. The arc length of a curve is independent of a particular parameter representation of that curve. Further, the arc length of a curve is invariant under a rigid motion of the curve in the plane (Stoker [1969]). By a *rigid motion* in the plane we mean a combination of a rotation and a translation.

By applying the fundamental theorem of calculus to

$$s(a, t) = \int_a^t [\dot{x}^2(\tau) + \dot{y}^2(\tau)]^{1/2} \mathrm{d}\tau \tag{2.1.13}$$

we obtain an expression for the element of arc length $\mathrm{d}s$

$$\mathrm{d}s = [\dot{x}^2(t) + \dot{y}^2(t)]^{1/2} \mathrm{d}t. \tag{2.1.14}$$

If we interpret the parameter $t$ as time, then

$$\dot{s} = \frac{\mathrm{d}s}{\mathrm{d}t} = [\dot{x}^2(t) + \dot{y}^2(t)]^{1/2} \tag{2.1.15}$$

expresses the *speed of motion* along the curve. From Eq. 2.1.15 we see that the condition in Eq. 2.1.7c simply states that the speed of motion along the curve as a function of the parameter $t$ shall nowhere be equal to zero. If the arc length $s$ is used to parametrize a curve, then

Eq. 2.1.15 becomes

$$\dot{s} = \left[\dot{x}^2(s) + \dot{y}^2(s)\right]^{1/2} = 1 \qquad (2.1.16)$$

everywhere along the curve. Due to the invariance properties of arc length, which we have just mentioned, it constitutes the natural parameter of a curve.

In practice we want to define interpretable measures for similarity or dissimilarity between shapes for which the computational efforts of matching pairs of shapes, by means of their parametric representations, can be kept within acceptable limits. Therefore we consider it appropriate, though not in all situations ideal, to require that the speed of motion along a curve, Eq. 2.1.15, is constant, i.e. that $\dot{s}$ is independent of $t$.

Almost all of the parametric representations for simple closed curves, that will be introduced in this and the next section, are periodic functions. In view of the Fourier expansion of such representations, it is convenient to choose for the fundamental parameter interval $[a, b]$ an interval of length $2\pi$. If the parametric representation of a curve is essentially periodic, then the representation is defined, for any real value of the parameter, as the periodic extension of the representation on $[0, 2\pi]$. For example, $z(t) = z(t + k \cdot 2\pi)$ for any value of $t \in [0, 2\pi]$ and for any $k \in \mathbb{Z}$, where $\mathbb{Z}$ indicates the *set of integer numbers*, cf. Figure 2.1.

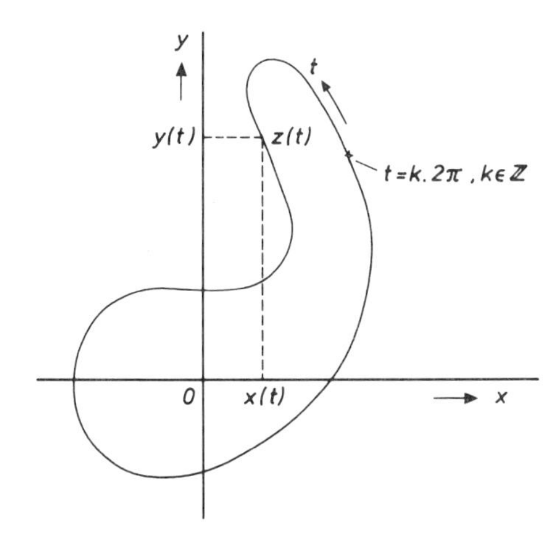


**Figure 2.1.** A contour represented by the position function $z = z(t) = x(t) + iy(t)$. Note that the contour has counterclockwise positive sense and that $z(t)$ has a fundamental parameter interval of length $2\pi$.

Given the requirements of a constant speed parametrization and choosing an arbitrary parameter interval of length $2\pi$, we find for the speed of motion along a rectifiable curve:

$$\dot{s} = \frac{L}{2\pi}, \tag{2.1.17a}$$

where[1]

$$L = \int_{2\pi} [\dot{x}^2(t) + \dot{y}^2(t)]^{1/2} \mathrm{d}t. \tag{2.1.17b}$$

For simple closed curves, $L$ is the perimeter of the curve. For a curve with a constant speed parametrization, there exists a linear relation between arc length $s$ and parameter $t$. The parameter $t$ is in fact a *normalized arc length parameter*. With respect to the analysis of the shape of object contours $\gamma$ we assume that such contours meet the following conditions:

- $\gamma \in \boldsymbol{\Gamma}_{\mathrm{ps}}$, (2.1.18a)
- a parametric representation of $\gamma$ satisfies the condition of a constant speed parametrization, (2.1.18b)
- a parametric representation is $2\pi$-periodic or, if arc length is used as a parameter, $L$-periodic. (2.1.18c)

The position function $z$ has frequently been used for shape representation, most times in the context of the Fourier expansion of $z$. One of the earliest references is Granlund [1972], who used $z$ to represent character outlines. To give some impression of the application of the position function in shape analysis and representation we further mention Richard and Hemami [1974] and Wallace and Wintz [1980] (airplane silhouettes), Young, Walker and Bowie [1974], Sychra et al. [1976], Chen and Shi [1980] and Proffitt [1982] (cell boundaries), Tai, Li and Chiang [1982] (particle analysis) and Giardina and Kuhl [1977], Burkhardt [1979], Kuhl and Giardina [1982], and Crimmins [1982] (general shape analysis applications).

[1] Throughout this book we use the notation $\int_{2\pi}$ to indicate integration over an interval of length $2\pi$.

As we remarked earlier, $z$ is the direct representation of a planar curve. This means in image analysis that $z$ can be measured directly in the image plane and in computer graphics that the curve can be generated directly from $z$. Therefore $z$ is also frequently encountered in studies on contour estimation and approximation. The position function $z$ can also be used as a basis for the computation of shape features, as will become clear in Chapters 3 and 4 and in Appendix B.

We continue this discussion of parametric contour representation for shape analysis purposes with representations that specify the distance between the contour and an appropriate reference position for the contour, $\zeta$. For example, $\zeta$ can be the centroid of the region enclosed by the contour or it can be the contour average of $z$. A discussion on appropriate translation normalization parameters will be presented in Section 4.3.

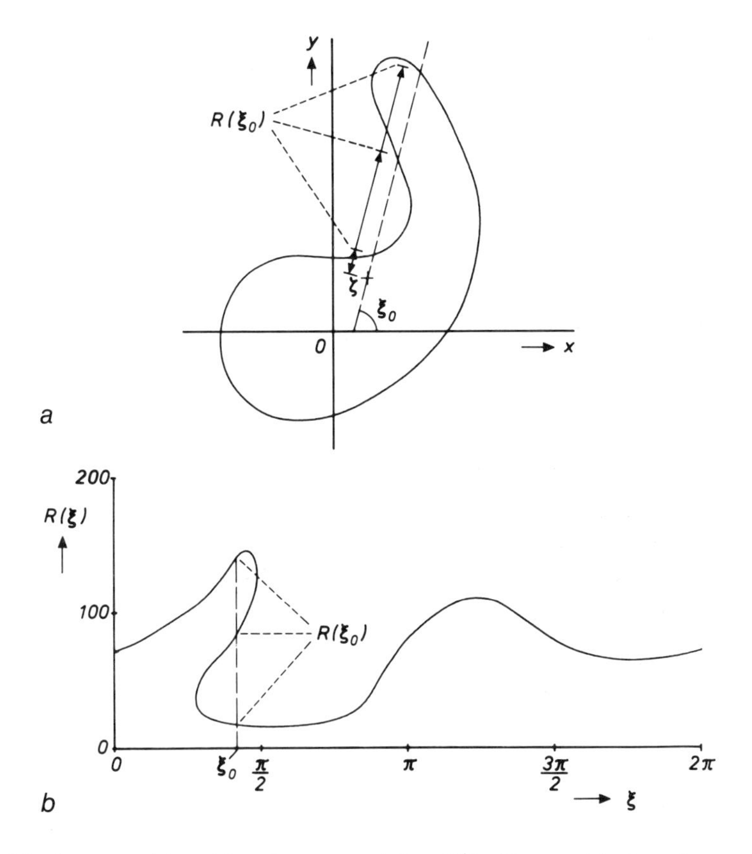


**Figure 2.2.** A contour ($a$) and its polar representation $R(\xi)$ with respect to the reference position $\zeta$ ($b$). Note that $R(\xi)$ is a multiple valued function.

In the following discussion of contour representations that all specify the distance between the contour and a reference position in the plane, we will show that these contour representations have undesirable properties or that they do not provide any advantage over the position function $z$.

The polar representation of a contour, $R(\xi)$, is likely to be one of the earliest propositions for shape representation. According to Rosenfeld it had already been proposed for that purpose in the 1950s (Barrow and Popplestone [1971]). $R(\xi)$ specifies the distance between the contour and reference position $\zeta$ as a function of the angle of revolution $\xi$. See Figure 2.2 for an illustration of $R(\xi)$.

Though $R(\xi)$ constitutes an information-preserving contour representation it has the serious drawback that, for non-holomorphic shapes, it is not a single-valued representation. This is illustrated in Figure 2.2. Despite this fact, $R(\xi)$ has been used frequently to represent the shape of fine particles (e.g. Schwarcz and Shane [1969], Ehrlich and Weinberg [1970], Beddow and Philip [1975], Meloy [1977a], Meloy [1977b] and Beddow et al. [1977], Luerkens, Beddow and Vetter [1982a]). $R(\xi)$ has also been used for chromosome analysis (Rutovitz [1970]) and in the context of robot vision (Kammenos [1978]). Recently, some propositions have been published to transform $R(\xi)$ into a single-valued representation (Gotoh [1979], Luerkens, Beddow and Vetter [1982b]). We will return to the latter propositions shortly.

Another contour representation that specifies the distance between the contour and a reference position $\zeta$ is the *radial distance function* $r$, which is closely related to $z$ and defined as

$$r(t) = |z(t) - \zeta|, \quad t \in [0, 2\pi]. \tag{2.1.19}$$

See Figure 2.3 for an illustration of $r$.

Though the radial distance function $r$ was already proposed for shape representation by Searle [1970], not many references reporting its use for that purpose can be found. The real-valued radial distance function $r$ is a single-valued function of the parameter $t$, as opposed to the polar representation $R(\xi)$. However, $r$ is not an information-preserving contour representation, a property that $R(\xi)$ possesses. Since phase information is not present in $r$, a contour cannot be reconstructed from it. The most important drawback of $r$ with respect to shape representation

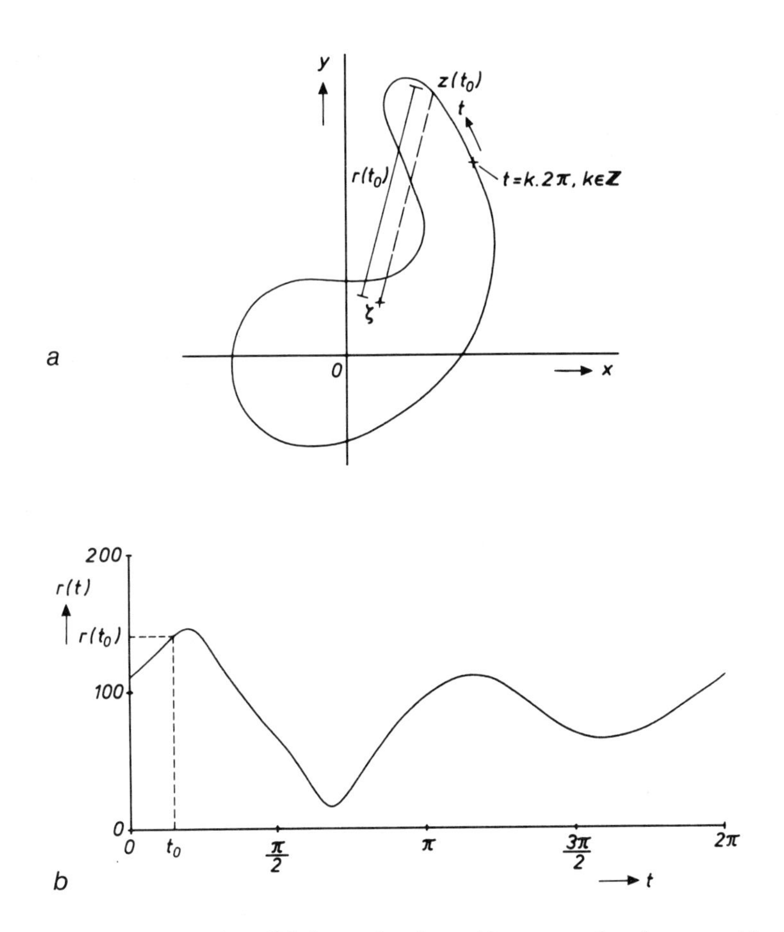


**Figure 2.3.** A contour ($a$) and its radial distance function $r$ with respect to the reference position $\zeta$($b$).

is the fact that there exist contours with different shapes that have identical radial distance functions. An example of such an ambiguity is shown in Figure 2.4.

This ambiguity makes $r$ a less suitable candidate for shape representation. Freeman [1978a] proposes to avoid such ambiguities by providing the radial distance function with a sign. The *signed radial distance function* $r'$ has negative sign if the angle of revolution $\xi$ changes in clockwise direction upon tracing the contour in counterclockwise direction and positive sign otherwise. Formally $r'$ is defined as

$$r'(t) = \operatorname{sgn}(\dot{\xi}(t)) \cdot r(t), \tag{2.1.20}$$

where the *sign function* sgn (·) is defined as

$$\operatorname{sgn}(x) = \begin{cases} +1, & x \geqslant 0, & (2.1.21a) \\ -1, & x < 0. & (2.1.21b) \end{cases}$$

If the parameter $t$ is a constant speed parameter, i.e. Eq. 2.1.17a is satisfied, then it can be shown that $r'(t)$ is a contour representation that preserves shape information.

**Theorem 2.1.**
Let $r'(t)$, $0 \leqslant t \leqslant 2\pi$, be the signed radial distance function of an arbitrary contour, where $t$ is a constant speed parameter. Then the position function of that contour can be reconstructed from $r'(t)$ up to a rigid motion in the plane.

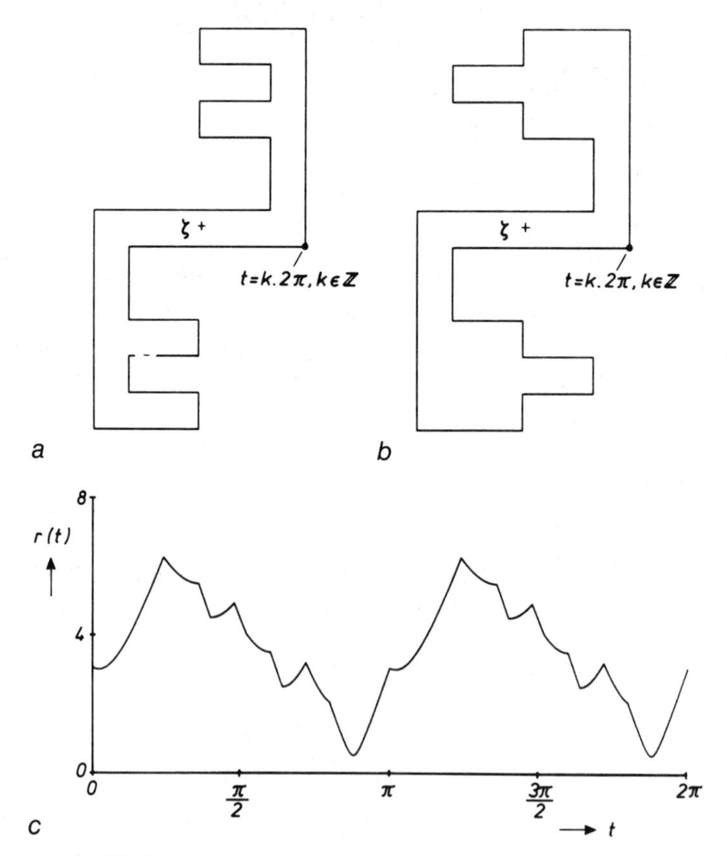


**Figure 2.4.** (*a*) and (*b*) display two contours of different shape which have identical radial distance functions $r(t)$, shown in (*c*).

*Proof*

We denote the angle of revolution as a function of the normalized arc length parameter $t$, as $\xi(t)$. Then we can write, cf. Eq. 2.1.19

$$\begin{aligned} z(t) - \zeta &= |z(t) - \zeta| e^{i\xi(t)} \\ &= r(t) e^{i\xi(t)}. \end{aligned} \tag{2.1.22}$$

Taking derivatives with respect to $t$ on both sides yields

$$\dot{z}(t) = \dot{r}(t) e^{i\xi(t)} + i r(t) e^{i\xi(t)} \cdot \dot{\xi}(t). \tag{2.1.23}$$

Employing the constant speed property of $t$, Eqs. 2.1.15 and 2.1.17a, we find

$$\dot{z}(t) \cdot \bar{\dot{z}}(t) = \dot{r}^2(t) + r^2(t)\dot{\xi}^2(t) = \left(\frac{L}{2\pi}\right)^2. \tag{2.1.24}$$

From this equation we can solve $\dot{\xi}(t)$

$$\dot{\xi}(t) = \pm\, r^{-1}(t) \cdot \sqrt{\left(\frac{L}{2\pi}\right)^2 - \dot{r}^2(t)}\,. \tag{2.1.25}$$

Eq. 2.1.25 shows that when $r(t) = 0$, which happens when the contour passes through the reference position $\zeta$, $\dot{\xi}(t)$ is undefined. Further we remark that, due to the continuity of $z(t)$ and to the constant speed property of $t$, the absolute maximum of $\dot{r}^2(t)$ is $(L/2\pi)^2$.

From Eq. 2.1.20 it follows immediately that

$$r(t) = |r'(t)|\,. \tag{2.1.26}$$

It follows from Eqs. 2.1.25 and 2.1.26 that, given $r'(t)$, we can find $\dot{\xi}(t)$. Assuming an initial angle of revolution $\xi(0)$, we can find $\xi(t)$ through integration, i.e.

$$\xi(t) = \xi(0) + \int_0^t \dot{\xi}(\tau)\mathrm{d}\tau. \tag{2.1.27}$$

If we choose a reference position $\zeta$ in the plane, Eq. 2.1.22 shows that we obtain $z(t)$ from $r(t)$, $\xi(t)$ and $\zeta$. This completes the proof of the theorem. □

Despite its preservation of shape information, $r'(t)$ has an undesirable property in relation to shape representation. We illustrate this in Figure 2.5.

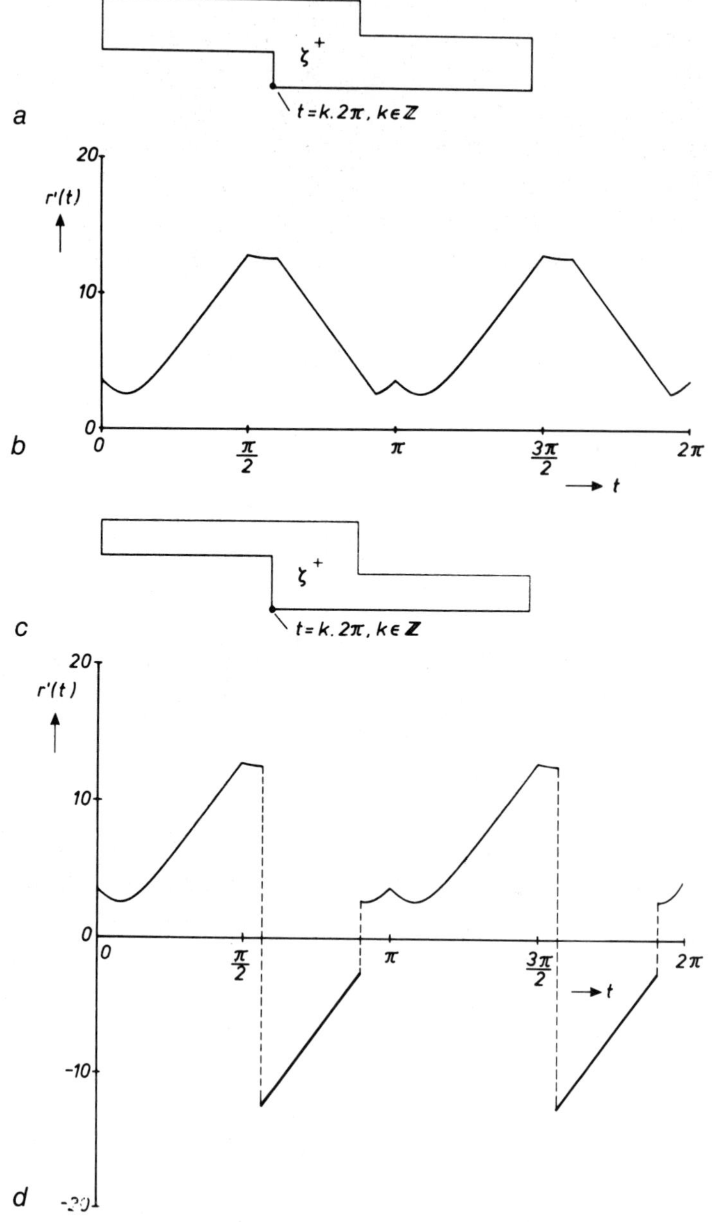


**Figure 2.5.** ($a$) and ($c$) display two rather similar contours which have considerably different signed radial distance functions $r'(t)$, shown in ($b$) and ($d$), respectively.

Figure 2.5 shows that two rather similar contours can have considerably different signed radial distance functions. This fact makes it unlikely that useful similarity measures can be formulated on the basis of the signed radial distance function $r'(t)$.

We now return to the propositions by Gotoh [1979] and by Luerkens, Beddow and Vetter [1982b] to transform the polar representation $R(\xi)$ into a single-valued representation.

The representation proposed by Gotoh [1979] is in fact similar to the function $r'(t)$. Gotoh proposes a *signed polar representation* $R'(\xi')$, where $\xi'$ is the accumulated angle of revolution as the contour is traced. As a result $R'(\xi')$ does not have a fixed period. The sign of $R'(\xi')$ is also determined by $\dot{\xi}(t)$. Formally $R'(\xi')$ is defined as

$$R'(\xi'(t)) = \operatorname{sgn}(\dot{\xi}(t)) \cdot R(\xi(t)), \tag{2.1.28}$$

where

$$\xi'(t) = \int_0^t \dot{\xi}(\tau)\mathrm{d}\tau. \tag{2.1.29}$$

It is easily verified that $R'(\xi'(t))$ can be obtained from $r'(t)$ by a parameter transformation and vice versa. $R'(\xi')$ is a shape information-preserving contour representation, but it has the same undesirable property as we observed in $r'(t)$, which makes it less suited for shape representation.

Finally, Luerkens, Beddow and Vetter [1982b] proposed to use the pair $(r(t), \dot{\xi}(t))$ for shape representation, instead of the multiple-valued polar representation $R(\xi)$. Indeed, the pair $(r(t), \dot{\xi}(t))$ constitutes a shape information-preserving contour representation, as can be verified from Eqs. 2.1.22 and 2.1.27. Further, it does not suffer from the disadvantages that we observed in $r'(t)$ and $R'(\xi')$. This shows that, in combination with appropriate additional information, the radial distance function $r(t)$ can be useful for shape representation. However, the advantages of such a representation over the position function $z$ are not clear. Therefore we will not consider shape representations based on the distance between the contour and an appropriate reference position $\zeta$ any further.

## 2.2 Tangent, normal and curvature

In this section we introduce representations for the shape of a contour, based on the derivatives of the position function $z$ or, equivalently, on the derivatives of the position vector function $\boldsymbol{z}$. It will be shown that these representations are information-preserving, i.e. that they allow full reconstruction of the shape of the contour, though information about the position and/or the orientation of the contour may be lost. However, absolute position and orientation are frequently unimportant features for the shape analysis problem. If necessary, such features can be estimated from the position function of a contour. We will use vector- and complex-valued representations alternatingly.

The *tangent vector function* $\dot{\boldsymbol{z}}$ of a contour $\gamma$, with position vector function $\boldsymbol{z}$, is defined as

$$\dot{\boldsymbol{z}}(t) = \frac{\mathrm{d}\boldsymbol{z}(t)}{\mathrm{d}t} = \big(\dot{x}(t), \dot{y}(t)\big) = \dot{x}(t)\boldsymbol{u}_x + \dot{y}(t)\boldsymbol{u}_y. \tag{2.2.1}$$

The corresponding complex-valued *tangent function* $\dot{z}$ is defined as

$$\dot{z}(t) = \frac{\mathrm{d}z(t)}{\mathrm{d}t} = \dot{x}(t) + \mathrm{i}\dot{y}(t). \tag{2.2.2}$$

Combining Eq. 2.1.15 and Eq. 2.2.1, we observe that

$$\left|\dot{\boldsymbol{z}}(t)\right| = \dot{s}(t), \tag{2.2.3}$$

i.e. if we interpret $t$ as time, then the length of $\dot{\boldsymbol{z}}$ expresses the speed of motion along the curve. Using the arc length $s$ as a parameter, we found in Eq. 2.1.16 that the speed of motion, and thus the length of the tangent vector, is always one. The unit tangent vector function $\boldsymbol{p}(s)$ and the tangent vector function $\dot{\boldsymbol{z}}(t)$ are related as

$$\boldsymbol{p}(s) \equiv \dot{\boldsymbol{z}}(s) = \dot{\boldsymbol{z}}(t)\,\frac{\mathrm{d}t}{\mathrm{d}s}, \tag{2.2.4}$$

using the chain rule of differentiation with $t = t(s)$.

If the parameter $t$ is a constant speed parameter, or, equivalently, a normalized arc length parameter (cf. Eq. 2.1.17a), then $\boldsymbol{p}(s)$ and $\dot{\boldsymbol{z}}(t)$

are related as

$$p(s) = \frac{2\pi}{L}\dot{z}(t). \qquad (2.2.5)$$

The *tangent angle function* $\theta$ of a contour gives, for each value of the parameter, the angle of inclination of the tangent vector $\dot{z}$ with the positive $x$-axis, as illustrated in Figure 2.6.

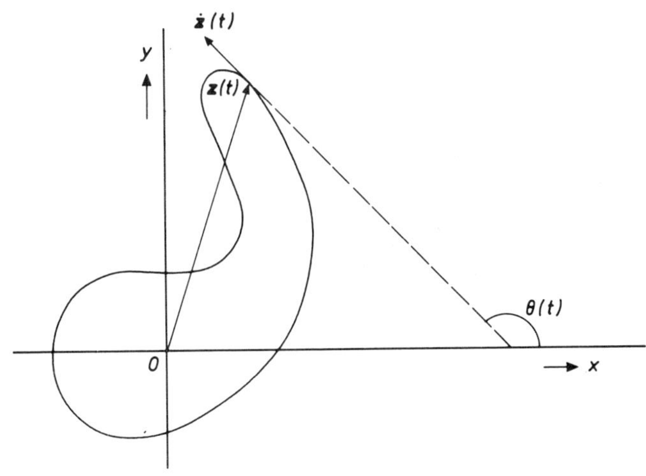


**Figure 2.6.** Illustration of the tangent vector $\dot{z}(t)$ and the tangent angle $\theta(t)$ at the position vector $z(t)$ on the contour.

Clearly, the tangent angle $\theta(t)$ is the argument of the tangent vector $\dot{z}(t)$ and can be found from Eq. 2.2.1 as

$$\theta(t) = \arctan\left(\frac{\dot{y}(t)}{\dot{x}(t)}\right) \qquad (2.2.6)$$

and can be solved, without ambiguity, from the signs of $\dot{x}(t)$ and $\dot{y}(t)$ or, equivalently, from the direction cosines of the tangent vector $\dot{z}(t)$

$$\cos\theta(t) = \frac{\dot{x}(t)}{[\dot{x}^2(t) + \dot{y}^2(t)]^{1/2}}, \qquad (2.2.7a)$$

$$\sin\theta(t) = \frac{\dot{y}(t)}{[\dot{x}^2(t) + \dot{y}^2(t)]^{1/2}}. \qquad (2.2.7b)$$

Combining Eqs. 2.1.15, 2.2.1 and 2.1.7a,b we can rewrite $\dot{\boldsymbol{z}}$ as

$$\dot{\boldsymbol{z}}(t) = \frac{\mathrm{d}s}{\mathrm{d}t} \cdot \left[\cos\theta(t)\boldsymbol{u}_x + \sin\theta(t)\boldsymbol{u}_y\right] \tag{2.2.8}$$

or similarly, by combining Eqs. 2.1.15, 2.2.2 and 2.1.7a,b, we find for $\dot{z}$ the expression

$$\dot{z}(t) = \frac{\mathrm{d}s}{\mathrm{d}t}\,\mathrm{e}^{\mathrm{i}\theta(t)}. \tag{2.2.9}$$

These two equations again show clearly that $\dot{\boldsymbol{z}}$, or $\dot{z}$, expresses the velocity along the curve. Interpreting the parameter $t$ again as time, the second derivative of the position (vector) function expresses the acceleration along the contour. For this reason we call

$$\ddot{\boldsymbol{z}}(t) = \frac{\mathrm{d}^2\boldsymbol{z}(t)}{\mathrm{d}t^2} = \left(\ddot{x}(t), \ddot{y}(t)\right) = \ddot{x}(t)\boldsymbol{u}_x + \ddot{y}(t)\boldsymbol{u}_y \tag{2.2.10}$$

the *acceleration vector function* and

$$\ddot{z}(t) = \frac{\mathrm{d}^2 z(t)}{\mathrm{d}t^2} = \ddot{x}(t) + \mathrm{i}\ddot{y}(t) \tag{2.2.11}$$

the *acceleration function*. These functions express the rate of change of the velocity along the curve as a function of $t$. When the arc length $s$ is used as the parameter, the second derivative of $\boldsymbol{z}$ with respect to arc length is called, for reasons that will soon become clear, the *curvature vector function* $\boldsymbol{k}(s)$

$$\boldsymbol{k}(s) \equiv \ddot{\boldsymbol{z}}(s) = \frac{\mathrm{d}\boldsymbol{p}(s)}{\mathrm{d}s}. \tag{2.2.12}$$

The relation between $\boldsymbol{k}(s)$ and $\ddot{\boldsymbol{z}}(t)$ can be found through the chain rule of differentiation

$$\boldsymbol{k}(s) = \ddot{\boldsymbol{z}}(t)\left(\frac{\mathrm{d}t}{\mathrm{d}s}\right)^2 + \dot{\boldsymbol{z}}(t)\,\frac{\mathrm{d}^2 t}{\mathrm{d}s^2}, \tag{2.2.13}$$

with $t = t(s)$. In Section 2.1, we discussed the conditions for the invertibility of the functional relationship between the parameters $s$ and $t$, which we assume to be valid. This means that we may write

$$\left(\frac{dt}{ds}\right) = \left(\frac{ds}{dt}\right)^{-1} \tag{2.2.14}$$

and (cf. Abramowitz and Stegun [1972], p. 11)

$$\frac{d^2t}{ds^2} = -\frac{d^2s}{dt^2}\left(\frac{ds}{dt}\right)^{-3}. \tag{2.2.15}$$

With these two relations and the definition of the unit tangent vector function $\boldsymbol{p}(s)$ in Eq. 2.2.4, we can rewrite Eq. 2.2.13 as

$$\ddot{\boldsymbol{z}}(t) = \left(\frac{ds}{dt}\right)^2 \boldsymbol{k}(s) + \frac{d^2s}{dt^2}\boldsymbol{p}(s). \tag{2.2.16}$$

To explain the meaning of this equation we reformulate Eq. 2.1.16 as

$$|\dot{\boldsymbol{z}}(s)|^2 = \dot{\boldsymbol{z}}(s) \cdot \dot{\boldsymbol{z}}(s) = 1 \tag{2.2.17}$$

and differentiate to obtain

$$\dot{\boldsymbol{z}}(s) \cdot \ddot{\boldsymbol{z}}(s) \equiv \boldsymbol{p}(s) \cdot \boldsymbol{k}(s) = 0. \tag{2.2.18}$$

Interpreting this result in the context of Eq. 2.2.16 we find that, at any point along the contour, $\ddot{\boldsymbol{z}}(t)$ can be decomposed into two components that are perpendicular to each other: a tangential component of length $d^2s/dt^2$ and a normal component. If we use arc length as a parameter in Eq. 2.2.8 we obtain an expression for $\boldsymbol{p}(s)$, cf. Eq. 2.2.4. We substitute this expression into Eq. 2.2.12 to obtain

$$\begin{aligned}\boldsymbol{k}(s) &= \frac{d}{ds}[\cos\theta(s)\boldsymbol{u}_x + \sin\theta(s)\boldsymbol{u}_y] \\ &= [-\sin\theta(s)\boldsymbol{u}_x + \cos\theta(s)\boldsymbol{u}_y]\cdot\frac{d\theta(s)}{ds} \\ &= \boldsymbol{n}(s)\frac{d\theta(s)}{ds}.\end{aligned} \tag{2.2.19}$$

In Eq. 2.2.19 $\boldsymbol{n}(s)$ is a unit vector perpendicular to $\boldsymbol{p}(s)$. If the sense of $\boldsymbol{n}(s)$ is kept the same everywhere along the curve, then it may be chosen arbitrarily. Usually $\boldsymbol{n}(s)$ is obtained by a rotation of $\boldsymbol{p}(s)$ over $+\pi/2$ radians, which can also be observed by comparing Eqs. 2.2.8 and 2.2.19. At any point along a contour $\gamma$, with position vector function $\boldsymbol{z}(t)$ and arbitrary parameter $t$, the pair of orthogonal unit vectors $\boldsymbol{p}(t)$ and $\boldsymbol{n}(t)$ can be expressed in terms of the tangent angle function $\theta(t)$ as

$$\boldsymbol{p}(t) = \cos\theta(t)\boldsymbol{u}_x + \sin\theta(t)\boldsymbol{u}_y, \tag{2.2.20a}$$

$$\boldsymbol{n}(t) = -\sin\theta(t)\boldsymbol{u}_x + \cos\theta(t)\boldsymbol{u}_y, \tag{2.2.20b}$$

under the condition that $\theta(t)$ is defined in that point. The pair of vectors $(\boldsymbol{p}, \boldsymbol{n})$ is usually called the moving frame of a contour (see Figure 2.7).

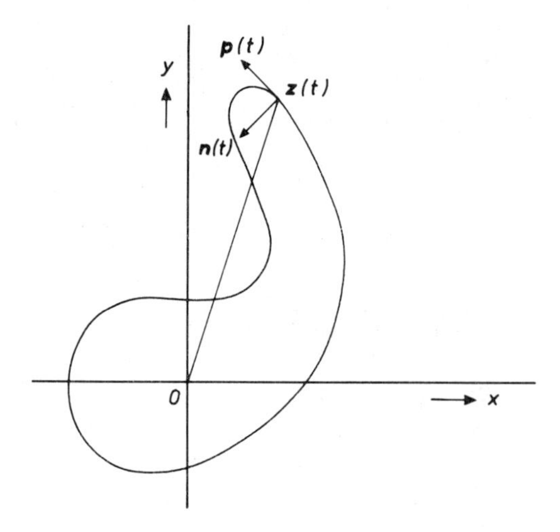


**Figure 2.7.** Illustration of the moving frame along a curve. This figure shows the moving window formed by the pair of orthogonal unit vectors $\{\boldsymbol{p}, \boldsymbol{n}\}$ at position $\boldsymbol{z}(t)$.

If $t$ is the normalized arc length parameter, then the tangential component in Eq. 2.2.16 is zero everywhere along the contour and $\ddot{\boldsymbol{z}}(t)$ can be expressed as

$$\ddot{\boldsymbol{z}}(t) = \left(\frac{L}{2\pi}\right)^2 \frac{\mathrm{d}\theta(s)}{\mathrm{d}s}\,\boldsymbol{n}(s). \tag{2.2.21}$$

The *curvature* $K$ at any point along the contour is defined as the rate of change in tangent direction of a contour, as a function of arc length. In Section 2.1 we remarked that, because of its invariance properties, arc length is the natural parameter of a curve. With this in mind we observe from the definition of curvature that curvature is also an invariant property of a contour, since it is, similar to arc length, invariant under a rigid motion of the curve and under a change in the choice of a parameter. Commonly, the curvature function $K$ is expressed in formula as

$$K(s) = \frac{d\theta(s)}{ds} = \frac{\dot{\theta}}{\dot{s}}. \tag{2.2.22}$$

With this definition of the curvature function $K$ we can rewrite the last expression in Eq. 2.2.19 to obtain

$$\boldsymbol{k}(s) = K(s)\boldsymbol{n}(s). \tag{2.2.23}$$

This expression explains why $\boldsymbol{k}$ is called the curvature vector function. Substitution of this expression for $\boldsymbol{k}$ into Eq. 2.2.16 yields

$$\ddot{\boldsymbol{z}}(t) = \left(\frac{ds}{dt}\right)^2 K(s)\boldsymbol{n}(s) + \frac{d^2s}{dt^2}\boldsymbol{p}(s). \tag{2.2.24}$$

When $t$ is the normalized arc length parameter this expression reduces to, cf. Eq. 2.2.21

$$\ddot{\boldsymbol{z}}(t) = \left(\frac{L}{2\pi}\right)^2 K(s)\boldsymbol{n}(s). \tag{2.2.25}$$

Furthermore, by substitution of Eqs. 2.1.15 and 2.2.6 into Eq. 2.2.2 we derive the following expression for the curvature along a contour, with arbitrary parameter $t$, as

$$K(t) = \frac{\dot{x}(t)\ddot{y}(t) - \ddot{x}(t)\dot{y}(t)}{\left[\dot{x}^2(t) + \dot{y}^2(t)\right]^{3/2}}. \tag{2.2.26}$$

Using this expression for the curvature function $K$, it is straightforward to find the influence of scaling upon $K$. We denote a pure *scaling* of a contour $\gamma$ by a positive real factor $\beta$ as $\mathcal{S}_\beta\gamma$. The position functions of

$\gamma$ and $\mathcal{S}_\beta \gamma$ are related as

$$\mathcal{S}_\beta z(t) = \beta \cdot z(t). \tag{2.2.27}$$

From the definitions of the tangent function $\dot{z}$, Eq. 2.2.2, and the acceleration function $\ddot{z}$, Eq. 2.2.11, we see that scale information is preserved in these representations. Therefore $\mathcal{S}_\beta \dot{z}$ and $\mathcal{S}_\beta \ddot{z}$ are related to $\dot{z}$ and $\ddot{z}$, respectively, in the same way as $\mathcal{S}_\beta z$ is related to $z$, Eq. 2.2.27

$$\mathcal{S}_\beta \dot{z}(t) = \beta \cdot \dot{z}(t), \tag{2.2.28a}$$

$$\mathcal{S}_\beta \ddot{z}(t) = \beta \cdot \ddot{z}(t). \tag{2.2.28b}$$

With the aid of Eq. 2.2.26 the curvature function $\mathcal{S}_\beta K$ of the scaled contour $\mathcal{S}_\beta \gamma$ can be written as

$$\begin{aligned}
\mathcal{S}_\beta K(t) &= \frac{(\mathcal{S}_\beta \dot{x}(t))(\mathcal{S}_\beta \ddot{y}(t)) - (\mathcal{S}_\beta \ddot{x}(t))(\mathcal{S}_\beta \dot{y}(t))}{\left[(\mathcal{S}_\beta \dot{x}(t))^2 + (\mathcal{S}_\beta \dot{y}(t))^2\right]^{3/2}} \\
&= \frac{1}{\beta} \cdot \frac{\dot{x}(t)\ddot{y}(t) - \ddot{x}(t)\dot{y}(t)}{\left[\dot{x}^2(t) + \dot{y}^2(t)\right]^{3/2}} \\
&= \frac{1}{\beta} K(t).
\end{aligned} \tag{2.2.29}$$

This derivation shows that a scaling of contour by a factor $\beta \in \mathbb{R}^+$ leads to a scaling of the curvature function by a factor $1/\beta$.

The definition of the curvature function in Eq. 2.2.22 still merits some discussion. From the definition of $\theta(t)$ in Eqs. 2.2.6 and 2.2.7a, b, it can be seen that the tangent angle function can only assume values in a range of length $2\pi$, usually in the interval $[-\pi, \pi]$ or $[0, 2\pi]$. Therefore $\theta(t)$ in general contains discontinuities of size $2\pi$. The *cumulative angular function* $\varphi(t)$, defined by Zahn and Roskies [1972] as the net amount of angular bend between the starting position $z(0)$ and position $z(t)$ on the curve, does not suffer from this problem. The functions $\varphi(t)$ and $\theta(t)$ are related as

$$\theta(t) = \left[\varphi(t) + \theta(0)\right] \bmod 2\pi. \tag{2.2.30}$$

See Figure 2.8 for an illustration.

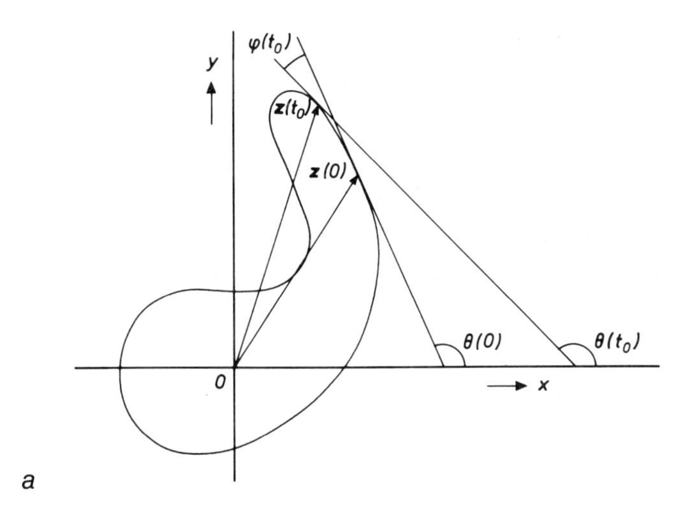


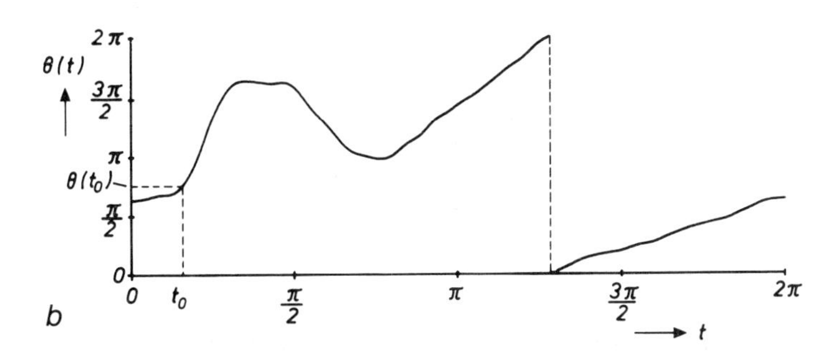


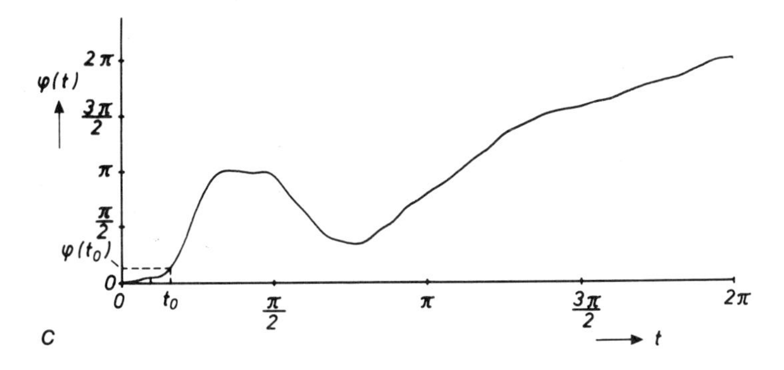


**Figure 2.8.** Illustration of the tangent angle function $\theta(t)$ $(b)$ and the cumulative angular function $\varphi(t)$ $(c)$ of a contour $(a)$.

From the definition of $\varphi(t)$ also follows $\varphi(0) \equiv 0$. For smooth simple closed curves $\gamma$, i.e. $\gamma \in \boldsymbol{\Gamma}_s$, $\varphi(t)$ is a continuous function. In relation to shape representation and analysis $\varphi(t)$ has one serious drawback: unlike $\theta(t)$, $\varphi(t)$ is not a periodic function. For simple closed curves $\varphi$ has the property

$$\varphi(t + 2\pi) = \varphi(t) + 2\pi. \tag{2.2.31}$$

In order to be able to perform Fourier analysis, Zahn and Roskies [1972] introduced a $2\pi$-periodic variant of the cumulative angular function. Using the normalized arc length parameter $t$, the *periodic cumulative angular function* $\psi(t)$ is defined as

$$\psi(t) = \varphi(t) - t. \tag{2.2.32}$$

When Eqs. 2.2.31 and 2.2.32 are combined the periodicity of $\psi$ follows:

$$\psi(t + 2\pi) = \psi(t) \tag{2.2.33}$$

for all values of $t$.

The curvature function $K$, defined earlier in Eq. 2.2.22, can now be redefined as

$$K = \frac{\mathrm{d}\varphi(s)}{\mathrm{d}s} = \frac{\dot{\varphi}}{\dot{s}}. \tag{2.2.34}$$

The inverse relationship also exists between $K$ and $\varphi$, i.e.

$$\varphi(s) = \int_0^s K(\sigma)\mathrm{d}\sigma, \quad \forall s \in \mathbb{R}, \tag{2.2.35}$$

where $\sigma$ also denotes arc length. Note that a similar inverse relationship between $K$ and $\theta$ in general does not exist.

For a variety of reasons, the curvature function, with arc length as a parameter, is very important in the study of shape. One is that this function completely determines a contour up to a rigid motion in the plane (Stoker [1969], Guggenheimer [1963]); the curvature function is the natural or intrinsic equation of a curve. The sign of $K$ determines the convex and concave parts of the curve: for $K > 0$ the curve is *convex*, for $K < 0$ the curve is *concave* and for $K = 0$ the curve is said to have a *point of inflection*.

The radius of curvature $\varrho$ at position $z(t)$ on a curve is defined as

$$\varrho(t) = \frac{1}{K(t)}. \tag{2.2.36}$$

Figure 2.9 illustrates the concept of the *radius of curvature* $\varrho(t)$. The endpoint of the vector $\boldsymbol{P}(t)$ in this figure is called the *centre of curvature* and the circle, defined by the equation

$$[\boldsymbol{z}(t) - \boldsymbol{P}(t)] \cdot [\boldsymbol{z}(t) - \boldsymbol{P}(t)] - \varrho^2(t) = 0, \tag{2.2.37}$$

is called the *osculating circle* to the curve at $\boldsymbol{z}(t)$.

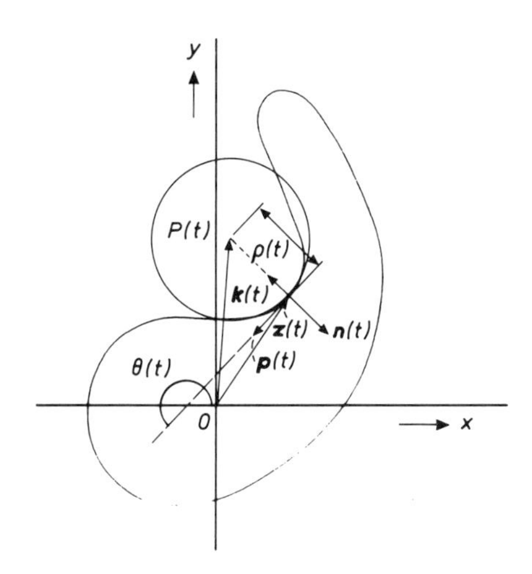


**Figure 2.9.** This figure illustrates the concepts of radius of curvature $\varrho(t)$, center of curvature $\boldsymbol{P}(t)$ and osculating circle along a curve. The moving window $\{\boldsymbol{p}(t), \boldsymbol{n}(t)\}$ and the curvature vector function $\boldsymbol{k}(t)$ at position $\boldsymbol{z}(t)$ are also shown.

So far, in this section little has been said about the properties concerning the preservation of shape information by the contour representations that we introduced. In Stoker [1969] a theorem, stating the unique reconstructability of a regular curve, up to a rigid motion in the plane, from its continuous curvature function, with arc length as a parameter, is presented. Zahn and Roskies [1972] contains a similar theorem for the tangent angle function $\theta(s)$, and a corollary that follows from it for the cumulative angular function $\varphi(s)$. We now restate the theorem in Stoker [1969] in a slightly more general form and describe its proof. The shape information-preserving properties of the other contour representations, introduced in this section, follow from this theorem as corollaries.

**Theorem 2.2.** (cf. e.g. Stoker [1969].)
Let $K(s)$, $a \leqslant s \leqslant b$, be integrable on $[a, b]$, i.e. $K \in \mathbf{L}^1[a, b]$ (cf. Appendix A). Then there exists one and only one smooth curve, up to a rigid motion, for which $K(s)$ is the curvature function and $s$ the arc length.

*Proof*

Without loss of generality, we may assume that $a \leqslant 0 \leqslant b$. In view of Eqs. 2.2.30 and 2.2.34 it is natural to define

$$\varphi(s) = \int_0^s K(\sigma)\mathrm{d}\sigma, \quad \forall s \in [a, b]$$

where we use $\varphi(0) \equiv 0$. Assuming a tangent angle $\theta(0)$ at $s = 0$ we find the tangent angle function $\theta(s)$ by Eq. 2.2.30. Once $\theta(s)$ has been determined, the unit tangent vector function $\boldsymbol{p}(s) \equiv \dot{\boldsymbol{z}}(s)$ is also known through Eq. 2.2.8.

Assuming a position vector $\boldsymbol{z}(0)$ at $s = 0$, we obtain the position vector function $\boldsymbol{z}(s)$ from $\dot{\boldsymbol{z}}(s)$ by the integral

$$\boldsymbol{z}(s) = \boldsymbol{z}(0) + \int_0^s \dot{\boldsymbol{z}}(\sigma)\mathrm{d}\sigma, \quad \forall s \in [a, b]. \tag{2.2.38}$$

It is now to be shown that $\boldsymbol{z}(s)$, as defined in Eq. 2.2.38, is a smooth curve for which $s$ is the arc length and $K(s)$ is the curvature. From the relation between $K$ and $\varphi$ and from $K \in \mathbf{L}^1[a, b]$ it follows that $\varphi \in \mathbf{AC}[a, b]$ (cf. Riesz and Sz.-Nagy [1955], pp. 50-52, Janssen and Van der Steen [1984], pp. 165-170). Through Eqs. 2.2.30 and 2.2.8 and Definition A.10 it is straightforward to show that then also $\dot{z} \in \mathbf{AC}[a, b]$. With this result and Eqs. 2.2.1, 2.2.2, 2.2.17 and 2.2.38 it follows that the function $\boldsymbol{z}(s)$ represents a smooth curve (cf. Eqs.

2.1.7a-2.1.7c). From Eqs. 2.2.12, 2.2.19 and 2.2.22 it follows that $K(s)$ is the curvature function of the curve represented by $\boldsymbol{z}(s)$ in Eq. 2.2.38. Finally it is readily seen that any pair of curves $\gamma_1$ and $\gamma_2$, represented by $\boldsymbol{z}_1(s)$ and $\boldsymbol{z}_2(s)$ respectively, that have the same arc length $s$ and curvature $K(s)$, differ at most by a rigid motion, as follows. By a rigid motion, i.e. by an appropriate translation followed by a rotation, the two curves can be brought together such that $\boldsymbol{z}_1(0) = \boldsymbol{z}_2(0)$ and $\dot{\boldsymbol{z}}_1(0) = \dot{\boldsymbol{z}}_2(0)$. Given the equality of the curvature functions of both curves, it follows from Eq. 2.2.35 that $\varphi_1(s) = \varphi_2(s)$, $\forall s \in [a, b]$. Further, from $\dot{\boldsymbol{z}}_1(0) = \dot{\boldsymbol{z}}_2(0)$ and Eqs. 2.2.9 and 2.2.30 we find that $\theta_1(s) = \theta_2(s)$, $\forall s \in [a, b]$. Hence we find from Eq. 2.2.9 that $\dot{\boldsymbol{z}}_1(s) = \dot{\boldsymbol{z}}_2(s)$, $\forall s \in [a, b]$ and with the equality $\boldsymbol{z}_1(0) = \boldsymbol{z}_2(0)$ we can finally state

$$\begin{aligned}\boldsymbol{z}_1(s) &= \boldsymbol{z}_1(0) + \int_0^s \dot{\boldsymbol{z}}_1(\sigma)\mathrm{d}\sigma \\ &= \boldsymbol{z}_2(0) + \int_0^s \dot{\boldsymbol{z}}_2(\sigma)\mathrm{d}\sigma \\ &= \boldsymbol{z}_2(s), \qquad \forall s \in [a, b], \end{aligned} \tag{2.2.39}$$

which completes the proof of the theorem.

□

From this theorem it is clear that the two invariants $K$ and $s$ form a complete set of invariants for a plane curve, since they determine it uniquely up to rigid motions. Note that the theorem excludes piecewise smooth curves. This is caused by the fact that curvature is not defined at corners in the curve. If we want curvature also to be meaningful as a representation for piecewise smooth curves, we will have to revert to generalized functions, as we will see later.

**Corollary 2.1.**
Let $\boldsymbol{k}(s)$, $a \leqslant s \leqslant b$, be an arbitrary two-dimensional vector function, bounded and continuous in its component functions, except possibly in a finite number of points. Then there exists one and only one smooth curve, up to a rigid motion, for which $\boldsymbol{k}(s)$ is the curvature vector function and $s$ the arc length. □

For the representations of a curve at the level of the first derivative of the position vector function $\boldsymbol{z}(s)$, i.e. $\dot{\boldsymbol{z}}$, $\theta$, $\varphi$ and $\psi$, the corresponding theorems not only hold for smooth curves, but also for piecewise smooth curves. It follows from the definition of a piecewise smooth curve in Section 2.1 that these representations are bounded and that they are continuous, except possibly in a finite number of points. Therefore these representations are integrable functions. We now formulate the following corollaries.

**Corollary 2.2.**
Let $\boldsymbol{p}(s)$, $a \leqslant s \leqslant b$, be an arbitrary two-dimensional unit vector function, i.e. $|\boldsymbol{p}(s)| \equiv 1$. Let $\boldsymbol{p}(s)$ be piecewise absolutely continuous in its component functions. Then there exists one and only one piecewise smooth curve, up to a translation, for which $\boldsymbol{p}(s)$ is the unit tangent vector function and $s$ the arc length.

□

**Corollary 2.3.**
Let $\theta(s)$, $a \leqslant s \leqslant b$, be an arbitrary piecewise absolutely continuous function, defined on a range of length $2\pi$. Then there exists one and only one piecewise smooth curve, up to a translation, for which $\theta(s)$ is the tangent angle function and $s$ is the arc length.

□

As mentioned earlier, the latter corollary has also been formulated by Zahn and Roskies [1972].

**Corollary 2.4.**
Let $f(s)$, $a \leqslant s \leqslant b$, be an arbitrary piecewise absolutely continuous function, with $f(0) = 0$. Then there exists one and only one curve for which $f(s)$ is the cumulative angular function $\varphi(s)$ and $s$ the arc length. And there exists one and only one curve for which $f(s)$ is the periodic cumulative angular function $\psi(s)$ and $s$ is the arc length.

□

Corollary 2.4 has also been stated by Zahn and Roskies [1972] in a slightly different form. The proofs of the Corollaries 2.1-2.4 are similar to the proof of Theorem 2.2 and, for the main part, they can be derived from it.

Two remarks are still in order. First we note that both Theorem 2.2 and Corollaries 2.1-2.4 have been formulated for curve representations with arc length as a parameter. If the curve representations are parametrized by an arbitrary parameter $t$, such that the functional relationship $t = \chi(s)$ between arc length $s$ and the arbitrary parameter $t$ is fully known and this relationship is invertible, then the corresponding curve representation with $s$ as a parameter can always be found. Therefore Theorem 2.2 and Corollaries 2.1-2.4 also hold for curve representations with an arbitrary parameter $t$, provided that this parameter satisfies the conditions just mentioned. In the sequel we will use the normalized arc length parameter, introduced in Section 2.1.

Secondly, we note that in Theorem 2.2 and Corollaries 2.1-2.4 neither closure conditions nor simplicity conditions are enforced on the curves. They hold for arbitrary, not necessarily simple or closed, curves. If we require the curves to be simple and/or closed, then this leads to additional constraints on the individual curve representations.

We conclude from the foregoing that we have introduced in this and in the previous section a number of curve representations that contain, under the condition that the relation between parameter and arc length is known, complete shape information. This fact makes these curve representations candidates for use in shape analysis and classification. The curve representations that we refer to are:

$z$ – position function
$\dot{z}$ – tangent function
$\ddot{z}$ – acceleration function
$\theta$ – tangent angle function
$\varphi$ – cumulative angular function
$\psi$ – periodic cumulative angular function
$K$ – curvature function.

Earlier in this section we already defined the scaling operator $\mathcal{S}_\beta$, by which a contour $\gamma$ is scaled by a factor $\beta \in \mathbb{R}^+$. The following operators are also important in the analysis of the shape of plane curves, represented by a parametric function:

$\mathcal{D}_\zeta$ – *translation* (or *displacement*) over $\zeta \in \mathbb{C}$
$\mathcal{R}_\alpha$ – *rotation* over $\alpha \in \mathbb{R}$
$\mathcal{T}_\tau$ – *backward shift* of the representation funtion parameter over $\tau \in \mathbb{R}$
$\mathcal{M}_x$ – *mirror reflection* about the $x$-axis.

**Table 2.1.** Representations of the (mirror-)similarity operators in the function spaces of the contour representations.

| Contour representation | Operator | | | | |
|---|---|---|---|---|---|
| | $\mathcal{S}_\beta$ | $\mathcal{D}_\zeta$ | $\mathcal{R}_\alpha$ | $\mathcal{T}_\tau$ | $\mathcal{M}_x$ |
| $z$ | $\beta z(t)$ | $z(t)+\zeta$ | $e^{i\alpha}z(t)$ | $z(t-\tau)$ | $\bar{z}(-t)$ |
| $\dot{z}$ | $\beta\dot{z}(t)$ | $\dot{z}(t)$ | $e^{i\alpha}\dot{z}(t)$ | $\dot{z}(t-\tau)$ | $-\bar{\dot{z}}(-t)$ |
| $\ddot{z}$ | $\beta\ddot{z}(t)$ | $\ddot{z}(t)$ | $e^{i\alpha}\ddot{z}(t)$ | $\ddot{z}(t-\tau)$ | $\bar{\ddot{z}}(-t)$ |
| $\theta$ | $\theta(t)$ | $\theta(t)$ | $\{\theta(t)+\alpha\}$ mod $2\pi$ | $\theta(t-\tau)$ | $\{-\theta(-t)+\pi\}$ mod $2\pi$ |
| $\varphi$ | $\varphi(t)$ | $\varphi(t)$ | $\varphi(t)$ | $\varphi(t-\tau)-\varphi(-\tau)$ | $-\varphi(-t)$ |
| $\psi$ | $\psi(t)$ | $\psi(t)$ | $\psi(t)$ | $\psi(t-\tau)-\psi(-\tau)$ | $-\psi(-t)$ |
| $K$ | $\beta^{-1}K(t)$ | $K(t)$ | $K(t)$ | $K(t-\tau)$ | $K(-t)$ |

**Table 2.2** Variance (•) or invariance (○) of the contour representations for the (mirror-)similarity operators.

| Contour representation | Operator | | | | |
|---|---|---|---|---|---|
| | $\mathcal{S}_\beta$ | $\mathcal{D}_\zeta$ | $\mathcal{R}_\alpha$ | $\mathcal{T}_\tau$ | $\mathcal{M}_x$ |
| $z$ | • | • | • | • | • |
| $\dot{z}$ | • | ○ | • | • | • |
| $\ddot{z}$ | • | ○ | • | • | • |
| $\theta$ | ○ | ○ | • | • | • |
| $\varphi$ | ○ | ○ | ○ | • | • |
| $\psi$ | ○ | ○ | ○ | • | • |
| $K$ | • | ○ | ○ | • | • |

We will call the operators $\mathcal{S}_\beta$, $\mathcal{D}_\zeta$, $\mathcal{R}_\alpha$ and $\mathcal{T}_\tau$ collectively *similarity operators* or *equiform operators*, since they do not affect the shape of a contour, as defined earlier. For the same reason $\mathcal{M}_x$ is called a *mirror-similarity operator*.

The representations of these (mirror-)similarity operators in each of the function spaces, defined by the aforementioned contour representations, are listed in Table 2.1. The operator representations can be derived from the definitions of the contour representations in a straightforward manner. For the scaling operator $\mathcal{S}_\beta$ we already listed some of its representations in Eqs. 2.2.28a,b and 2.2.29. Note in Table 2.1 that, in order to maintain the convention that the positive sense of a contour is counterclockwise, the sign of parameter $t$ is changed upon application of the mirror-reflection operator $\mathcal{M}_x$.

From Table 2.1 we immediately find the variance or invariance of the individual contour representations for the (mirror-)similarity operators. A survey of these properties can be found in Table 2.2. The properties in the Tables 2.1 and 2.2 are important for the determination of the conditions that are satisfied by the representations of similar or mirror-similar contour pairs. We will return to this topic in Section 2.3.

Simple closed polygons constitute an important class of piecewise smooth simple closed curves for shape representation, approximation and analysis purposes. In the following we introduce notations for simple closed polygons and derive expressions for the representation of such contours. First an expression for the position function $z$ of a polygon will be derived in this illustration, and subsequently expressions for the representations $\dot{z}$, $\ddot{z}$, $\theta$, $\varphi$, $\psi$ and $K$.

A simple closed polygon with $N$ vertices is completely specified by the ordered set of its complex-valued vertices

$$\{z(t_n)\} = \{x(t_n) + \mathrm{i}y(t_n)\}, \quad n = 0, \ldots, N-1. \tag{2.2.40}$$

The periodicity of the position function of the polygon is expressed for the vertices by the equation

$$z(t_n) = z(t_{n+pN}), \quad \forall p \in \mathbb{Z}. \tag{2.2.41}$$

We define *first* and *second order discrete differences* as

$$\Delta z(t_n) = z(t_{n+1}) - z(t_n) \tag{2.2.42}$$

and
$$\Delta^2 z(t_n) = \Delta z(t_n) - \Delta z(t_{n-1})$$
$$= z(t_{n+1}) - 2z(t_n) + z(t_{n-1}) \tag{2.2.43}$$

and *normalized first* and *second order discrete differences* as

$$\Delta z^*(t_n) = \frac{\Delta z(t_n)}{|\Delta z(t_n)|} \tag{2.2.44}$$

and

$$\Delta^2 z^*(t_n) = \Delta z^*(t_n) - \Delta z^*(t_{n-1}). \tag{2.2.45}$$

For notational convenience we defined the second order discrete differences as central differences. Without loss of generality we can make the starting point of the parametric representations coincide with the vertex $z(t_0)$, i.e. $t_0 = 0$. We choose $t$ to be a constant speed parameter. Then for polygons the relation between arc length $s$ and $t$ is

$$s = \frac{L}{2\pi}\, t, \tag{2.2.46}$$

where $L$ is the perimeter of the polygon. The arc length $s_n$ at the vertex $z(t_n)$ is given by the equations

$$s_n = \begin{cases} \sum_{m=0}^{n-1} |\Delta z(t_m)|, & n \geqslant 1, \quad (2.2.47a) \\ 0, & n = 0, \quad (2.2.47b) \\ -\sum_{m=n}^{-1} |\Delta z(t_m)|, & n \leqslant -1 \quad (2.2.47c) \end{cases}$$

and the perimeter can be expressed as $L = s_N$. From Eqs. 2.2.46 and 2.2.47a-c we find the equation

$$|\Delta z(t_n)| = s_{n+1} - s_n = \frac{L}{2\pi}(t_{n+1} - t_n). \tag{2.2.48}$$

The terminology that we introduced for simple closed polygons is illustrated in Figure 2.10.

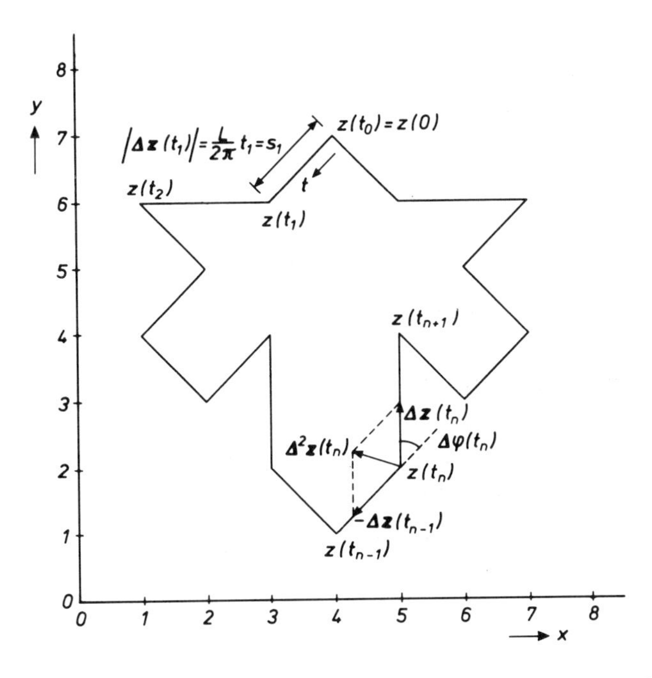


**Figure 2.10.** A simple closed polygon with $N = 16$ vertices. Normalized discrete differences are indicated as vectors.

Based on Eqs. 2.2.40-2.2.48 we find for the position function $z$ between the vertices $z(t_n)$ and $z(t_{n+1})$ the expression

$$z(t) = \Delta z^*(t_n)\,\frac{L}{2\pi}\,(t - t_n) - z(t_n), \qquad t_n \leqslant t \leqslant t_{n+1}. \tag{2.2.49}$$

Through differentiation of $z$ we obtain an expression for the tangent function $\dot{z}$ between the vertices $z(t_n)$ and $z(t_{n+1})$

$$\dot{z}(t) = \Delta z^*(t_n)\,\frac{L}{2\pi}, \qquad t_n < t < t_{n+1}. \tag{2.2.50}$$

Note that the tangent function of a polygon is formally undefined at the vertices, though we may decide to use left or right limits at the vertices.

The acceleration function $\ddot{z}$ can only be defined as a distribution and not as an ordinary function. We will not go into the details of the theoretical aspects of distributions (cf. Lighthill [1962], Zemanian [1965]), but use distributions as if they were ordinary functions.

The *Dirac delta functional* $\delta(t)$ is defined as the generalized function that satisfies

$$\int_{-\infty}^{\infty} f(t)\delta(t)\mathrm{d}t = f(0) \qquad (2.2.51)$$

for all continuous functions $f(t)$. It can be identified with the derivative of the *Heaviside unit step function* $H(t)$, i.e.

$$\delta(t) = \frac{\mathrm{d}H(t)}{\mathrm{d}t} \qquad (2.2.52)$$

where $H(t)$ is defined as

$$H(t) = \begin{cases} 0, & t<0 \\ 1, & t \geqslant 0. \end{cases} \qquad (2.2.53)$$

The tangent function can now be rewritten as a combination of functions $H(t)$

$$\dot{z}(t) = \sum_{p \in \mathbb{Z}} \Delta z^*(t_p) \frac{L}{2\pi} [H(t-t_p) - H(t-t_{p+1})] \qquad (2.2.54)$$

which can be further simplified to

$$\dot{z}(t) = \frac{L}{2\pi} \sum_{p \in \mathbb{Z}} [\Delta z^*(t_p) - \Delta z^*(t_{p-1})] H(t-t_p)$$

$$= \frac{L}{2\pi} \sum_{p \in \mathbb{Z}} \Delta^2 z^*(t_p) H(t-t_p). \qquad (2.2.55)$$

We can find an expression for $\ddot{z}$ through generalized differentiation of

$\dot{z}$ in Eq. 2.2.55

$$\ddot{z}(t) = \frac{L}{2\pi} \sum_{p \in \mathbb{Z}} \Delta^2 z^*(t_p)\delta(t - t_p). \tag{2.2.56}$$

Due to its periodicity, $\ddot{z}(t)$ can be expressed as a periodic distribution

$$\ddot{z}(t) = \frac{L}{2\pi} \sum_{n=0}^{N-1} \Delta^2 z^*(t_n)\delta_{2\pi}(t - t_n) \tag{2.2.57}$$

where $\delta_{2\pi}(t)$ is defined as

$$\delta_{2\pi}(t) = \sum_{q \in \mathbb{Z}} \delta(t - 2\pi q). \tag{2.2.58}$$

It is straightforward to show that $\ddot{z}$ specifies the polygon up to a rigid motion in the plane. By integration of both sides of Eq. 2.2.56 and by inversion of integration and summation we obtain through Eq. 2.2.52 the expression for $\dot{z}$ in Eq. 2.2.55. Assuming an initial direction $\Delta z^*(t_0)$ we can derive the expression for $\dot{z}$ in Eq. 2.2.50. Further, assuming an initial position $z(t_0)$ on the polygon, each point on it can be reconstructed by integration of both sides of Eq. 2.2.50.

The derivation of the tangent angle function $\theta$ of a simple closed polygon is as follows. We observe that the normalized discrete differences $z^*(t_n)$ can be decomposed as

$$\Delta z^*(t_n) = \Delta x^*(t_n) + i\Delta y^*(t_n) \tag{2.2.59}$$

where

$$\Delta x^*(t_n) = \frac{x(t_{n+1}) - x(t_n)}{|\Delta z(t_n)|} \tag{2.2.60a}$$

and

$$\Delta y^*(t_n) = \frac{y(t_{n+1}) - y(t_n)}{|\Delta z(t_n)|}. \tag{2.2.60b}$$

Through Eq. 2.2.6, which defines $\theta$, and Eq. 2.2.50 we find an expression for the tangent angle function of a simple closed polygon

$$\theta(t) = \arctan\left(\frac{\Delta y^*(t_n)}{\Delta x^*(t_n)}\right), \qquad t_n < t < t_{n+1}. \tag{2.2.61}$$

We can solve $\theta(t)$, without ambiguity from the signs of $\Delta x^*(t)$ and $\Delta y^*(t)$ or, equivalently, from the direction cosines of the tangent function $\dot{z}(t)$

$$\cos\theta(t) = \Delta x^*(t_n) \tag{2.2.62a}$$

and

$$\sin\theta(t) = \Delta y^*(t_n), \qquad t_n < t < t_{n+1}. \tag{2.2.62b}$$

In order to measure the amount of angular change at vertex $z(t_n)$ we define the *discrete tangent angle difference*

$$\Delta\theta(t_n) = \theta(t_n) - \theta(t_{n-1}). \tag{2.2.63}$$

Recalling that the tangent angle function $\theta$ assumes values in a range of length $2\pi$, we obtain the amount of *angular change* $\Delta\varphi(t_n)$ at vertex $z(t_n)$ by a mapping of $\Delta\theta(t_n)$

$$\Delta\varphi(t_n) = \begin{cases} \Delta\theta(t_n) + 2\pi, & -2\pi < \Delta\theta(t_n) < -\pi, \quad (2.2.64a) \\ \Delta\theta(t_n), & -\pi < \Delta\theta(t_n) < \pi, \quad (2.2.64b) \\ \Delta\theta(t_n) - 2\pi, & \pi < \Delta\theta(t_n) < 2\pi. \quad (2.2.64c) \end{cases}$$

See Figure 2.11 for an illustration of $\Delta\varphi(t_n)$. The angular change $\Delta\varphi(t_n)$ takes on values in the range $(-\pi, \pi)$, where positive values are obtained at convex angles and negative values at concave angles. Note that the values $-2\pi$, $-\pi$, $\pi$ and $2\pi$ are not included in the ranges in Eqs. 2.2.64a-c. The exclusion of $-\pi$ and $\pi$ is obvious, while the simplicity of the polygon leads to the exclusion of $-2\pi$ and $2\pi$. The cumulative angular function $\varphi$ of an arbitrary simple closed polygon can now be expressed as

$$\varphi(t) = \begin{cases} \sum_{m=1}^{n} \Delta\varphi(t_m), & n \geqslant 1, \quad (2.2.65a) \\ 0, & n = 0, \quad (2.2.65b) \\ -\sum_{m=n+1}^{0} \Delta\varphi(t_m), & n \leqslant -1, \quad t_n < t < t_{n+1}. \quad (2.2.65c) \end{cases}$$

The periodic cumulative angular function $\psi$ of a simple closed polygon is found by applying its definition, in Eq. 2.2.32, to the results in Eqs. 2.2.65a-c.

To provide insight into the derivation of an expression for the curvature of a simple closed polygon, as a distribution, we rewrite Eqs. 2.2.65a-c as a combination of Heaviside unit step functions

$$\varphi(t) = \sum_{p \in \mathbb{Z}} \Delta\varphi(t_p) H(t - t_p) - \sum_{p=-\infty}^{0} \Delta\varphi(t_p). \quad (2.2.66)$$

We can find an expression for the curvature $K$ of a simple closed polygon, using the definition of $K$ in Eq. 2.2.34, through generalized differentiation of $\varphi(t)$ in Eq. 2.2.66, giving

$$K(t) = \sum_{p \in \mathbb{Z}} \Delta\varphi(t_p)\delta(t - t_p)\frac{\mathrm{d}t}{\mathrm{d}s}. \quad (2.2.67)$$

Since the parameter $t$ has been defined in Eq. 2.2.46 as a normalized arc length parameter, Eq. 2.2.67 can be rewritten as

$$K(t) = \frac{2\pi}{L} \sum_{p \in \mathbb{Z}} \Delta\varphi(t_p)\delta(t - t_p). \quad (2.2.68)$$

Finally, due to its periodicity, $K(t)$ can be expressed as a periodic distribution

$$K(t) = \frac{2\pi}{L} \sum_{n=0}^{N-1} \Delta\varphi(t_n)\delta_{2\pi}(t - t_n). \quad (2.2.69)$$

We now show that a polygon is completely specified, up to a rigid motion in the plane, by the expression that we found for $K(t)$. We first note that for a simple closed polygon with $N$ vertices

$$\sum_{n=0}^{N-1} \Delta\varphi(t_n) = 2\pi. \tag{2.2.70}$$

Using this equation and at least one period of $K(t)$, we can find the perimeter $L$. Through Eq. 2.2.34 and Eq. 2.2.46 we find $\varphi(t)$ from $K(t)$

$$\varphi(t) = \int_0^t K(\tau)\frac{ds}{d\tau}d\tau$$

$$= \frac{L}{2\pi}\int_0^t K(\tau)d\tau. \tag{2.2.71}$$

In Eq. 2.2.71 we have employed the convention $\psi(0) = 0$. For polygons, $\varphi(t)$ is undefined for $t = 0$ and the convention becomes $\varphi(t) = 0$, $0 = t_0 < t < t_1$. Rewriting Eq. 2.2.71 as

$$\varphi(t) = \frac{L}{2\pi}\left[\int_{-\infty}^{t} K(\tau)d\tau - \int_{-\infty}^{0} K(\tau)d\tau\right] \tag{2.2.72}$$

we can find the expression in Eq. 2.2.66 through substitution of Eq. 2.2.68 into Eq. 2.2.72, the inversion of summation and integration and finally the application of Eq. 2.2.52. The expressions in Eqs. 2.2.65a-c are equivalent to the one in Eq. 2.2.66. From Eqs. 2.2.65a-c we can find the individual angular changes $\Delta\varphi(t_n)$ at the vertices in a straightforward manner. Since the relation between arc length and the parameter $t$ is known and the values of $t_n$ in at least one period are known, the lengths of the sides of the polygon are known. If we use this information and define an initial orientation $\theta(t)$, $0 = t_0 < t < t_1$, and an initial position $z(t_0)$, we can reconstruct the polygon from $\Delta\varphi(t_n)$. This shows that the polygon can be reconstructed up to a rigid motion from $K(t)$.

Through this derivation we have shown that it is possible to find information-preserving expressions for $\dot{z}$, $\ddot{z}$, $\theta$, $\varphi$, $\psi$ and $K$ of a polygon. If we combine this result with the information-preserving properties of the same representations for smooth curves (cf. Theorem 2.1 and Corollaries 2.1-2.4), we may conclude that we can find information-preserving expressions for $\dot{z}$, $\ddot{z}$, $\theta$, $\varphi$, $\psi$ and $K$ of any piecewise smooth curve with a finite number of corners.

Applications of polygonal curves occur frequently in the literature, e.g. in connection with contour coding (Freeman [1961a], Freeman [1970], Saghri and Freeman [1981]), polygonal contour approximation (Montanari [1970], Ramer [1972], Pavlidis and Horowitz [1974], McClure and Vitale [1975], Ellis and Eden [1976], Pavlidis [1977b], Williams [1981]), shape property measurement (Freeman [1961b], Zahn and Roskies [1972], Wilson and Farrior [1976], Persoon and Fu [1977], Kuhl and Giardina [1982], Sarvarayudu and Sethi [1983]) and shape classification (Pavlidis and Ali [1975], Davis [1977a], Davis [1979], Kashyap and Oommen [1982]).

Reports in the literature on the use of $z$ for shape representation have already been reviewed in Section 2.1. Concerning the use of the remaining representations an account of the literature now follows.

The contour representations $\dot{z}$ and $\theta$ have rarely been used explicitly for shape representation. The tangent function $\dot{z}$ appears naturally in the derivation of the curvature function (Young, Walker and Bowie [1974], Bennett and MacDonald [1975], Groen [1977], van Otterloo [1978]) or is used as an intermediate representation for the derivation of $\theta$ (Ozaki et al. [1982]). However, $\dot{z}$ is not identified in these references as an information-preserving shape, and therefore potentially useful, contour representation. We already saw in Eqs. 2.2.30 and 2.2.32 that the tangent angle function $\theta$, the cumulative angular function $\varphi$ and the periodic cumulative angular function $\psi$ are closely related. The earliest report that we found on the use of these functions for shape representation is Cosgriff [1960]. Having no access to this report, it remains unclear whether Cosgriff suggested the use of $\theta$ or of $\psi$. Brill [1968] and Zahn and Roskies [1972], who use $\psi$ for shape representation, make conflicting statements on this issue. As a result of the problem of discontinuities of size $2\pi$ in $\theta$, that are not shape-related (see Figure 2.8), $\theta$ is not a popular shape representation, although Perkins [1978] reports on its use in the context of shape matching in an industrial vision system. This problem is overcome by both $\varphi$ and $\psi$, where $\psi$ has the advantage over $\varphi$ of being a periodic function. Usually not $\psi$ itself but its Fourier coefficients (Brill [1968], Barrow and Popplestone [1971], Zahn and Roskies [1972], Fong, Beddow and Vetter [1979], Beddow [1980], Strackee and Nagelkerke [1983]) or its Walsh coefficients (Dinstein and Silberberg [1980], Sethi and Sarvarayudu [1980], Sarvarayudu [1982], Sarvarayudu and Sethi [1983]) are used for shape representation. Martin and Aggarwal [1979] discuss the use of $\psi$ for curve segmentation.

Another approach to overcome the discontinuity problems in $\theta$ is its mapping into a *slope density function*, i.e. a distribution of the occurrence of each value of $\theta$ along the contour (Sklansky and Davison [1971], Sklansky and Nahin [1972], Nahin [1974], Ozaki et al. [1982]). The slope density function is not an information-preserving shape representation. Related to this method is the Hough transform technique, which computes the frequency of occurrence of $(r, \theta)$-pairs along the contour (Hough [1962], Sklansky [1978], Shapiro [1978], Ballard [1981], Davis [1982]).

The chain code, that was introduced and later generalized by Freeman to represent contours (Freeman [1961a], Freeman [1978b]), constitutes in fact a sampled, quantized and coded approximation of $\theta$.

Just as the tangent function $\dot{z}$, the acceleration function $\ddot{z}$ has only been mentioned in connection with the derivation of the formula for the curvature function $K$ from the position function $z$ (Young, Walker and Bowie [1974], Groen [1977], van Otterloo [1978], Anderson and Bezdek [1984]). We have found no reference to its explicit use for shape representation.

The concept of curvature plays an important role in many approaches to shape analysis (Pavlidis [1977a]). The observation of the importance of curvature for human shape perception dates back to the work of Attneave (Attneave [1954], Attneave and Arnoult [1956]). Especially the perceptually dominant role of points of high absolute curvature has become apparent from these studies (see also Zusne [1970]). From a mathematical point of view, evidence for the importance of curvature extrema has been obtained by McClure. McClure [1975] showed that in piecewise linear spline approximation with free knots, using a minimum integral square error criterion, the distribution of the knots follows the curvature of the curve (see also Pavlidis [1978]).

In the foregoing we saw that curvature is, apart from scaling, a mathematical shape invariant. This fact has given curvature the status of *intrinsic equation* of a curve (cf. e.g. Duda and Hart [1973]). In view of these facts it is not surprising that applications of curvature can already be found in the early pattern recognition literature (e.g. Cosgriff [1960], Ledley [1964], Zahn [1966]).

In practice it is hard to obtain a reliable estimate of the curvature function $K$ of a curve since it involves second derivatives of the position function $z$ (cf. Eq. 2.2.26). Many authors have dealt with this estimation problem, (e.g. Ledley [1964], Young, Walker and Bowie [1974], Bennett and MacDonald [1975], Bowie and Young [1977a], van Otterloo [1978], Wallace, Mitchell and Fukunaga [1981], Kasvand and Otsu [1982], Smeulders [1983], Anderson and Bezdek [1984]). To obtain curvature estimates that are less sensitive to noise, a variant of the curvature function has been defined, which essentially consists of a mapping of the angle between a leading and a trailing vector on the curve to a measure of curvature (cf. Figure 2.11). In the latter tech-

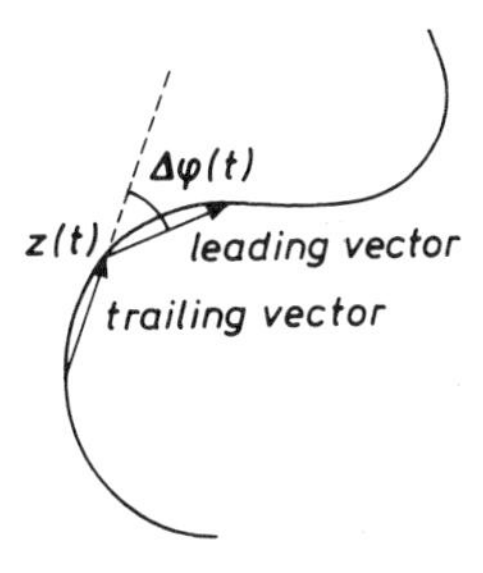


**Figure 2.11.** Example of a leading and a trailing vector on a curve, at position $z(t)$, spanning equal arc length. The angle $\Delta\varphi(t)$ is used to obtain a measure of curvature at $z(t)$.

niques the perceptually significant concept of a *corner* plays an important role. Details of such methods can be found in Rosenfeld and Johnston [1973], Freeman and Davis [1977], Freeman [1979] and Pineda and Horaud [1983]. Examples of the many applications of contour curvature in image analysis and shape analysis, with some references, are the following:

- corner detection and critical point detection (Rosenfeld and Johnston [1973], Freeman and Davis [1977], Sankar and Sharma [1978],

Kitchen and Rosenfeld [1982], Pineda and Horaud [1983], Hung and Kasvand [1983], Anderson and Bezdek [1984]),

- curve partitioning (Ledley [1964], Ledley [1972], Bowie and Young [1977b], Davis [1977a], Davis and Rosenfeld [1977], Perkins [1978], Freeman [1978a], Nevins [1979], Rutkowski, Peleg and Rosenfeld [1981], Fischler and Bolles [1983]),

- polygonal contour approximation (Shirai [1973], Davis [1977a], Davis and Rosenfeld [1977]),

- segmentation of overlapping objects (Eccles, McQueen and Rosen [1977], Dessimoz [1978], Dessimoz [1980], Bengtsson et al. [1981], Kailay, Sadananda and Das [1981], Smeulders [1983], Segen [1984]),

- shape matching (Freeman [1979], Davis [1979], Wallace, Mitchell and Fukunaga [1981], Grogan and Mitchell [1983]).

Closely related to curvature is the concept of *bending energy*. The bending energy in a thin elastic beam is proportional to the integrated squared curvature along the beam (cf. Landau and Lifschitz [1970]). Freeman and Glass [1969] used this property to compute a minimum energy curve in a tolerance region. Young, Walker and Bowie [1974] proposed bending energy as a shape feature, while Chang [1976] and Perkins [1978] used bending energy to match arcs of curves.

Further discussions about the role of curvature in shape analysis can be found in Pavlidis [1977a] and Pavlidis [1978].

## 2.3 Geometric similarity and geometric mirror-similarity

In Section 1.1 we already indicated that we consider the internal structure of an object, such as its brightness, colour, texture, etc., not to be part of its shape. We also mentioned that we consider the shape of an object to be invariant under translation, scaling and rotation. If we use parametric contour representations to render the shape of an object, the choice of a particular parameter $t$ and a starting point $t = 0$ on the contour is regarded not to affect the shape. We recall here our assumptions concerning the choice of a parameter $t$, expressed in Eqs. 2.1.18b, c, namely that the parameter is either the arc length or the normalized arc length. In the latter case the period of the periodic contour representations is normalized to $2\pi$. For these reasons we have called the operators $\mathcal{D}_\zeta$, $\mathcal{S}_\beta$, $\mathcal{R}_\alpha$ and $\mathcal{T}_\tau$, that perform a translation, a scaling, a rotation and a shift of the parametric starting point, respectively, collectively similarity operators or equiform operators. Consequently, also the order of application of the similarity operators does not affect the shape of an object.

It will be clear from the foregoing that in this thesis we are concerned with the geometrical aspects of shape. We emphasize that information about the absolute position, size and orientation of an object may contain valuable information in many applications. For example, in optical character recognition, information about character orientation is needed to be able to discriminate between numerals of the class '6' and numerals of the class '9'. If our objective is to estimate the motion of an object in a sequence of images, information about its position, size and orientation is indispensable (Richard and Hemami [1974], Wallace and Mitchell [1980]). It will become clear later on that, in many cases, such information can be obtained directly from the parametric contour representation of the object or from a similarity measurement. In view of the invariance properties of shape, we will require similarity measurement itself to be invariant for position, size, orientation and position of the parametric starting point. The formulation of similarity measures that satisfy these requirements is the subject of Chapter 4.

Before we proceed with a formal definition of geometric similarity, it is important to note that the shape of an object may vary with the level of magnification and resolution at which it is observed. There are numerous examples to illustrate this statement: coastlines (Mandelbrot [1967]), cell boundaries, snowflakes, fine particles, etc.; all will change in perceived geometrical shape when observed with a finer resolution. A detailed discussion of the mathematical modelling of such phenomena is given by Mandelbrot (Mandelbrot [1977], Mandelbrot [1982a]). Applications of such models can be found, for example, in computer graphics, where they are used for the computer rendering of curves and surfaces at variable levels of resolution (Carpenter [1980], Fournier, Fussel and Carpenter [1982], Mandelbrot [1982b], Kajiya [1983], Pentland [1983]). Though there exist some references in the literature on particle analysis (e.g. Kaye [1978], Flook [1978]), the consequences of the dependence of shape on resolution for digital shape analysis largely remain to be studied.

We now give a formal definition of geometric similarity.

**Definition 2.4.** *Geometric similarity.*
A contour $\gamma_1$ is said to be geometrically similar to a contour $\gamma_2$ iff $\gamma_1$ can be mapped into $\gamma_2$ by a sequence of translation, scaling and rotation operations.

□

An example of a pair of geometrically similar contours is given in Figure 2.12. Geometric similarity, as defined here, is an equivalence relation, i.e. it is reflexive, symmetric and transitive. Therefore it may be used to partition the set $\boldsymbol{\Gamma}_{\mathrm{ps}}$ into equivalence classes of geometrically similar contours (Richard and Hemami [1974]).

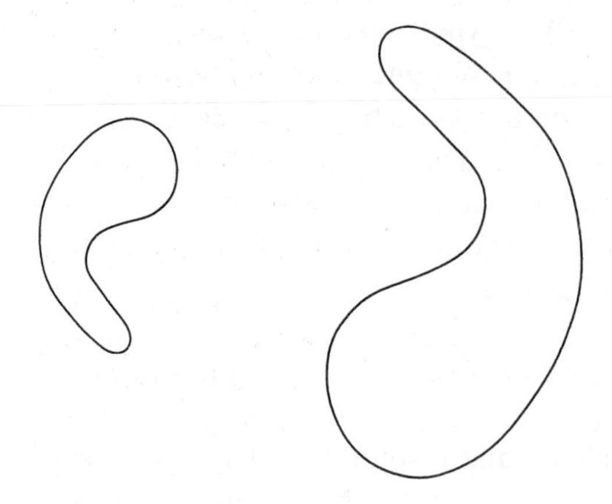

**Figure 2.12.** Example of a pair of geometrically similar contours.

In the remainder of this chapter we will use $f$ as a generic symbol for any of the parametric contour representations $z$, $\dot{z}$, $\ddot{z}$, $\theta$, $\varphi$, $\psi$ and $K$. Since these contour representations preserve shape information, we can collectively formulate the conditions that must be satisfied by the representations of a pair of contours in order to render geometrically similar contours.

**Theorem 2.3.**
Two contours $\gamma_1$ and $\gamma_2$, with contour representations $f_1$ and $f_2$ respec-

tively, are geometrically similar iff there exist scalars $\zeta \in \mathbb{C}$, $\beta \in \mathbb{R}^+$, $\alpha, \tau \in \mathbb{R}$ such that

$$f_2 = \mathcal{T}_\tau \mathcal{R}_\alpha \mathcal{S}_\beta \mathcal{D}_\zeta f_1. \tag{2.3.1}$$

*Proof*

In Section 2.2 we have shown that the contour representations $z$, $\dot{z}$, $\ddot{z}$, $\theta$, $\varphi$, $\psi$ and $K$ are information-preserving, i.e. given a particular contour representation from any of these types, there is one and only one contour, possibly up to a rigid motion and/or a scaling in the plane, that is described by this representation.

On the other hand, using the formulas in Section 2.2 the contour representations $\dot{z}$, $\ddot{z}$, $\theta$, $\varphi$, $\psi$ and $K$ can be determined uniquely from a contour representation $z$, being the direct representation of contour $\gamma$. Here we assume $z$ to possess sufficient differentiability properties.

Thus there exists a one-to-one correspondence, possibly up to a rigid motion and/or a scaling, between a contour $\gamma$ and each of its contour representations $z$, $\dot{z}$, $\ddot{z}$, $\theta$, $\varphi$, $\psi$ and $K$. From this, from Definition 2.4 and from the invariance of the shape of a contour under a starting point shift in its parametric representation it follows immediately that $\gamma_1$ and $\gamma_2$ are geometrically similar if Eq. 2.3.1 is satisfied.

If no scalars $\zeta$, $\beta$, $\alpha$ and $\tau$ can be found for which Eq. 2.3.1 is satisfied then it also follows from the one-to-one correspondence between a contour and its contour representations and from Definition 2.4 that $\gamma_1$ and $\gamma_2$ cannot be geometrically similar.

□

A survey of the formulation of this condition for geometric similarity, in terms of the individual contour representations, can be found in Table 2.3. To derive these formulations we have used the representations of the similarity operators in the function spaces of the individual contour representations, given in Table 2.1, and the variance or invariance properties of the individual contour representations for the similarity operators, given in Table 2.2.

Another important concept in shape analysis, which is closely related to similarity, is that of mirror-similarity. In two-dimensional shape analysis the planar shapes have usually been obtained as a result of a

**Table 2.3.** Necessary and sufficient conditions that the individual contour representations must satisfy, for some $\zeta \in \mathbb{C}$, $\beta \in \mathbb{R}^+$, $\alpha, \tau \in \mathbb{R}$ and $\forall t \in [0, 2\pi]$, in order to render a pair of geometrically similar contours.

| Contour representation | Condition for geometric similarity |
|---|---|
| $z$ | $z_2(t) = \beta e^{i\alpha} \{z_1(t-\tau) + \zeta\}$ |
| $\dot{z}$ | $\dot{z}_2(t) = \beta e^{i\alpha} \dot{z}_1(t-\tau)$ |
| $\ddot{z}$ | $\ddot{z}_2(t) = \beta e^{i\alpha} \ddot{z}_1(t-\tau)$ |
| $\theta$ | $\theta_2(t) = \{\theta_1(t-\tau) + \alpha\} \bmod 2\pi$ |
| $\varphi$ | $\varphi_2(t) = \varphi_1(t-\tau) - \varphi_1(-\tau)$ |
| $\psi$ | $\psi_2(t) = \psi_1(t-\tau) - \psi_1(-\tau)$ |
| $K$ | $K_2(t) = \beta^{-1} K_1(t-\tau)$ |

projection of a three-dimensional structure onto the plane of analysis. In many applications, the relative position and orientation of the objects with respect to the plane of projection or, equivalently, the point of observation with respect to the objects may vary. For example, thin industrial parts may land on a conveyor belt with either one or the other side up (Dessimoz [1980]). The same holds for biological cells that have been prepared on a glass plate for microscopic analysis. In airplane recognition, an airplane may have any position and orientation with respect to the point of observation (Richard and Hemami [1974], Wallace and Wintz [1980]). In order to be able to determine the orientation of an object with respect to the point of observation or to reduce the size of a library of plane projections of three-dimensional prototypes we need the concept of geometric mirror-similarity.

**Definition 2.5.** *Geometric mirror-similarity.*
A contour $\gamma_1$ is said to be geometrically mirror-similar to a contour $\gamma_2$ iff $\gamma_1$ becomes geometrically similar to $\gamma_2$ upon a mirror-reflection of $\gamma_1$ with respect to an arbitrary axis in the plane. □

An example of two geometrically mirror-similar contours is given in Figure 2.13. Geometric mirror-similarity is a symmetric binary relation between contours. It is not an equivalence relation in itself. Equivalence classes of geometrically similar contours, of which the elements are also geometrically mirror-similar, consist of geometrically mirror-symmetric contours. The concept of geometric mirror-symmetry will be defined in the next section. Contours that are geometrically mirror-similar but not geometrically similar are called *enantiomorphic* versions of the same shape (Weyl [1952], Shubnikov and Koptsik [1974]), i.e. there exists a 'left' and a 'right' version of that shape.

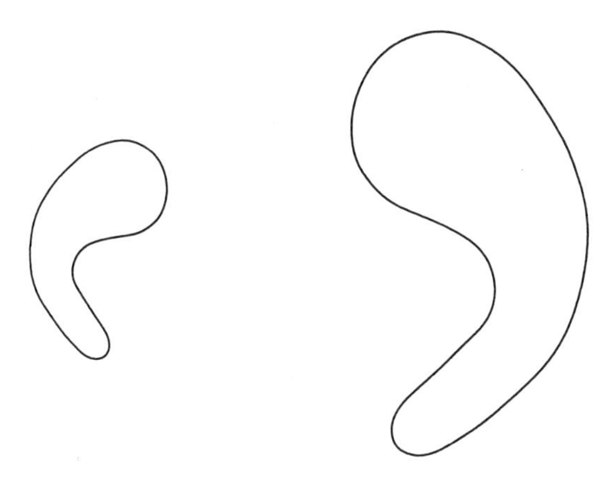

**Figure 2.13.** Example of a pair of geometrically mirror-similar contours.

We now formulate for the parametric contour representations, collectively indicated by the symbol $f$, the conditions that must be satisfied by the representations of a pair of contours in order to render geometrically mirror-similar contours. Without loss of generality, we choose the $x$-axis as the arbitrary axis, mentioned in Definition 2.5, about which mirror-reflection takes place. The reason for this choice is the analytical convenience it provides.

**Theorem 2.4.**
Two contours $\gamma_1$ and $\gamma_2$, with contour representations $f_1$ and $f_2$ respec-

tively, are geometrically mirror-similar iff there exist scalars $\zeta \in \mathbb{C}$, $\beta \in \mathbb{R}^+$, $\alpha, \tau \in \mathbb{R}$ such that

$$f_2 = \mathcal{M}_x \mathcal{T}_\tau \mathcal{R}_\alpha \mathcal{S}_\beta \mathcal{D}_\zeta f_1. \tag{2.3.2}$$

□

The proof of Theorem 2.4 is similar to that of Theorem 2.5.

**Remark.**
The order of the operators $\mathcal{D}_\zeta$, $\mathcal{S}_\beta$, $\mathcal{R}_\alpha$, $\mathcal{T}_\tau$ and $\mathcal{M}_x$ is immaterial for establishing the validity of Eqs. 2.3.1 or 2.3.2. However, the values of $\zeta$, $\beta$, $\alpha$ and $\tau$, for which this validity is established, are dependent upon the order of the (mirror-)similarity operators in these equations (cf. Table 2.1).

□

**Remark.**
For the contour representations $\dot{z}$, $\ddot{z}$, $\theta$, $\varphi$, $\psi$ and $K$, the validity of Eqs. 2.3.1 or 2.3.2 *almost everywhere* (denoted as a.e., cf. Definition A.3) is already a sufficient condition for geometric (mirror-)similarity.

**Table 2.4.** Necessary and sufficient conditions that the individual contour representations must satisfy, for some $\zeta \in \mathbb{C}$, $\beta \in \mathbb{R}^+$, $\alpha, \tau \in \mathbb{R}$ and $\forall t \in [0, 2\pi]$, in order to render a pair of geometrically mirror-similar contours.

| Contour representation | Condition for geometric mirror-similarity |
|---|---|
| $z$ | $z_2(t) = \beta e^{-i\alpha}\{\bar{z}_1(-t+\tau) + \bar{\zeta}\}$ |
| $\dot{z}$ | $\dot{z}_2(t) = -\beta e^{-i\alpha}\bar{\dot{z}}_1(-t+\tau)$ |
| $\ddot{z}$ | $\ddot{z}_2(t) = \beta e^{-i\alpha}\bar{\ddot{z}}_1(-t+\tau)$ |
| $\theta$ | $\theta_2(t) = \{-\theta_1(-t+\tau) + \pi - \alpha\} \bmod 2\pi$ |
| $\varphi$ | $\varphi_2(t) = -\varphi_1(-t+\tau) + \varphi_1(\tau)$ |
| $\psi$ | $\psi_2(t) = -\psi_1(-t+\tau) + \psi_1(\tau)$ |
| $K$ | $K_2(t) = \beta^{-1}K_1(-t+\tau)$ |

In this section, and in Section 2.4 on symmetry, the distinction between pointwise equality and equality a.e. for the contour representations is handled loosely or ignored altogether.

In Chapter 4, discussing the measurement of similarity and symmetry, it will become clear that, as a result of the mathematical form of dissimilarity and dissymmetry measures, there is no need here for a distinction between pointwise equality and equality a.e..

□

A survey of the formulations of this condition for geometric mirror-similarity, in terms of the individual contour representations, can be found in Table 2.4.

## 2.4 Symmetry in plane objects

Symmetry is an important feature that an object or a set of objects may exhibit. In the world that surrounds us we encounter various types of symmetry, each type with a different reason for occurring. This is one of the reasons why symmetry theory has found widespread application in various fields of science. To quote Weyl [1952]: 'Symmetry, as wide or as narrow as you may define its meaning, is one idea by which man through the ages has tried to comprehend and create order, beauty and perfection'. Because we find it aesthetically pleasing, symmetry is found in many works of art. A striking example of fascination by symmetry in graphic arts is found in the work of Escher (Escher et al. [1972]). A nice introduction to symmetry theory is given by Weyl [1952], while Shubnikov and Koptsik [1974] give a comprehensive description of this field. Both works give many examples of symmetry in diverse fields of art and science. Hargittai [1986] constitutes a recent survey of the widespread use of symmetry.

Animals, living on the surface of the earth, almost always consist of two mirror-equal halves, arranged relative to one another as an object and its mirror-image. See Figure 2.14 for an illustration of this phenomenon. The imaginary plane that divides such creatures into two mirror-parts is called the *symmetry plane* and is denoted by the symbol $\boldsymbol{m}$. The reason for the occurrence of this type of symmetry in animals is probably the fact that for animals the directions forward and backward and up and down are essentially different, while movements to left and right are executed with the same frequency.

Many man-made objects also contain a symmetry plane ***m***. This can have functional reasons, for instance in cars, bicycles, airplanes, armchairs, or it can have aesthetic reasons, for instance in ornaments, works of art, tools, musical instruments.

Associated with the concept of symmetry is an imaging operation, by means of which the figure can be made to coincide with itself. For a figure that contains a symmetry plane ***m*** the imaging operation consists of a reflection of the figure in the symmetry plane, assuming that the plane reflects on both sides. We observe that the symmetry plane ***m*** occupies a specific position in a figure, as opposed to the arbitrary

**Figure 2.14.** A butterfly of the species *Troides helena* (cf. D'Abrera [1975]), with an imaginary symmetry plane ***m***. Note that the symmetry plane ***m*** does not only apply to the shape of the butterfly but also to its color.

position of the mirror to perform the mirror-reflection operation in the definition of mirror-similarity, Definition 2.5. Any operation of making objects coincide with themselves is called a *symmetry operation* or *symmetry transformation*. Auxiliary geometric elements, such as points, lines or planes, by means of which symmetry operations are effected are called *symmetry elements*. Strictly speaking a symmetry element is the locus of points that remain in place when a specific symmetry operation is performed. Every figure that possesses at least one nontrivial symmetry element is symmetric by definition. If a figure contains a symmetry plane $\boldsymbol{m}$, then we say that this figure has *symmetry* $\boldsymbol{m}$ or that it is *mirror-symmetric*.

Another well-known symmetry element is a symmetry axis, i.e. a line such that, when a figure is *rotated* about it, the figure comes into coincidence with itself several times. The number of coincidences in a complete rotation over an angle $2\pi$ is called the *order of the axis* and is indicated by the symbol $\boldsymbol{n}$. So $\boldsymbol{n}$ serves two purposes: to indicate the type of symmetry and to specify the order of this symmetry. The *elementary angle of rotation* is the smallest angle for which the figure

**Figure 2.15.** Ornament that occurs in Asmat woodcarving (cf. Gerbrands [1967]). The ornament has twofold rotational symmetry, i.e. $\boldsymbol{n} = \mathbf{2}$.

comes into coincidence with itself and is $2\pi/\boldsymbol{n}$ for a figure with a symmetry axis of order $\boldsymbol{n}$. Symmetry axes can be of any order, from $\mathbf{1}$ to $\boldsymbol{\infty}$. Infinitely many symmetry axes of order $\mathbf{1}$ are trivially present in any figure. We will not treat a symmetry axis of order $\mathbf{1}$ as a genuine symmetry element. A figure with a symmetry axis of order $\infty$ can be made to coincide with itself for any angle of rotation, since the elementary angle of rotation is infinitely small.

If we study symmetry in the two-dimensional plane then a mirror-symmetry plane becomes a mirror-symmetry line and an axis of rotational symmetry becomes a point of rotational symmetry.

Of all simple closed contours that bound simply connected two-dimensional figures, only a circle has a symmetry point of order $\infty$. If a figure contains a symmetry axis or point of order $\boldsymbol{n}$, then we say that this figure has *symmetry* $\boldsymbol{n}$ or that it has *$\boldsymbol{n}$-fold rotational symmetry*. See Figure 2.15 for an example of a figure with symmetry $\boldsymbol{n} = \mathbf{2}$.

In the living world, species with symmetry $\boldsymbol{n}$ but without mirror-symmetry are not frequently encountered. In man-made objects, however, symmetry $\boldsymbol{n}$ without mirror-symmetry is rather common, especially in technological objects such as machine parts, rotating about a specific axis, or the vanes of a windmill. Many ornaments also have this type of symmetry.

Figures can have a symmetry plane $\boldsymbol{m}$ combined with a rotation axis of order $\boldsymbol{n}$ that lies in $\boldsymbol{m}$. Such figures are said to have *symmetry* $\boldsymbol{n \cdot m}$, where the dot indicates that $\boldsymbol{n}$ lies in $\boldsymbol{m}$. It is easily verified that a figure with symmetry $\boldsymbol{n \cdot m}$ has $\boldsymbol{n}$ distinct symmetry planes, all coinciding at the symmetry axis of order $\boldsymbol{n}$. However, the axis of order $\boldsymbol{n}$ and one symmetry plane $\boldsymbol{m}$ can be considered as generating symmetry elements, the other $\boldsymbol{n}-\mathbf{1}$ symmetry planes as arising from these generating elements. See Figure 2.16 for an example of a figure with symmetry $\boldsymbol{n \cdot m} = \mathbf{5}\cdot\boldsymbol{m}$.

Symmetries of the types $\mathbf{2}\cdot\boldsymbol{m}$, $\mathbf{3}\cdot\boldsymbol{m}$, $\mathbf{4}\cdot\boldsymbol{m}$ and $\mathbf{6}\cdot\boldsymbol{m}$ are widespread in the plant and animal world. Symmetry of the type $\mathbf{5}\cdot\boldsymbol{m}$ is often encountered in the fruits of plants, for instance in apples, and sometimes in animals, for instance in various starfish (Shubnikov and Koptsik [1974]). In man-made objects symmetry of the type $\boldsymbol{n \cdot m}$ is also common, for example in vases, lamps, tables, rotating machine parts, ornaments.

The complete set of all symmetry elements in a figure determines its *symmetry class*. The complete set of symmetry operations that is provided by the symmetry class of a figure is called the *symmetry group* of that figure.

In Chapter 1 we already mentioned that we are concerned in this book with the geometrical aspects of shape. As a consequence we restrict ourselves to the geometrical aspects of symmetry and refrain from considering the physical aspects of symmetry such as colour and internal structure. Furthermore, we only pay attention to the types of symmetry that may occur in a single object. We do not study, for instance, the types of symmetry that may occur in unbounded structures such as bands or network patterns.

**Figure 2.16** A picture of a flower of Sedum acre. This flower has symmetry $\boldsymbol{n \cdot m = 5 \cdot m}$

In Section 1.1 we remarked that the plane figures that we study are usually the result of a projection of three-dimensional objects onto the two-dimensional plane of analysis. If a three-dimensional object has symmetry $\boldsymbol{m}$, $\boldsymbol{n}$ or $\boldsymbol{n \cdot m}$, then we can only observe this symmetry in a two-dimensional projection if the plane of projection is perpendicular to the symmetry elements in the three-dimensional object. As we remarked earlier, in two dimensions the mirror-symmetry element is a line $\boldsymbol{m}$ and the element of $\boldsymbol{n}$-fold rotational symmetry is a point of order $\boldsymbol{n}$. The study of the geometrical aspects of symmetry in single plane figures can be restricted to the symmetry classes ***m***, ***n*** and ***n·m***. In Section 1.2 we argued that the study of the shape of simply connected plane figures can be performed through the analysis of their contours. Therefore we now present definitions of geometric symmetry of the types ***m***, ***n*** and ***n·m*** in the plane in terms of the contours of simply connected two-dimensional figures.

**Definition 2.6.** *Geometric symmetry **m** or geometric mirror-symmetry.* A contour $\gamma$ is said to have geometric symmetry ***m*** or geometric mirror-symmetry iff there exists a line ***m*** in the plane such that, when $\gamma$ is mirror-reflected about ***m***, it coincides with itself.

□

By necessity, the symmetry line ***m*** of a geometrically mirror-symmetric contour $\gamma$ has two intersections with $\gamma$ and passes through the centroid of the interior of $\gamma$. In the literature on shape analysis, symmetry ***m*** is also called axial symmetry (Zahn and Roskies [1972]) and lateral symmetry (Davis [1977b], Chaudhuri and Dutta Majumder [1980]) or bilateral symmetry (Wallace and Wintz [1980]). Recalling the concept of enantiomorphism, introduced in Section 2.3, it is easy to verify that a necessary and sufficient condition for the existence of enantiomorphic versions, or a 'left' and a 'right' version, of a two-dimensional shape is that it has no symmetry line ***m***. Thus we observe that the set of equivalence classes of similar shapes is divided into mirror-symmetric classes and pairs of enantiomorphic classes.

**Definition 2.7.** *Geometric symmetry **n** or **n**-fold geometric rotational symmetry.*
A contour $\gamma$ is said to have geometric symmetry ***n*** or ***n***-fold geometric rotational symmetry iff there exists a point of order ***n*** in the plane such that, when $\gamma$ is rotated about this point, it coincides with itself after each rotation over an angle $2\pi/\boldsymbol{n}$.

□

By necessity, the symmetry point of order ***n*** of an ***n***-fold geometrically rotationally symmetric contour $\gamma$ coincides with the centroid of the interior of $\gamma$. In the literature on shape analysis symmetry ***n*** is also called rotational symmetry of degree ***n*** (Granlund [1972]) and ***n***-fold axial symmetry (Santisteban, García and Carrascosa [1981]).

**Definition 2.8.** *Geometric symmetry **n·m** or **n**-fold geometric compositional symmetry.*
A contour $\gamma$ is said to have geometric symmetry ***n·m*** or ***n***-fold geometric compositional symmetry iff $\gamma$ has both symmetry ***n*** and symmetry ***m***.

□

We call this type of symmetry ***n***-fold compositional symmetry because it is composed of ***n***-fold rotational symmetry and mirror-symmetry. As we remarked earlier, the dot in ***n·m*** indicates that the symmetry element ***n*** is coincident with the symmetry element ***m***. In a two-dimensional figure, that has both symmetry ***n*** and symmetry ***m***, it is obviously guaranteed that the symmetry point ***n*** lies on the symmetry line ***m***.

Based on the Definitions 2.6 and 2.7 we now formulate the conditions that must be satisfied by the parametric contour representations $z$, $\dot{z}$, $\ddot{z}$, $\theta$, $\varphi$, $\psi$ and $K$, indicated by the generic symbol $f$, of a contour $\gamma$ in order that $\gamma$ has geometric symmetry ***m*** or geometric symmetry ***n***, respectively.

**Theorem 2.5.**
A contour $\gamma$, with contour representation $f$, has geometric symmetry ***m*** or geometric mirror-symmetry iff there exist scalars $\zeta \in \mathbb{C}$, $\alpha$, $\tau \in \mathbb{R}$ such that

$$\mathcal{D}_\zeta f = \mathcal{M}_x \mathcal{T}_\tau \mathcal{R}_\alpha \mathcal{D}_\zeta f. \tag{2.4.1}$$

*Proof*

It is easily verified that a contour $\gamma$ has geometric symmetry ***m*** iff $\gamma$ is geometrically mirror-similar with itself (cf. Definitions 2.6 and 2.7). As a result of this equivalence the validity of this theorem follows immediately from Theorem 2.4. We remark that no scaling of the contour is performed if we wish to establish the mirror-similarity of the contour with itself.

□

A survey of the formulations of this condition for geometric symmetry ***m***, in terms of the individual contour representations, can be found in Table 2.5.

**Table 2.5.** Necessary and sufficient conditions that the individual contour representations must satisfy, for some $\zeta \in \mathbb{C}$, $\alpha, \tau \in \mathbb{R}$ and $\forall t \in [0, 2\pi]$, in order to render a contour with symmetry $\boldsymbol{m}$.

| Contour representation | Condition for geometric symmetry $\boldsymbol{m}$ |
|---|---|
| $z$ | $z(t) + \zeta = e^{-i\alpha}\{\bar{z}(-t+\tau) + \bar{\zeta}\}$ |
| $\dot{z}$ | $\dot{z}(t) = -e^{-i\alpha}\bar{\dot{z}}(-t+\tau)$ |
| $\ddot{z}$ | $\ddot{z}(t) = e^{-i\alpha}\bar{\ddot{z}}(-t+\tau)$ |
| $\theta$ | $\theta(t) = \{-\theta(-t+\tau) + \pi - \alpha\} \bmod 2\pi$ |
| $\varphi$ | $\varphi(t) = -\varphi(-t+\tau) + \varphi(\tau)$ |
| $\psi$ | $\psi(t) = -\psi(-t+\tau) + \psi(\tau)$ |
| $K$ | $K(t) = K(-t+\tau)$ |

The values of the scalars $\zeta$, $\alpha$ and $\tau$, for which the representation $f$ of a geometrically mirror-symmetric contour satisfies Eq. 2.4.1, can be interpreted as follows. The translation operator $\mathfrak{D}_\zeta$ assures that the symmetry line $\boldsymbol{m}$ passes through the origin. This is the reason why the translation operator $\mathfrak{D}_\zeta$ appears on both sides of Eq. 2.4.1, which however does not lead to a loss in generality. In practice, $-\zeta$ usually corresponds to the position of the centroid of the interior of $\gamma$, which coincides with the symmetry line $\boldsymbol{m}$ of a geometrically mirror-symmetric contour. The angle between the symmetry line $\boldsymbol{m}$ and the positive $x$-axis equals $-\alpha/2$ or $-\alpha/2 \pm \pi$. If the starting point on the contour is parametrically shifted over $-\tau/2$, or $-\tau/2 \pm \pi$, then it coincides with the symmetry line $\boldsymbol{m}$.

**Theorem 2.6.**
A contour $\gamma$, with contour representation $f$, has geometric symmetry $\boldsymbol{n}$ or $\boldsymbol{n}$-fold geometric rotational symmetry iff there exists a scalar $\zeta \in \mathbb{C}$ such that

$$\mathfrak{D}_\zeta f = \mathfrak{T}_{2\pi/n}\mathfrak{R}_{2\pi/n}\mathfrak{D}_\zeta f. \tag{2.4.2}$$

*Proof*

The translation operator $\mathcal{D}_\zeta$ on both sides of Eq. 2.4.2 allows for the determination of the point in the plane about which the contour must be rotated.

If a contour, that has $\boldsymbol{n}$-fold geometric rotational symmetry, is rotated about this point over $2\pi/\boldsymbol{n}$, then it comes into coincidence with itself (cf. Definition 2.7), which can be considered as similarity with itself upon this rotation. Since the corresponding contour representations are parametrized by a normalized arc length parameter, with a fundamental parameter interval of length $2\pi$, the rotation $\mathcal{R}_{2\pi/\boldsymbol{n}}$ must be compensated by a starting point shift $\mathcal{T}_{2\pi/\boldsymbol{n}}$ in these contour representations, thus leading to Eq. 2.4.2.

Because of the one-to-one relation between a contour and the contour representation $f$, possibly up to a rigid motion in the plane (cf. the proof of Theorem 2.3), Eq. 2.4.2 will be valid iff $f$ represents a contour that has geometric rotational symmetry $\boldsymbol{n}$.

□

A survey of the formulations of this condition for geometric symmetry $\boldsymbol{n}$, in terms of the individual contour representations, can be found in Table 2.6.

The translation operator $\mathcal{D}_\zeta$ in Eq. 2.4.2 causes the centroid of the interior of the $\boldsymbol{n}$-fold geometrically rotationally symmetric contour to coincide with the origin. So if $\zeta$ is the complex value for which Eq. 2.4.2 is satisfied, then $-\zeta$ corresponds with the position of the centroid. As a result of the application of $\mathcal{D}_\zeta$, subsequent rotations are about the centroid. This is again the reason why, similar to Eq. 2.4.1, the translation operator $\mathcal{D}_\zeta$ appears on both sides of Eq. 2.4.2. It is easily verified that if Eq. 2.4.2 is valid, then also

$$\mathcal{D}_\zeta f = \mathcal{T}_{m(2\pi/\boldsymbol{n})}\mathcal{R}_{m(2\pi/\boldsymbol{n})}\mathcal{D}_\zeta f, \qquad \forall m \in \mathbb{Z}. \tag{2.4.3}$$

holds. Note in Table 2.6 that we used the property of the cumulative angular function $\varphi$ of a contour with geometric symmetry $\boldsymbol{n}$ that

$$\varphi\left(m\frac{2\pi}{\boldsymbol{n}}\right) = m\frac{2\pi}{\boldsymbol{n}}, \qquad \forall m \in \mathbb{Z}. \tag{2.4.4}$$

**Table 2.6.** Necessary and sufficient conditions that the individual contour representations must satisfy, for some $\zeta$ and $\forall t \in [0, 2\pi]$, in order to render a contour with symmetry $\boldsymbol{n}$.

| Contour representation | Condition for geometric symmetry $\boldsymbol{n}$ |
|---|---|
| $z$ | $z(t) + \zeta = \mathrm{e}^{\mathrm{i}\frac{2\pi}{n}} \left\{ z\left(t - \frac{2\pi}{n}\right) + \zeta \right\}$ |
| $\dot{z}$ | $\dot{z}(t) = \mathrm{e}^{\mathrm{i}\frac{2\pi}{n}} \dot{z}\left(t - \frac{2\pi}{n}\right)$ |
| $\ddot{z}$ | $\ddot{z}(t) = \mathrm{e}^{\mathrm{i}\frac{2\pi}{n}} \ddot{z}\left(t - \frac{2\pi}{n}\right)$ |
| $\theta$ | $\theta(t) = \left\{ \theta\left(t - \frac{2\pi}{n}\right) + \frac{2\pi}{n} \right\} \bmod 2\pi$ |
| $\varphi$ | $\varphi(t) = \varphi\left(t - \frac{2\pi}{n}\right) - \varphi\left(- \frac{2\pi}{n}\right) = \varphi\left(t - \frac{2\pi}{n}\right) + \frac{2\pi}{n}$ |
| $\psi$ | $\psi(t) = \psi\left(t - \frac{2\pi}{n}\right)$ |
| $K$ | $K(t) = K\left(t - \frac{2\pi}{n}\right)$ |

**Theorem 2.7.**
A contour $\gamma$, with contour representation $f$, has geometric symmetry $\boldsymbol{n \cdot m}$ or $\boldsymbol{n}$-fold geometric compositional symmetry iff there exist scalars $\zeta \in \mathbb{C}$, $\alpha, \tau \in \mathbb{R}$ such that both Eq. 2.4.1 and Eq. 2.4.2 are satisfied.

□

The validity of Theorem 2.7 follows immediately from Definition 2.8 and Theorems 2.5 and 2.6.

## 2.5 Concluding remarks

In Sections 2.1 and 2.2 we introduced a number of parametric contour representations that preserve shape information. These are:

$z$ – position function
$\dot{z}$ – tangent function
$\ddot{z}$ – acceleration function
$\theta$ – tangent angle function
$\varphi$ – cumulative angular function
$\psi$ – periodic cumulative angular function
$K$ – curvature function.

The relations between these representations have been described. We have introduced the symbol $f$ to collectively indicate the aforementioned contour representations. For simple closed contours all these representations, except for $\varphi$, are periodic if we move continuously along the contour. We have decided to use the normalized arc length parameter $t$ in the parametric contour representations, normalizing their period to $2\pi$.

We have also introduced the similarity operators or equiform operators $\mathcal{S}_\beta$, $\mathcal{D}_\zeta$, $\mathcal{R}_\alpha$ and $\mathcal{T}_\tau$, to perform scaling, translation, rotation and parametric starting point shift on contours, respectively, and the mirror-similarity operator $\mathcal{M}_x$ to perform a mirror-reflection about the $x$-axis.

In Section 2.1 we argued why we consider contour representations that specify, in some way or another, the distance between a contour and a contour-dependent reference position, not suitable for shape analysis purposes.

In Section 2.3 we have expressed what we consider to be the shape of a two-dimensional object. Based on this we have defined geometric similarity and geometric mirror-similarity between contours. In Tables 2.3 and 2.4 we have listed the necessary and sufficient conditions that the individual contour representations must satisfy in order to render pairs of geometrically similar and geometrically mirror-similar contours. We identified geometric similarity as an equivalence relation that generates equivalence classes of geometrically similar contours. These equivalence classes are divided by the concepts of geometric mirror-similarity and geometric mirror-symmetry into mirror-symmetric classes and pairs of enantiomorphic classes.

In Section 2.4 we defined the types of geometric symmetry that may occur in single two-dimensional shapes. These are:

- geometric symmetry ***m*** or geometric mirror-symmetry,
- geometric symmetry ***n*** or ***n***-fold geometric rotational symmetry,
- geometric symmetry ***n·m*** of ***n***-fold geometric compositional symmetry, being the combination of symmetry ***n*** and symmetry ***m***.

In Tables 2.5 and 2.6 we have listed the necessary and sufficient conditions that the individual contour representations must satisfy in order to render a contour with symmetry ***m*** and symmetry ***n***, respectively.

As formulated in Sections 2.3 and 2.4, geometric similarity, geometric mirror-similarity and the three types of geometric symmetry are mathematical abstractions. For various reasons they are not likely to occur in practice. In order to have the disposal of means to establish the extent of similarity or mirror-similarity that exists between shapes, we will have to define appropriate measures. Likewise we will have to determine appropriate measures to establish the extent with which a particular type of symmetry is present in a shape. The conditions for geometric similarity, geometric mirror-similarity and the three types of geometric symmetry, in terms of the contour representations, will serve to set boundary conditions for such measures. These topics will be dealt with in Chapter 4.

## References

Abramowitz, M. and I.A. Stegun, Eds. [1972]
*Handbook of Mathematical Functions with Formulas, Graphs, and Mathematical Tables*, Washington, DC: U.S. Dept. of Commerce, National Bureau of Standards.

Ahlfors, L.V. [1953]
*Complex Analysis: An Introduction to the Theory of Analytic Functions of a Complex Variable*, New York: McGraw-Hill Book Co., Inc.

Anderson, I.M. and J.C. Bezdek [1984]
'Curvature and Tangential Deflection of Discrete Arcs: A Theory Based on the Commutator of Scatter Matrix Pairs and Its Application to Vertex Detection in Planar Shape Data', IEEE Trans. Patt. Anal. and Mach. Intell. **PAMI-6**: 27-40.

Attneave, F. [1954]
'Some Informational Aspects of Visual Perception', Psychol. Rev. **61**: 183-193.

Attneave, F. and M.D. Arnoult [1956]
'The Quantitative Study of Shape and Pattern Perception', Psychol. Bull. **53**: 452-471. Reprinted in: *Pattern Recognition*, L. Uhr (Ed.): 123-141, New York: John Wiley and Sons, Inc., 1966.

Ballard, D.H. [1981]
'Generalizing the Hough Transform to Detect Arbitrary Shapes', Patt. Recogn. **13**: 111-122.

Barrow, H.G. and R.J. Popplestone [1971]
'Relational Descriptions in Picture Processing'. In: *Machine Intelligence 6*, B. Meltzer and D. Michie (Eds.): 377-396, Edinburgh: Edinburgh University Press.

Beddow, J.K. [1980]
'Particle Morphological Analysis'. In: *Advanced Particulate Morphology*, J.K. Beddow and T.P. Meloy (Eds.): 1-84, Boca Raton, FL: C.R.C. Press.

Beddow, J.K. and G. Philip [1975]
'On the Use of a Fourier Analysis Technique for Describing the Shape of Individual Particles', Planseeber. für Pulvermetall. **23**: 3-14.

Beddow, J.K., G.C. Philip, A.F. Vetter and M.D. Nasta [1977]
'On Relating Some Particle Profile Characteristics to the Profile Fourier Coefficients', Powd. Technol. **18**: 19-25.

Bengtsson, E., O. Eriksson, J. Holmquist, T. Jarkrans, B. Nordin and B. Stenkvist [1981]
'Segmentation of Cervical Cells: Detection of Overlapping Cell Nuclei', Comp. Graph. and Im. Proc. **16**: 382-394.

Bennett, J.R. and J.S. MacDonald [1975]
'On the Measurement of Curvature in a Quantized Environment', IEEE Trans. Comp. **C-24**: 803-820.

Bowie, J.E. and I.T. Young [1977a]
'An Analysis Technique for Biological Shape – II', Acta Cytol. **21**: 455-464.

Bowie, J.E. and I.T. Young [1977b]
'An Analysis Technique for Biological Shape – III', Acta Cytol. **21:** 739-746.

Brill, E.L. [1968]
'Character Recognition via Fourier Descriptors', Proc. WESCON, Session 25, Los Angeles, CA: Paper 25/3.

Burkhardt, H. [1979]
*Transformationen zur Lageinvarianten Merkmalgewinnung* (Habilitationsschrift), Düsseldorf, Germany: VDI-Verlag, Fortschritt-Berichte der VDI-Zeitschriften, Reihe 10 (Angewandte Informatik), Nr. 7.

Carpenter, L.C. [1980]
'Computer Rendering of Fractal Curves and Surfaces'. In: Siggraph '80 Conference Proceedings Supplement: 9-15, New York: Association for Computing Machinery (ACM).

Chang, T.L. [1976]
'On Similarity Measures for Orientated Line Patterns', Proc. Third Intl. Joint Conf. on Patt. Recogn., Coronado, CA: 208-210.

Chaudhuri, B.B. and D. Dutta Majumder [1980]
'Recognition and Fuzzy Description of Sides and Symmetries of Figures by Computer', Int. Journ. of Syst. Sci. **11**: 1435-1445.

Chen, C.-J. and Shi Q.-Y. [1980]
'Shape Features for Cancer Cell Recognition', Proc. Fifth Intl. Conf. on Patt. Recogn., Miami Beach, FL: 579–581.

Churchill, R.V., J.W. Brown and R.F. Verhey [1974]
*Complex Variables and Applications*, Third Edition, Tokyo: McGraw-Hill Kogakusha, Ltd.

Cosgriff, R.L. [1960]
'Identification of Shape', Rep. 820-11, ASTIA AD 254 792, Ohio State Univ. Res. Foundation, Columbus, OH.

Courant, R. and F. John [1965]
*Introduction to Calculus and Analysis, Volume One*, New York: John Wiley and Sons, Inc.

Crimmins, T.R. [1982]
'A Complete Set of Fourier Descriptors for Two-Dimensional Shapes', IEEE Trans. Syst., Man and Cybern. **SMC-12**: 848-855.

D'Abrera, B. [1975]
*Birdwing Butterflies of the World*, Melbourne, Australia: Lansdowne Press.

Davis, L.S. [1977a]
'Understanding Shape: Angles and Sides', IEEE Trans. Comp. **C-26**: 236-242.

Davis, L.S. [1977b]
'Understanding Shape: Symmetry', IEEE Trans. Syst., Man and Cybern. **SMC-7**: 204-212.

Davis, L.S. [1979]
'Shape Matching Using Relaxation Techniques', IEEE Trans. Patt. Anal. and Mach. Intell. **PAMI-1**: 60-72.

Davis, L.S. [1982]
'Hierarchical Generalized Hough Transforms and Line-Segment Based Generalized Hough Transforms', Patt. Recogn. **15**: 277-285.

Davis, L.S. and A. Rosenfeld [1977]
'Curve Segmentation by Relaxation Labeling', IEEE Trans. Comp. **C-26**: 1053-1057.

De Boor, C. [1978]
*A Practical Guide to Splines*, New York: Springer-Verlag.

Dessimoz, J.-D. [1978]
'Visual Identification and Location in a Multi-Object Environment by Contour Tracking and Curvature Description', Proc. Eighth Intern. Symp. on Industr. Robots, Stuttgart, Germany: 764-777.

Dessimoz, J.-D. [1980]
*Traitement des Contours en Reconnaissance de Formes Visuelles: Application en Robotique* (Ph.D. Thesis, No. 387, École Polytechnique Fédérale de Lausanne, Lausanne, Switzerland).

Dinstein, I. and T. Silberberg [1980]
'Shape Discrimination with Walsh Descriptors', Proc. Fifth Intl. Conf. on Patt. Recogn., Miami Beach, FL: 1055–1061.

Duda, R.O. and P.E. Hart [1973]
*Pattern Classification and Scene Analysis*, New York: John Wiley and Sons, Inc.

Eccles, M.J., M.P.C. McQueen and D. Rosen [1977]
'Analysis of the Digitized Boundaries of Planar Objects', Patt. Recogn. **9**: 31-41.

Ehrlich, R. and B. Weinberg [1970]
'An Exact Method for Characterization of Grain Shape', Journ. of Sedim. Petrol. **40**: 205-212.

Ellis, Jr., J.R. and M. Eden [1976]
'On the Number of Sides Necessary for Polygonal Approximation of Black-and-White Figures in a Plane', Inform. and Contr. **30**: 169-186.

Escher, M.C., J.L. Locher, C.H.A. Broos and H.S.M. Coxeter [1972]
*The World of M.C. Escher*, New York: H.N. Abrams.

Fischler, M.A. and R.C. Bolles [1983]
'Perceptual Organization and Curve Partitioning', Proc. IEEE Comp. Soc. Conf. on Comp. Vision and Patt. Recogn., Washington, DC: 38–46.

Flook, A.G. [1978]
'The Use of Dilation Logic on the Quantimet to Achieve Fractal Dimension Characterisation of Textured and Structured Profiles', Powd. Technol. **21**: 295-298.

Fong, S.-T., Beddow, J.K. and A.F. Vetter [1979]
'A Refined Method of Particle Shape Representation', Powd. Technol. **22**: 17-21.

Fournier, A., D. Fussel and L.C. Carpenter [1982]
'Computer Rendering of Stochastic Models', Comm. ACM **25**: 371-384.

Freeman, H. [1961a]
'On the Encoding of Arbitrary Geometric Configurations', IRE Trans. Electr. Comp. **EC-10**: 260-268.

Freeman, H. [1961b]
'A Technique for the Classification and Recognition of Geometric Patterns', Proc. Third Intl. Congr. on Cybern., Namur, Belgium: 348–368.

Freeman, H. [1970]
'Boundary Encoding and Processing'. In: *Picture Processing and Psychopictorics*, B.S. Lipkin and A. Rosenfeld (Eds.): 241-266, New York: Academic Press.

Freeman, H. [1978a]
'Shape Description via the Use of Critical Points', Patt. Recogn. **10**: 159-166.

Freeman, H. [1978b]
'Application of the Generalized Chain Coding Scheme to Map Data and Image Processing', Proc. IEEE Comp. Soc. Conf. on Patt. Recogn. and Map Data Proc., Chicago, IL: 220–226.

Freeman, H. [1979]
'Use of Incremental Curvature for Describing and Analyzing Two-Dimensional Shape', Proc. IEEE Comp. Soc. Conf. on Patt. Recogn. and Image Proc., Chicago, IL: 437-444.

Freeman, H. and L.S. Davis [1977]
'A Corner-Finding Algorithm for Chain-Coded Curves', IEEE Trans. Comp. **C-26**: 297-303.

Freeman, H. and J.M. Glass [1969]
'On the Quantization of Line-Drawing Data', IEEE Trans. Syst., Sci. and Cybern. **SSC-5**: 70-79.

Gerbrands, A.A. [1967]
*Wow-Ipits. Eight Asmat Woodcarvers of New Guinea*, The Hague and Paris: Mouton & Co., Publishers.

Giardina, C.R. and F.P. Kuhl [1977]
'Accuracy of Curve Approximation by Harmonically Related Vectors with Elliptical Loci', Comp. Graph. and Im. Proc. **6**: 277-285.

Gotoh, K. [1979]
'Shape Characterization of Two-Dimensional Forms', Powd. Technol. **23**: 131-134.

Granlund, G.H. [1972]
'Fourier Preprocessing for Hand Print Character Recognition', IEEE Trans. Comp. **C-21**: 195-201.

Groen, F.C.A. [1977]
*An Analysis of DNA Based Measurement Methods Applied to Human Chromosome Classification* (Ph.D. Thesis, Delft University of Technology, Delft, The Netherlands), Pijnacker, The Netherlands: Dutch Efficiency Bureau.

Grogan, T.A. and O.R. Mitchell [1983]
'Partial Shape Recognition Using Fourier-Mellin Transform Methods', Digest of Technical Papers Presented at the Topical Meeting on Signal Recovery and Synthesis with Incomplete Information and Partial Constraints, Incline Village, NV: ThA19–1–ThA19-4.

Guggenheimer, H.W. [1963]
*Differential Geometry*, New York: McGraw-Hill Book Co., Inc.

Hargittai, I. [1986]
*Symmetry: Unifying Human Understanding*, Elmsford, NY: Pergamon Press.

Hough, P.V.C. [1962]
'Method and Means for Recognizing Complex Patterns', US Patent 3,069,654, Dec. 18, 1962.

Hung, S.H.Y. and T. Kasvand [1983]
'Critical Points on a Perfectly 8- or 6-Connected Thin Binary Line', Patt. Recogn. **16**: 297-306.

Janssen, A.J.E.M. and P. van der Steen [1984]
*Integration Theory*, Lecture Notes in Mathematics **1078**, Berlin: Springer-Verlag.

Kailay, B.C., R. Sadananda and J.R. Das [1981]
'An Algorithm for Segmenting Juxtaposed Objects', Patt. Recogn. **13**: 347-351.

Kajiya, J.T. [1983]
'New Procedures for Ray Tracing Procedurally Defined Objects', ACM Trans. Graph. **2**: 161-181.

Kammenos, P. [1978]
'Performances of Polar Coding for Visual Localization of Planar Objects', Proc. Eighth Intern. Symp. on Industr. Robots, Stuttgart, Germany: 143-154.

Kashyap, R.L. and B.J. Oommen [1982]
'A Geometrical Approach to Polygonal Dissimilarity and the Classification of Closed Boundaries', IEEE Trans. Patt. Anal. and Mach. Intell. **PAMI-4**: 649-654.

Kasvand, T. and N. Otsu [1982]
'Regularization of Piece-Wise Linear Digitized Plane Curves for Shape Analysis and Smooth Reconstruction', Proc. Sixth Intl. Conf. on Patt. Recogn., Munich, Germany: 468-471.

Kaye, B.H. [1978]
'Specification of the Ruggedness and/or Texture of a Fine Particle Profile by Its Fractal Dimension', Powd. Technol. **21**: 1-16.

Kitchen, L. and A. Rosenfeld [1982]
'Grey-Level Corner Detection', Patt. Recogn. Lett. **1**: 95-102.

Kuhl, F.P. and C.R. Giardina [1982]
'Elliptic Fourier Features of a Closed Contour', Comp. Graph. and Im. Proc. **18**: 236-258.

Landau, L.D. and E.M. Lifschitz [1970]
*Theory of Elasticity*, Second Edition, Oxford: Pergamon Press.

Ledley, R.S. [1964]
'High-Speed Automatic Analysis of Biomedical Pictures', Science **146**: 216-223.

Ledley, R.S. [1972]
'Analysis of Cells', IEEE Trans. Comp. **C-21**: 740-753.

Lighthill, M.J. [1962]
*Introduction to Fourier Analysis and Generalised Functions*, Cambridge, England: Cambridge University Press.

Luerkens, D.W., J.K. Beddow and A.F. Vetter [1982a]
'Morphological Fourier Descriptors', Powd. Technol. **31**: 209-215.

Luerkens, D.W., J.K. Beddow and A.F. Vetter [1982b]
'A Generalized Method of Morphological Analysis (the (R,S) Method)', Powd. Technol. **31**: 217-220.

McClure, D.E. [1975]
'Nonlinear Segmented Function Approximation and Analysis of Line Patterns', Quart. of Appl. Math. **33**: 1-37.

McClure, D.E. and R.A. Vitale [1975]
'Polygonal Approximation of Plane Convex Bodies', Journ. Math. Anal. and Appl. **51**: 326-358.

Mandelbrot, B.B. [1967]
'How Long is the Coast of Britain: Statistical Self-Similarity and Fractional Dimension', Science **155**: 636-638.

Mandelbrot [1977]
*Fractals: Form, Chance and Dimension*, San Francisco, CA: W.H. Freeman and Co.

Mandelbrot, B.B. [1982a]
*The Fractal Geometry of Nature*, San Francisco, CA: W.H. Freeman and Co.

Mandelbrot, B.B. [1982b]
'Comment on Computer Rendering of Fractal Stochastic Models', Comm. ACM **25**: 581-584.

Martin, W.N. and J.K. Aggarwal [1979]
'Computer Analysis of Dynamic Scenes Containing Curvilinear Figures', Patt. Recogn. **11**: 167-178.

Meloy, T.P. [1977a]
'A Hypothesis for Morphological Characterization of Particle Shape and Physiochemical Properties', Powd. Technol. **16**: 233-253.

Meloy, T.P. [1977b]
'Fast Fourier Transforms Applied to Shape Analysis of Particle Silhouettes to Obtain Morphological Data', Powd. Technol. **17**: 27-35.

Montanari, U. [1970]
'A Note on Minimal Length Polygonal Approximation to a Digitized Contour', Comm. ACM **13**: 41-47.

Nahin, P.J. [1974]
'Silhouette Descriptor for Image Pre-Processing and Recognition', Patt. Recogn. **6**: 85-95.

Nevins, A.J. [1979]
'An Orientation Free Study of Handprinted Characters', Patt. Recogn. **11**: 155-164.

Ozaki, H., S. Waku, A. Mohri and M. Takata [1982]
'Pattern Recognition of a Grasped Object by Unit-Vector Distribution', IEEE Trans. Syst., Man and Cybern. **SMC-12**: 315-324.

Pavlidis, T. [1977a]
*Structural Pattern Recognition*, New York: Springer-Verlag.

Pavlidis, T. [1977b]
'Polygonal Approximations by Newton's Method', IEEE Trans. Comp. **C-26**: 800-807.

Pavlidis, T. [1978]
'A Review of Algorithms for Shape Analysis', Comp. Graph. and Im. Proc. **7**: 243-258.

Pavlidis, T. and F. Ali [1975]
'Computer Recognition of Handwritten Numerals by Polygonal Approximations', IEEE Trans. Syst., Man and Cybern. **SMC-5**: 610-614.

Pavlidis, T. and S.L. Horowitz [1974]
'Segmentation of Plane Curves', IEEE Trans. Comp. **C-23**: 860-870.

Pentland, A. [1983]
'Fractal-Based Description of Natural Scenes', Proc. IEEE Comp. Soc. Conf. on Comp. Vision and Patt. Recogn., Washington, DC: 201-209.

Perkins, W.A. [1978]
'A Model-Based Vision System for Industrial Parts', IEEE Trans. Comp. **C-27**: 126-143.

Persoon, E. and K.-S. Fu [1977]
'Shape Discrimination Using Fourier Descriptors', IEEE Trans. Syst., Man and Cybern. **SMC-7**: 170-179.

Pineda, J.-C. and P. Horaud [1983]
'An Improved Method for High-Curvature Detection with Applications to Automated Inspection', Sign. Proc. **5**: 117-125.

Proffitt, D. [1982]
'Normalization of Discrete Planar Objects', Patt. Recogn. **15**: 137-143.

Ramer, U. [1972]
'An Iterative Procedure for the Polygonal Approximation of Plane Curves', Comp. Graph. and Im. Proc. **1**: 244-256.

Richard, Jr., C.W. and H. Hemami [1974]
'Identification of Three-Dimensional Objects Using Fourier Descriptors of the Boundary Curve', IEEE Trans. Syst., Man and Cybern. **SMC-4**: 371-378.

Riesz, F. and B. Sz.-Nagy [1955]
*Functional Analysis*, New York: Frederick Ungar Publishing Company.

Rosenfeld, A. and E. Johnston [1973]
'Angle Detection on Digital Curves', IEEE Trans. Comp. **C-22**: 875-878.

Rutkowski, W.S., S. Peleg and A. Rosenfeld [1981]
'Shape Segmentation Using Relaxation', IEEE Trans. Patt. Anal. and Mach. Intell. **PAMI-3**: 368-375.

Rutovitz, D. [1970]
'Centromere Finding: Some Shape Descriptors for Small Chromosome Outlines'. In: *Machine Intelligence 5*, B. Meltzer and D. Michie (Eds.): 435-462, Edinburgh: Edinburgh University Press.

Saghri, J.A. and H. Freeman [1981]
'Analysis of the Precision of Generalized Chain Codes for the Representation of Planar Curves', IEEE Trans. Patt. Anal. and Mach. Intell. **PAMI-3**: 533-539.

Sankar, P.V. and C.U. Sharma [1978]
'A Parallel Procedure for the Detection of Dominant Points on a Digital Curve', Comp. Graph. and Im. Proc. **7**: 403-412.

Santisteban, A., N. García and J.L. Carrascosa [1981]
'Digital Analysis of Axially Symmetric Images: Application to Viral Structures', Proc. of the Second Scandinavian Conference on Image Analysis, Helsinki, Finland: 450-455.

Sarvarayudu, G.P.R. [1982]
*Shape Analysis Using Walsh Functions* (Ph.D. Thesis, Dept. of Electron. and Electr. Commun. Engin., Indian Institute of Technology, Kharagpur, India).

Sarvarayudu, G.P.R. and I.K. Sethi [1983]
'Walsh Descriptors for Polygonal Curves', Patt. Recogn. **16**: 327-336.

Schwarcz, H.P. and K.C. Shane [1969]
'Measurement of Particle Shape by Fourier Analysis', Sedimentology **13**: 213-231.

Searle, N.H. [1970]
'Shape Analysis by Use of Walsh Functions'. In: *Machine Intelligence 5*, B. Meltzer and D. Michie (Eds.): 395-410, Edinburgh: Edinburgh University Press.

Segen, J. [1984]
'Locating Randomly Oriented Objects from Partial View'. In: *Intelligent Robots: Third International Conference on Robot Vision and Sensory Controls RoViSeC3*, D.P. Casasent and E.L. Hall (Eds.), Proc. SPIE **449**: 676-684.

Sethi, I.K. and G.P.R. Sarvarayudu [1980]
'Boundary Approximation Using Walsh Series Expansion for Numeral Recognition', Proc. Fifth Intl. Conf. on Patt. Recogn., Miami Beach, FL: 879-881.

Shapiro, S.D. [1978]
'Properties of Transforms for the Detection of Curves in Noisy Pictures', Comp. Graph. and Im. Proc. **8**: 219-236.

Shirai, Y. [1973]
'A Context Sensitive Line Finder for Recognition of Polyhedra', Artif. Intell. **4**: 95-119.

Shubnikov, A.V. and V.A. Koptsik [1974]
*Symmetry in Science and Art*, New York: Plenum Press.

Sklansky, J. [1978]
'On the Hough Technique for Curve Detection', IEEE Trans. Comp. **C-27**: 923-926.

Sklansky, J. and G.A. Davison, Jr. [1971]
'Recognition of Three-Dimensional Objects by Their Silhouettes', Journ. Soc. Photo-Opt. Instrum. Engin. **10**: 10-17.

Sklansky, J. and P.J. Nahin [1972]
'A Parallel Mechanism for Describing Silhouettes', IEEE Trans. Comp. **C-21**: 1233-1239.

Smeulders, A.W.M. [1983]
*Pattern Analysis of Cervical Specimens* (Ph.D. Thesis, University of Leyden, Leyden, The Netherlands).

Stoker, J.J. [1969]
*Differential Geometry*, New York: John Wiley and Sons, Inc.

Strackee, J. and N.J.D. Nagelkerke [1983]
'On Closing the Fourier Descriptor Presentation', IEEE Trans. Patt. Anal. and Mach. Intell. **PAMI-4**: 660-661.

Sychra, J.J., P.H. Bartels, J. Taylor, M. Bibbo and G.L. Wied [1976]
'Cytoplasmic and Nuclear Shape Analysis for Computerized Cell Recognition', Acta Cytol. **20**: 68-78.

Tai, H.T., C.C. Li and S.H. Chiang [1982]
'Application of Fourier Shape Descriptors to Classification of Fine Particles', Proc. Sixth Intl. Conf. on Patt. Recogn., Munich, Germany: 748–751.

van Otterloo, P. J. [1978]
*A Feasibility Study of Automated Information Extraction from Anthropological Pictorial Data* (Thesis, Dept. of Elec. Engin., Delft University of Technology, Delft, The Netherlands).

Wallace, T.P. and O.R. Mitchell [1980]
'Analysis of Three-Dimensional Movement Using Fourier Descriptors', IEEE Trans. Patt. Anal. and Mach. Intell. **PAMI-2**: 583-588.

Wallace, T.P., O.R. Mitchell and K. Fukunaga [1981]
'Three-Dimensional Shape Analysis Using Local Shape Analysis', IEEE Trans. Patt. Anal. and Mach. Intell. **PAMI-3**: 310-323.

Wallace, T.P. and P.A. Wintz [1980]
'An Efficient Three-Dimensional Aircraft Recognition Algorithm Using Normalized Fourier Descriptors', Comp. Graph. and Im. Proc. **13**: 99-126.

Weyl, H. [1952]
*Symmetry*, Princeton, NJ: Princeton University Press.

Williams, C.M. [1981]
'Bounded Straight-Line Approximation of Digitized Planar Curves and Lines', Comp. Graph. and Im. Proc. **16**: 370-381.

Wilson, Jr., H.B. and D.S. Farrior [1976]
'Computation of Geometrical and Inertial Properties for General Areas and Volumes of Revolution', Comp. Aid. Des. **8**: 257-263.

Yaglom, I.M. [1968]
*Complex Numbers in Geometry*, New York: Academic Press.

Young, I.T., J.E. Walker and J.E. Bowie [1974]
'An Analysis Technique for Biological Shape – I', Inform. and Contr. **25**: 357-370.

Zahn, C.T. [1966]
'Two-Dimensional Pattern Description and Recognition via Curvature Points', SLAC Report No. 70, Stanford Linear Accelerator Center, Stanford, CA.

Zahn, C.T. and R.Z. Roskies [1972]
'Fourier Descriptors for Plane Closed Curves', IEEE Trans. Comp. **C-21**: 269-281.

Zemanian, A.H. [1965]
*Distribution Theory and Transform Analysis: An Introduction to Generalized Functions, with Applications*, New York: McGraw-Hill Book Co., Inc.

Zusne, L. [1970]
*Visual Perception of Form*, New York: Academic Press.

# Chapter 3

# Fourier series expansions of parametric contour representations and their relation to similarity and symmetry

## 3.1 Applications of Fourier series in the context of shape analysis – a review

In the previous chapter we introduced a number of periodic contour representations that preserve shape information. In the literature on shape analysis, it is not the periodic contour representations themselves, but the Fourier coefficients generated by such contour representations which have been given most attention. Fourier coefficients have been proposed to serve various purposes in shape analysis procedures.

Granlund [1972], Zahn and Roskies [1972] and Tai, Li and Chiang [1982] use a limited set of combinations of Fourier coefficients directly in a multidimensional feature space to enable shape clustering and classification. The features are defined such that they are invariant for similarity transformations of contours.

Many authors use a sequence of Fourier coefficients as a representation of an object contour and define a metric on sequences of Fourier coefficients as a measure of dissimilarity between contours. To ensure that such a dissimilarity measure is invariant for similarity transformations of contours, some authors propose a combined normalization/optimization procedure (e.g. Richard and Hemami [1974], Persoon and Fu [1977], Kuhl and Giardina [1982], Watson and Shapiro [1982]). Others first perform a normalization of the Fourier coefficients to obtain a dissimilarity measure that is invariant for similarity transformations. These normalization procedures are directly based on the Fourier coefficients themselves (e.g. Persoon and Fu [1977], Burkhardt [1979], Wallace and Mitchell [1979], Wallace and Wintz [1980], Wallace [1981], Proffitt [1982]). A detailed discussion of contour normalization methods, including the ones just mentioned, will be given in Section 4.3.

For the interpolation between shapes, schemes have been proposed that interpolate between the Fourier coefficients of the position functions of given shapes (e.g. Bertrand, Queval and Maitre [1982]). Wallace and Mitchell [1980] use an interpolation procedure based on the Fourier coefficients of airplane silhouette representations to obtain greater accuracy in airplane orientation measurement.

Fourier coefficients have also been proposed to detect symmetry in objects (e.g. Granlund [1972], Zahn and Roskies [1972], Burkhardt [1979], Wallace and Wintz [1980], Crimmins [1982], Mitchell and Grogan [1984]).

In biology and in particle science, frequently not absolute shape, but rather those shape characteristics from which conclusions concerning biological and physical properties can be derived, are important. Many propositions for such shape characteristics have been formulated in terms of Fourier coefficients of periodic contour representations. We refer to Young, Walker and Bowie [1974], Sychra et al. [1976], Chen and Shi [1980] and Nguyen, Poulsen and Louis [1983] for Fourier-based shape characteristics in cell analysis, and to Schwarcz and Shane [1969], Ehrlich and Weinberg [1970], Beddow et al. [1977], Meloy [1977a], Meloy [1977b], Beddow and Meloy [1980], Luerkens, Beddow and Vetter [1982a] and Luerkens, Beddow and Vetter [1982b] for the definition of such characteristics in the context of particle analysis.

The accuracy with which a finite Fourier series approximates a particular periodic function constitutes an important subject in the theory of Fourier series. In the context of contour representation, the accuracy of approximation of finite Fourier series has been given comparatively little attention (e.g. Giardina and Kuhl [1977], Kuhl and Giardina [1982] and Etesami and Uicker [1985]). Some new results on this subject are presented in Dekking and van Otterloo [1986] and in Section 3.4.

Closely related to the accuracy of approximation by finite Fourier series is the subject of the rate of decay of the Fourier coefficients. In Young, Walker and Bowie [1974] the finiteness of the bending energy in a curve is related to the rate of decay of the Fourier coefficients generated by the position function of such a curve. Some comments on this paper relating to this topic can be found in Section 3.2. The characterization of the behavior of Fourier coefficients and of the convergence properties of Fourier series for various classes of functions plays a central role in the theory of Fourier series (e.g. Titchmarsh [1939], Zygmund [1959a], Zygmund [1959b], Lighthill [1962], Katznelson [1968], Edwards [1979], Edwards [1982]). In the following three sections we discuss some aspects of Fourier series theory in relation to contour representation.

In Section 3.2 we analyze the convergence properties of the Fourier series and the Fourier coefficients of the contour representations introduced in Chapter 2. To facilitate this discussion we introduce two new smoothness classes of contours. We show that the sequences of Fourier coefficients, generated by the representations of contours that belong to these smoothness classes, preserve shape information.

In Section 3.3 we study the consequences of (normalized) arc length parametrization upon Fourier series expansions of contour representations. We show that the condition of (normalized) arc length parametrization causes the Fourier sequences, generated by the contour representations $z$, $\dot{z}$ and $\ddot{z}$ of all contours, except for the circle, to contain an infinite number of nonzero coefficients. However, in practice we can only work with finite sequences.

Therefore, in Section 3.4 we derive upper bounds for the truncation errors resulting from finite Fourier series expansions.

In Section 3.5 we formulate conditions for geometric similarity and for geometric mirror-similarity in terms of pairs of sequences of Fourier coefficients.

Section 3.6 describes conditions for geometric symmetry $\boldsymbol{m}$ and for geometric symmetry $\boldsymbol{n}$ in terms of sequences of Fourier coefficients.

Finally, Section 3.7 contains a review of this chapter and some concluding remarks. In this section we also discuss Walsh sequency expansions, which have been proposed as alternative transform domain representations of contours.

## 3.2 Fourier series theory in relation to parametric contour representation

In the following we discuss some elements of Fourier series theory in fairly global terms and relate these to the representation of object contours. We analyze the relation between smoothness classes of contours and function class membership of the corresponding contour representations. Based on these results we establish the convergence properties of the Fourier series and Fourier coefficients generated by these contour representations. We first define some essential concepts.

**Definition 3.1.** *Fourier series.*
Let $f$ denote a Lebesgue-integrable, complex-valued function of period $2\pi$. The *exponential Fourier series*, generated by $f$, is given by

$$f(t) \sim \sum_{k \in \mathbb{Z}} \hat{f}(k)\mathrm{e}^{\mathrm{i}kt} \tag{3.2.1}$$

where the *complex Fourier coefficients* $\hat{f}(k)$ are given by the formula

$$\hat{f}(k) = \frac{1}{2\pi} \int_{2\pi} f(t)\mathrm{e}^{-\mathrm{i}kt}\mathrm{d}t. \tag{3.2.2}$$

□

The notation '$\sim$' in Eq. 3.2.1 means that the Fourier series on the righthand side is generated by $f$. This formulation of Fourier series does not make any presupposition about the convergence of the series.

**Definition 3.2.** *Partial Fourier sum of degree n.*
The *partial Fourier sum of degree n* of a $2\pi$-periodic, Lebesgue-integrable function $f$ is defined as

$$(S_n f)(t) = \sum_{|k| \leq n} \hat{f}(k)\mathrm{e}^{\mathrm{i}kt} \tag{3.2.3}$$

where $\hat{f}(k)$ is the Fourier coefficient with index $k$ generated by $f$, as defined by Eq. 3.2.2.

□

The theory of Fourier series has established various types of convergence of Fourier series to the functions that generate them, and various ways in which a Fourier series can represent a function. The type of convergence and the way of representation depends upon the function class to which a function belongs. Also the way of summing the Fourier series influences the convergence properties.

The function classes referred to are, for example, characterized by the integrability or differentiability properties of its members, or by the important property of bounded variation.

To facilitate the interpretation of the discussion that follows, a number of definitions and properties from mathematical analysis have been incorporated in Appendix A.

We shall analyze the convergence properties of the Fourier series and the Fourier coefficients of the contour representations $z$, $\dot{z}$, $\ddot{z}$, $\psi$ and $K$ of contours belonging to the classes $\boldsymbol{\Gamma}_{ps}$, $\boldsymbol{\Gamma}_{pr}$, $\boldsymbol{\Gamma}_s$ and $\boldsymbol{\Gamma}_r$. To enable this we will first establish to which function class the representations of contours from various classes belong. In the discussion about the convergence properties of Fourier series and Fourier coefficents, we shall make frequent use of the inclusion relations that were established for the contour classes in Eqs. 2.1.10 and 2.1.11. Mathematically, the class $\boldsymbol{\Gamma}_s$ poses very few restrictions on $\ddot{z}$ and $K$ (cf. Definition 2.2), which prevents us from deriving some useful properties. On the other hand, the class $\boldsymbol{\Gamma}_r$ is too restrictive for most real-world applications (cf. Definition 2.3). Therefore we will define in this section the class $\boldsymbol{\Gamma}_{wr}$ of *weakly regular simple closed curves*. In analogy with the classes $\boldsymbol{\Gamma}_{ps}$ and $\boldsymbol{\Gamma}_{pr}$ (cf. Section 2.1) we will also define the class $\boldsymbol{\Gamma}_{pwr}$ of *piecewise weakly regular simple closed curves*.

In the following we shall show that the position function $z$ of any simple closed contour $\gamma \in \boldsymbol{\Gamma}_{ps}$ will always satisfy a Lipschitz condition, i.e. $z \in \boldsymbol{\Lambda}$ (cf. Appendix A).

**Lemma 3.1.**
If $\gamma \in \boldsymbol{\Gamma}_{ps}$, then $z \in \boldsymbol{\Lambda}$ with Lipschitz constant $L/2\pi$.

*Proof*

If $\gamma \in \boldsymbol{\Gamma}_{ps}$, then it consists of a finite number of smooth contour segments (cf. Section 2.1). Along each smooth segment the position function $z$ is continuously differentiable (cf. Definition 2.2). Consequently, each smooth segment of $\gamma$ is rectifiable. Since the number of smooth segments is finite, $\gamma$ itself is rectifiable and thus (normalized) arc length can be used as a parameter for $z$ (cf. Kreyszig [1968], pp. 28-30).

To see that $z \in \boldsymbol{\Lambda}$, with Lipschitz constant $L/2\pi$, we note that $\forall \delta > 0$ (cf. Eq. 2.1.13)

$$|z(t+\delta) - z(t)| \leqslant s(t, t+\delta) = \int_t^{t+\delta} |\dot{z}(\tau)| \mathrm{d}\tau = \delta \frac{L}{2\pi} \tag{3.2.4}$$

since $z$ is parametrized according to normalized arc length. □

Lemma 3.1 is also valid for contours belonging to the classes $\boldsymbol{\Gamma}_{\mathrm{pr}}$, $\boldsymbol{\Gamma}_{\mathrm{s}}$ and $\boldsymbol{\Gamma}_{\mathrm{r}}$ since these contour classes are contained in $\boldsymbol{\Gamma}_{\mathrm{ps}}$ (cf. Eqs. 2.1.10 and 2.1.11).

**Proposition 3.1.** (Cf. Apostol [1974], p. 139.)
If a function $f$, defined on $[a, b]$, belongs to $\boldsymbol{\Lambda}$ (with Lipschitz constant $\lambda$), then $f \in \mathbf{AC}$ (cf. Appendix A).

*Proof*

This is true because for any $\varepsilon > 0$, taking $\delta = \varepsilon/\lambda$, we find for every $n$ disjoint open subintervals $(a_k, b_k)$ of $[a, b]$, $n = 1, 2, \ldots$, such that $\sum_{k=1}^{n}(b_k - a_k)$ is less than $\delta$, that

$$\sum_{k=1}^{n} |f(b_k) - f(a_k)| \leqslant \lambda \sum_{k=1}^{n} |b_k - a_k| < \lambda\delta = \varepsilon \tag{3.2.5}$$

and hence $f \in \mathbf{AC}$.

□

From Lemma 3.1 and Proposition 3.1 we may conclude that, if $\gamma \in \boldsymbol{\Gamma}_{\mathrm{ps}}$, then $z \in \mathbf{AC}$.

We now define the classes $\boldsymbol{\Gamma}_{\mathrm{wr}}$ and $\boldsymbol{\Gamma}_{\mathrm{pwr}}$ of weakly regular and piecewise weakly regular simple closed curves.

**Definition 3.3.** *Weakly regular simple closed curve.*
A curve $\gamma$, parametrized on $[0, 2\pi]$, is a weakly regular simple closed curve iff:

- $\gamma \in \boldsymbol{\Gamma}_{\mathrm{s}}$, (3.2.6a)
- $\ddot{z}$ exists and is continuous along $\gamma$ except in a finite number of points; at the latter points left and right limits of $\ddot{z}$ exist, (3.2.6b)
- $\ddot{z}(0) = \ddot{z}(2\pi)$. (3.2.6c)

□

At those points $t$ where $\ddot{z}$ does not exist we can define for example $\ddot{z}(t) = \frac{1}{2}\{\ddot{z}(t^+) + \ddot{z}(t^-)\}$. With this convention, Definition 3.3 implies that $|\ddot{z}(t)|$ is bounded by some positive value $|\ddot{z}|_{max} < \infty$ if $\gamma \in \boldsymbol{\Gamma}_{wr}$. From Eq. 2.2.25 it then follows that the absolute curvature function $|K(t)|$ is also bounded by $K_{max} = (2\pi/L)^2\ |\ddot{z}|_{max} < \infty$. Note that $\boldsymbol{\Gamma}_r \subset \boldsymbol{\Gamma}_{wr} \subset \boldsymbol{\Gamma}_s$ (cf. Definitions 2.2 and 2.3).

**Definition 3.4.** *Piecewise weakly regular simple closed curve.*
A curve $\gamma$, parametrized on $[0, 2\pi]$, is a piecewise weakly regular simple closed curve iff:

- $\gamma \in \boldsymbol{\Gamma}_{ps}$, (3.2.7a)

- at those points, where $\dot{z}$ does not exist, left and right limits of $\dot{z}$ exist, (3.2.7b)

- everywhere, where $\gamma$ is smooth, $\ddot{z}$ exists and is continuous except in a finite number of points; at the latter points left and right limits of $\ddot{z}$ exist, (3.2.7c)

- $\ddot{z}(0) = \ddot{z}(2\pi)$. (3.2.7d)

□

Note that $\boldsymbol{\Gamma}_{pr} \subset \boldsymbol{\Gamma}_{pwr} \subset \boldsymbol{\Gamma}_{ps}$ (cf. Eq. 2.1.11).

Another way to characterize the class $\boldsymbol{\Gamma}_{pwr}$ is to say that the curves in this class are weakly regular, except in a finite number of points. At those points where a curve is not weakly regular, Eq. 3.2.7b applies.

In the following we first analyze to which function classes the representations of contours of the class $\boldsymbol{\Gamma}_{wr}$ belong. We first show that $\dot{z} \in \boldsymbol{\Lambda}$ and $\psi \in \boldsymbol{\Lambda}$ if $\gamma \in \boldsymbol{\Gamma}_{wr}$, and then that $\ddot{z} \in \mathbf{L}^{\infty}$ and $K \in \mathbf{L}^{\infty}$. On the basis of these results we draw conclusions about the convergence properties of the corresponding Fourier series and Fourier coefficients.

**Lemma 3.2.**
If $\gamma \in \boldsymbol{\Gamma}_{wr}$, then $\dot{z} \in \boldsymbol{\Lambda}$ with Lipschitz constant $(L/2\pi)^2 K_{max}$.

*Proof*

The truth of this lemma can be verified as follows. If $\gamma \in \boldsymbol{\Gamma}_{wr}$, then

$$
\begin{aligned}
|\dot{z}(t+\delta) - \dot{z}(t)| \\
&= \frac{L}{2\pi}\left|e^{i\theta(t+\delta)} - e^{i\theta(t)}\right| \\
&= \frac{L}{2\pi}\left|2\sin\left\{\frac{\theta(t+\delta)-\theta(t)}{2}\right\}\right| \\
&\leqslant \frac{L}{2\pi}\left|\theta(t+\delta) - \theta(t)\right| \\
&= \frac{L}{2\pi}\left|\int_t^{t+\delta} \dot{\theta}(\tau)\mathrm{d}\tau\right| \\
&\leqslant \delta\left|\frac{L}{2\pi}\right|^2 K_{\max}
\end{aligned}
\tag{3.2.8}
$$

for all $t \in [0, 2\pi - \delta]$ and for all small $\delta > 0$.

**Lemma 3.3.**
If $\gamma \in \boldsymbol{\Gamma}_{\mathrm{wr}}$, then $\psi \in \boldsymbol{\Lambda}$ with Lipschitz constant $(L/2\pi) K_{\max} + 1$.

*Proof*

We recall from Eq. 2.2.32 that

$$\psi(t+\delta) - \psi(t) = \varphi(t+\delta) - \varphi(t) - \delta.$$

To see that the statement in this lemma is true we note that if $\gamma \in \boldsymbol{\Gamma}_{\mathrm{wr}}$, then (cf. Eqs. 2.2.34 and 2.2.35)

$$
\begin{aligned}
&|\psi(t+\delta) - \psi(t)| \\
&= \left|\int_t^{t+\delta} \dot{\varphi}(\tau)\mathrm{d}\tau + \delta\right| \leqslant \delta\left(\frac{L}{2\pi} K_{\max} + 1\right)
\end{aligned}
\tag{3.2.9}
$$

for all $t \in [0, 2\pi - \delta]$ and all small $\delta > 0$. □

From Definition 3.3 it follows immediately that $\ddot{z} \in \mathbf{L}^{\infty}$ if $\gamma \in \boldsymbol{\Gamma}_{\mathrm{wr}}$. This in turn means that the Riemann-Lebesgue lemma applies to $\hat{\ddot{z}}(k)$

$$\lim_{|k| \to \infty} \hat{\ddot{z}}(k) = 0, \tag{3.2.10}$$

i.e. $\hat{\ddot{z}}(k) = o(1)$ as $|k| \to \infty$ (cf. Edwards [1979], pp. 32 and 36).

From $\dot{z} \in \mathbf{\Lambda}$ it follows that $\dot{z}$ is integrable, i.e. $\dot{z} \in \mathbf{L}^{1}$. We then find the relation (cf. Edwards [1979], p. 32)

$$\begin{aligned} \hat{z}(k) &= \frac{1}{2\pi} \int_{2\pi} z(t)\mathrm{e}^{-\mathrm{i}kt}\mathrm{d}t \\ &= -\frac{1}{2\pi \mathrm{i}k}\, z(t)\mathrm{e}^{-\mathrm{i}kt}\Big|_{0}^{2\pi} + \frac{1}{2\pi \mathrm{i}k} \int_{2\pi} \dot{z}(t)\mathrm{e}^{-\mathrm{i}kt}\mathrm{d}t \\ &= \frac{1}{\mathrm{i}k}\, \hat{\dot{z}}(k). \end{aligned} \tag{3.2.11}$$

By applying Eq. 3.2.12 twice we obtain the relation

$$\hat{z}(k) = -\frac{1}{k^{2}}\, \hat{\ddot{z}}(k). \tag{3.2.12}$$

From Eqs. 3.2.10 and 3.2.12 we conclude that $\hat{z}(k) = o(k^{-2})$ as $|k| \to \infty$ if $\gamma \in \boldsymbol{\Gamma}_{\mathrm{wr}}$. Hence we find that $\hat{\dot{z}}(k) = o(|k|^{-1})$ as $|k| \to \infty$ if $\gamma \in \boldsymbol{\Gamma}_{\mathrm{wr}}$. As $\dot{z} \in \mathbf{AC}$, $S_n\dot{z}$ converges uniformly to $\dot{z}$ (cf. Appendix A).

Along the same lines analogous results can be derived for $\psi$ and $K$ if $\gamma \in \boldsymbol{\Gamma}_{\mathrm{wr}}$. From Definition 3.3 it follows immediately that $K \in \mathbf{L}^{\infty}$ if $\gamma \in \boldsymbol{\Gamma}_{\mathrm{wr}}$. The Riemann-Lebesgue lemma then yields

$$\lim_{|k| \to \infty} \hat{K}(k) = 0 \tag{3.2.13}$$

i.e. $\hat{K}(k) = o(1)$ as $|k| \to \infty$.

Wherever $\dot{\psi}$ exists the relation between $\psi$ and $K$ is defined as (cf. Eqs. 2.2.32 and 2.2.34)

$$\dot{\psi}(t) = \frac{L}{2\pi} K(t) - 1. \tag{3.2.14}$$

From this equation we derive (cf. Eq. 3.2.11)

$$\hat{\psi}(k) = \frac{L}{2\pi i k}\hat{K}(k), \qquad \forall k \in \mathbb{Z} - \{0\}. \tag{3.2.15}$$

From Eqs. 3.2.13 and 3.2.15 we find that $\hat{\psi}(k) = o(|k|^{-1})$ as $|k| \to \infty$ if $\gamma \in \boldsymbol{\Gamma}_{\text{wr}}$. Since $\psi \in \mathbf{AC}$ if $\gamma \in \boldsymbol{\Gamma}_{\text{wr}}$, $S_n\psi$ converges uniformly to $\psi$ (cf. Appendix A).

If $\gamma \in \boldsymbol{\Gamma}_{\text{wr}}$, we already found that $\ddot{z} \in \mathbf{L}^\infty$ and $K \in \mathbf{L}^\infty$ and that $\hat{\ddot{z}}(k) = o(1)$ and $\hat{K}(k) = o(1)$ as $|k| \to \infty$. It was shown by Hunt [1967], elaborating on the results of Carleson [1966], that[1]

$$\lim_{n \to \infty} (S_n f)(x) = f(x), \quad \text{a.e.} \tag{3.2.16}$$

if $f \in \mathbf{L}^p$, for any index $p > 1$. Thus, if $\gamma \in \boldsymbol{\Gamma}_{\text{wr}}$, then

$$\lim_{n \to \infty} (S_n \ddot{z})(t) = \ddot{z}(t), \quad \text{a.e.} \tag{3.2.17}$$

and

$$\lim_{n \to \infty} (S_n K)(t) = K(t), \quad \text{a.e.} \tag{3.2.18}$$

If a contour belongs to $\boldsymbol{\Gamma}_{\text{r}}$, then obviously the same properties are valid for the representations of this contour and its Fourier expansions that are valid when a contour belongs to $\boldsymbol{\Gamma}_{\text{wr}}$. Though $\ddot{z} \in \mathbf{C}$ and $K \in \mathbf{C}$ if $\gamma \in \boldsymbol{\Gamma}_{\text{r}}$, it is not necessarily true that $\ddot{z} \in \mathbf{BV}$ and $K \in \mathbf{BV}$. Therefore stronger convergence properties for $\hat{\ddot{z}}(k)$ and $\hat{K}(k)$ are generally not valid if $\gamma \in \boldsymbol{\Gamma}_{\text{r}}$.

Mere continuity of a function is not a particularly strong property in relation to the convergence of the Fourier series that it generates. This is illustrated by the fact that the Fourier series of a continuous function can diverge on an uncountable set (cf. Edwards [1979], pp. 162-164). On the other hand, Eqs. 3.2.17 and 3.2.18 are clearly valid if $\gamma \in \boldsymbol{\Gamma}_{\text{r}}$.

We now pay some attention to contours with finite bending energy. Young, Walker and Bowie [1974] considered a contour as a thin flexible rod and related the finiteness of the bending energy in this rod to the rate of convergence of $\hat{z}(k)$ as $|k| \to \infty$. Finiteness of bending energy

[1] a.e. stands for *almost everywhere* (cf. Definition A.3).

is considered to characterize the smoothness of contours. We will show that if a contour $\gamma \in \boldsymbol{\Gamma}_{wr}$ then $\gamma$ has finite bending energy. Through an example we will show that the converse is not necessarily true. Through this example it will also become clear that finiteness of bending energy is not a sufficient criterion to judge the smoothness of a curve.

Finiteness of bending energy is equivalent to square integrability of the curvature function, i.e. $K \in \mathbf{L}^2$. In the aforementioned paper it was shown that

$$\sum_{k \in \mathbb{Z}} k^4 |\hat{z}(k)|^2 < \infty \tag{3.2.19}$$

i.e. $\{k^2\hat{z}(k)\} \in \ell^2(\mathbb{Z})$ (cf. Appendix A), if $K \in \mathbf{L}^2$. Hence $\{k^2\hat{z}(k)\} \in \mathbf{c}_0(\mathbb{Z})$, i.e. $\{k^2\hat{z}(k)\} = o(1)$ as $|k| \to \infty$. Thus we find that the rate of convergence of $\hat{z}(k)$ is at least $o(k^{-2})$ as $|k| \to \infty$ if the bending energy in the contour is finite. (This result corrects a minor error in the conclusion drawn by Young, Walker and Bowie [1974].)

We now verify that the bending energy in a contour $\gamma$ is finite if $\gamma \in \boldsymbol{\Gamma}_{wr}$. We have already remarked that finiteness of bending energy is equivalent to $K \in \mathbf{L}^2$. If $\gamma \in \boldsymbol{\Gamma}_{wr}$, then $K \in \mathbf{L}^\infty$. Observing that $\mathbf{L}^\infty \subset \mathbf{L}^2$ (cf. Appendix A) and combining this with the facts just mentioned, we can immediately conclude that $\gamma \in \boldsymbol{\Gamma}_{wr}$ indeed implies finiteness of bending energy in $\gamma$. Incidentally this observation also implies that $\boldsymbol{\Gamma}_{wr}$ is a proper subset of the class of contours with finite bending energy. Furthermore we remark that the result previously obtained, i.e. $\hat{z}(k) = o(k^{-2})$ as $|k| \to \infty$ if $\gamma \in \boldsymbol{\Gamma}_{wr}$, is in accordance with the rate of decay of $\hat{z}(k)$ just found if the bending energy in the contour is finite.

An example of a curve with finite bending energy which does not belong to $\boldsymbol{\Gamma}_{wr}$ is the following. Consider the tangent angle function

$$\theta(t) = \frac{\pi}{2} - \left(\frac{\pi}{2}\right)^{1-\alpha} |t|^\alpha, \qquad 0 < \alpha < 1 \tag{3.2.20}$$

in the interval $-\pi/2 \leqslant t \leqslant \pi/2$. In this interval $\theta$ is continuous. This function does not define a closed contour, but this fact is immaterial for this discussion. The derivative of $\theta$ is given by

$$\dot{\theta}(t) = -\alpha\left(\frac{\pi}{2}\right)^{1-\alpha} \operatorname{sgn}(t)\, |t|^{\alpha-1} \tag{3.2.21}$$

which does not exist at $t = 0$ since $0 < \alpha < 1$. If $t$ represents the arc length along the curve, then $\dot{\theta}$ expresses its curvature $K$ (cf. Eq. 2.2.22). The curves, defined by $\theta$ and $\dot{\theta} = K$ in Eqs. 3.2.20 and 3.2.21, are not weakly regular since $\lim_{t\uparrow 0} K(t) = \infty$ and $\lim_{t\downarrow 0} K(t) = -\infty$.

In order for $K$ to belong to $\mathbf{L}^p[-\pi/2, \pi/2]$ we find from Eq. 3.2.21 the requirement that $\alpha > (p-1)/p$. Thus $K \in \mathbf{L}^2[-\pi/2, \pi/2]$ if $\frac{1}{2} < \alpha < 1$.

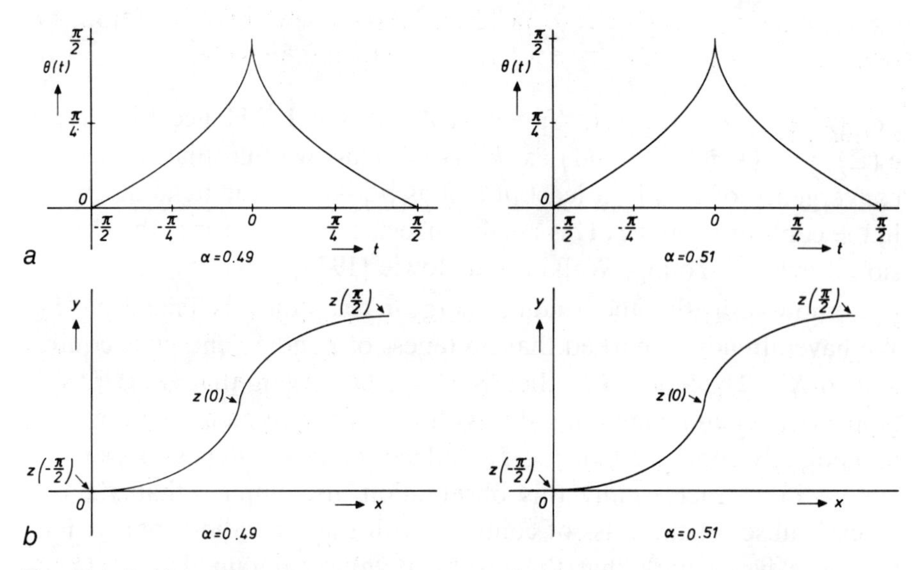


**Figure 3.1.** In (*a*) the tangent function $\theta(t) = \pi/2 - (\pi/2)^{1-\alpha}|t|^{\alpha}$ is shown for $\alpha = 0.49$ and $\alpha = 0.51$ In (*b*) the associated position functions $z(t)$, reconstructed according to Eq. 3.2.22, are displayed. Note that $z(t)$ does not have finite bending energy for $\alpha = 0.49$, whereas for $\alpha = 0.51$ it does.

This example also shows that finiteness of the bending energy is not a sufficient criterion to determine the smoothness of a curve. In Figure 3.1a we have shown $\theta(t)$ in the interval $-\pi/2 \leqslant t \leqslant \pi/2$ for $\alpha = 0.51$ and $\alpha = 0.49$. Figure 3.1b displays the reconstructed position functions

$$z(t) = \int_{-\pi/2}^{t} e^{i\theta(\tau)} d\tau \tag{3.2.22}$$

for these values of $\alpha$. Despite the continuity of $\theta$ we can see an irregularity in the reconstructed curves at $t = 0$. Therefore it seems that membership of the class $\boldsymbol{\Gamma}_{wr}$ is a better criterion for a curve to be smooth.

We now turn to an analysis of the function class membership of representations of contours that belong to $\boldsymbol{\Gamma}_{pwr}$. We show that $\dot{z} \in \mathbf{BV}$ and $\psi \in \mathbf{BV}$ if $\gamma \in \boldsymbol{\Gamma}_{pwr}$ and that $\ddot{z}$ and $K$ constitute distributions of order 0. Subsequently we consider the convergence properties of the corresponding Fourier series and Fourier coefficients.

In the following lemma we use the fact that, if $f \in \mathbf{\Lambda}$ in a certain interval, then $f \in \mathbf{AC}$ (cf. Proposition 3.1) and certainly $f \in \mathbf{BV}$ in that interval (cf. Appendix A or Apostol [1974], pp. 137-139).

**Lemma 3.4.**
If $\gamma \in \boldsymbol{\Gamma}_{pwr}$, then $\dot{z} \in \mathbf{BV}$ and $\psi \in \mathbf{BV}$.

*Proof*

From Lemmas 3.2 and 3.3 and the statement above it follows immediately that $\dot{z} \in \mathbf{BV}$ and $\psi \in \mathbf{BV}$ if $\gamma \in \boldsymbol{\Gamma}_{wr}$. If $\gamma \in \boldsymbol{\Gamma}_{pwr}$, then the number of points where $\dot{z}$ does not exist is finite (cf. Definition 3.4). If $\dot{z}$ does not exist at a certain point $t$, then the contribution to the total variation at $t$ is upperbounded by

$$|\dot{z}(t^+) - \dot{z}(t^-)| = \frac{L}{2\pi}\,|e^{i\theta(t^+)} - e^{i\theta(t^-)}| \leqslant \frac{L}{\pi}. \qquad (3.2.23)$$

Likewise the contribution at $t$ to the total variation in $\psi$ is upperbounded by $\pi$. Since both the number of such contributions and the contributions themselves are finite, the total variations of $\dot{z}$ and $\psi$ remain finite if $\gamma \in \boldsymbol{\Gamma}_{pwr}$.

□

Since $\dot{z}$ may contain jump discontinuities if $\gamma \in \boldsymbol{\Gamma}_{pwr}$, we assume $\ddot{z}$ to represent the distributional derivative of $\dot{z}$ (cf. Edwards [1982], p. 63). Likewise, if $\gamma \in \boldsymbol{\Gamma}_{pwr}$, then $\psi$ may contain jump discontinuities. Therefore $K$ is linearly related to the distributional derivative of $\psi$ (cf. Eq. 3.2.14). Then, as a result of Lemma 3.4, $\ddot{z}$ and $K$ constitute distributions of order 0 or Radon measures (cf. Edwards [1982], p. 72).

We are now in a position to make some statements about the behavior of the Fourier series and the Fourier coefficients of the contour representations $\dot{z}$, $\ddot{z}$, $\psi$ and $K$ if $\gamma \in \boldsymbol{\Gamma}_{\text{pwr}}$.

We just found that $\dot{z} \in \mathbf{BV}$ if $\gamma \in \boldsymbol{\Gamma}_{\text{pwr}}$. By Edwards [1979], pp. 33-34, we then find that

$$|k\hat{\dot{z}}(k)| \leqslant \frac{1}{2\pi} \operatorname{Var}(\dot{z}), \qquad \forall k \in \mathbb{Z} \tag{3.2.24}$$

where Var ($\dot{z}$) denotes the total variation of $\dot{z}$ in a fundamental parameter interval of length $2\pi$ (cf. Appendix A). Thus Eq. 3.2.24 yields $\hat{\dot{z}}(k) = O(|k|^{-1})$ as $|k| \to \infty$.

Combining Eqs. 3.2.11 and 3.2.24 we find that if $\gamma \in \boldsymbol{\Gamma}_{\text{pwr}}$, then

$$|k^2\hat{z}(k)| \leqslant \frac{1}{2\pi} \operatorname{Var}(\dot{z}), \qquad \forall k \in \mathbb{Z}. \tag{3.2.25}$$

Thus $\hat{z}(k) = O(k^{-2})$ as $|k| \to \infty$, so that the Fourier series of $z$ converges absolutely and uniformly to $z$.

In Eq. 3.2.24 we already observed that $\hat{\dot{z}}(k) = O(|k|^{-1})$ as $|k| \to \infty$ if $\gamma \in \boldsymbol{\Gamma}_{\text{pwr}}$. Since $\dot{z} \in \mathbf{BV}$ it follows from the Dirichlet-Jordan test (cf. Appendix A or Zygmund [1959a], p. 57) that $(S_n\dot{z})(t)$ converges to $\frac{1}{2}\{\dot{z}(t^+) - \dot{z}(t^-)\}$, $\forall t \in [0, 2\pi]$. Consequently, $(S_n\dot{z})$ converges to $\dot{z}$ at every point of continuity of $\dot{z}$. At every point where $\dot{z}$ has a jump discontinuity, $(S_n\dot{z})$ exhibits the well-known Gibbs phenomenon. The Gibbs phenomenon is a feature of the nonuniformity of the convergence of the sequence $(S_n\dot{z})$ in the neighborhood of a point of discontinuity. See Hewitt and Hewitt [1979] for a detailed account of this phenomenon.

Along the same lines we can perform the analysis for $\psi$ if $\gamma \in \boldsymbol{\Gamma}_{\text{pwr}}$, leading to analogous results.

We have established above that $\ddot{z}$ constitutes a distribution of order 0 if $\gamma \in \boldsymbol{\Gamma}_{\text{pwr}}$. As a result the Fourier coefficients of $\dot{z}$ and $\ddot{z}$ are related as $\hat{\ddot{z}}(k) = \mathrm{i}k\hat{\dot{z}}(k)$, $\forall k \in \mathbb{Z}$ (cf. Edwards [1982], p. 72). Since $\hat{\dot{z}}(k) = O(|k|^{-1})$ as $|k| \to \infty$ if $\gamma \in \boldsymbol{\Gamma}_{\text{pwr}}$, we then find that $\hat{\ddot{z}}(\mathrm{k}) = O(1)$ as $|k| \to \infty$ if $\gamma \in \boldsymbol{\Gamma}_{\text{pwr}}$. The Fourier series $S_n\ddot{z}$ converges distributionally to $\ddot{z}$ (cf. Edwards [1979], pp. 8-9).

Similarly, if $\gamma \in \boldsymbol{\Gamma}_{\text{pwr}}$ we find that $\hat{K}(k) = (2\pi \mathrm{i}k/L)\hat{\psi}(k)$, $\forall k \in \mathbb{Z} - \{0\}$, and, since $\hat{\psi}(k) = O(|k|^{-1})$ as $|k| \to \infty$, this yields $\hat{K}(k) = O(1)$ as $|k| \to \infty$. The Fourier series $S_nK$ converges distributionally to $K$.

**Table 3.1.** Contour representations and corresponding Fourier coefficients of an arbitrary closed $N$-sided polygon, specified by the ordered set of vertices $\{z(t_n): n = 0, \ldots, N-1\}$. The notations were introduced in Section 2.2.

| Contour representation | Fourier coefficients |
| --- | --- |
| $z(t) = \Delta z^*(t_n) \dfrac{L}{2\pi} \{t - t_n\} + z(t_n),$ <br> $t_n \leqslant t \leqslant t_{n+1}$ | $\hat{z}(0) = \dfrac{1}{2L} \sum_{n=0}^{N-1} \lvert\Delta z(t_n)\rvert \{z(t_{n+1}) + z(t_n)\}$ <br> $\hat{z}(k) = -\dfrac{L}{(2\pi k)^2} \sum_{n=0}^{N-1} \Delta^2 z^*(t_n) \mathrm{e}^{-\mathrm{i}kt_n}, \quad k \neq 0$ |
| $\dot{z}(t) = \Delta z^*(t_n) \dfrac{L}{2\pi},$ <br> $t_n < t < t_{n+1}$ | $\hat{\dot{z}}(0) = 0$ <br> $\hat{\dot{z}}(k) = -\dfrac{\mathrm{i}L}{(2\pi)^2 k} \sum_{n=0}^{N-1} \Delta^2 z^*(t_n) \mathrm{e}^{-\mathrm{i}kt_n}, \quad k \neq 0$ |
| $\ddot{z}(t) = \dfrac{L}{2\pi} \sum_{n=0}^{N-1} \Delta^2 z^*(t_n) \delta_{2\pi}(t - t_n)$ | $\hat{\ddot{z}}(0) = 0$ <br> $\hat{\ddot{z}}(k) = \dfrac{L}{(2\pi)^2} \sum_{n=0}^{N-1} \Delta^2 z^*(t_n) \mathrm{e}^{-\mathrm{i}kt_n}, \quad k \neq 0$ |
| $\psi(t) = \begin{cases} \sum_{m=1}^{n} \Delta\varphi(t_m) - t, & n \geqslant 1, \\ -t, & n = 0, \\ -\sum_{m=n+1}^{0} \Delta\varphi(t_m) - t, & n \leqslant -1, \end{cases}$ <br> $t_n < t < t_{n+1}$ | $\hat{\psi}(0) = \pi - \dfrac{1}{2\pi} \sum_{n=1}^{N} t_n \Delta\varphi(t_n)$ <br> $\hat{\psi}(k) = -\dfrac{\mathrm{i}}{2\pi k} \sum_{n=0}^{N-1} \Delta\varphi(t_n) \mathrm{e}^{-\mathrm{i}kt_n}, \quad k \neq 0$ |
| $K(t) = \dfrac{2\pi}{L} \sum_{n=0}^{N-1} \Delta\varphi(t_n) \delta_{2\pi}(t - t_n)$ | $\hat{K}(0) = \dfrac{2\pi}{L}$ <br> $\hat{K}(k) = \dfrac{1}{L} \sum_{n=0}^{N-1} \Delta\varphi(t_n) \mathrm{e}^{-\mathrm{i}kt_n}, \quad k \neq 0$ |

If $\gamma \in \boldsymbol{\Gamma}_{\mathrm{pr}}$ we cannot improve on the results obtained above for $\gamma \in \boldsymbol{\Gamma}_{\mathrm{pwr}}$. This is caused by the potential jump discontinuities in $\dot{z}$ and $\psi$ if $\gamma \in \boldsymbol{\Gamma}_{\mathrm{pr}}$, while $\ddot{z}$ and $K$ also constitute distributions of order 0 in this case.

As an illustrative example we have listed in Table 3.1 the expressions for the various contour representations of an arbitrary closed $N$-sided polygon. These expressions were previously derived in Section 2.2. Next to these contour representations we have listed in Table 3.1 the expressions for their Fourier coefficients. Note that polygons belong to $\boldsymbol{\Gamma}_{\mathrm{pr}}$.

In Figures 3.2b-f the magnitudes are shown of the first 125 Fourier coefficients generated by the contour representations $z$, $\dot{z}$, $\ddot{z}$, $\psi$ and $K$ of the polygon in Figure 3.2a.

We have now completed an investigation of the convergence properties of the Fourier series and the Fourier coefficients generated by the

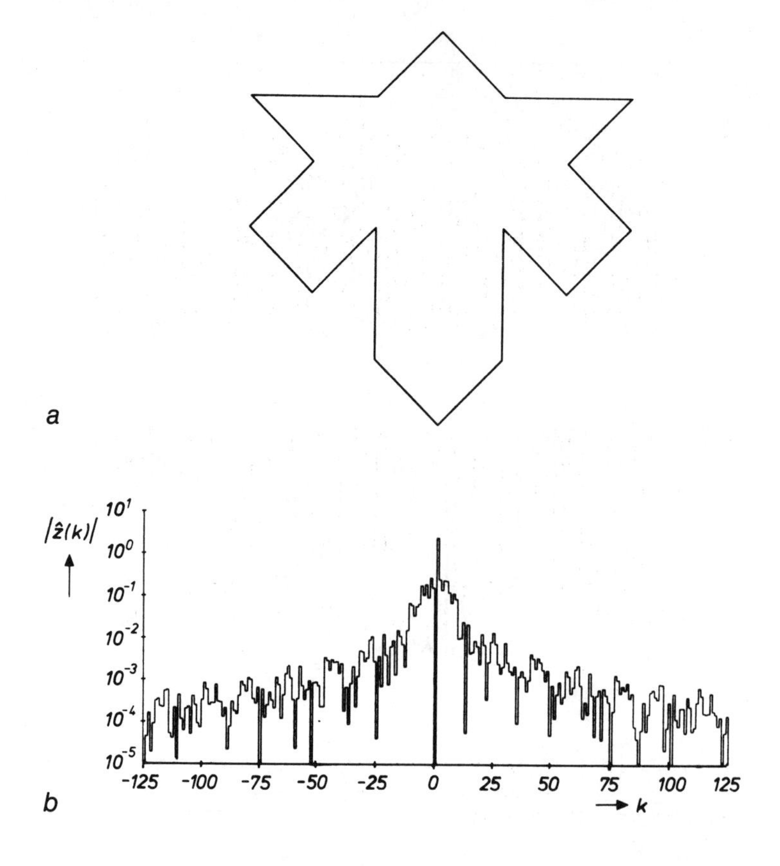

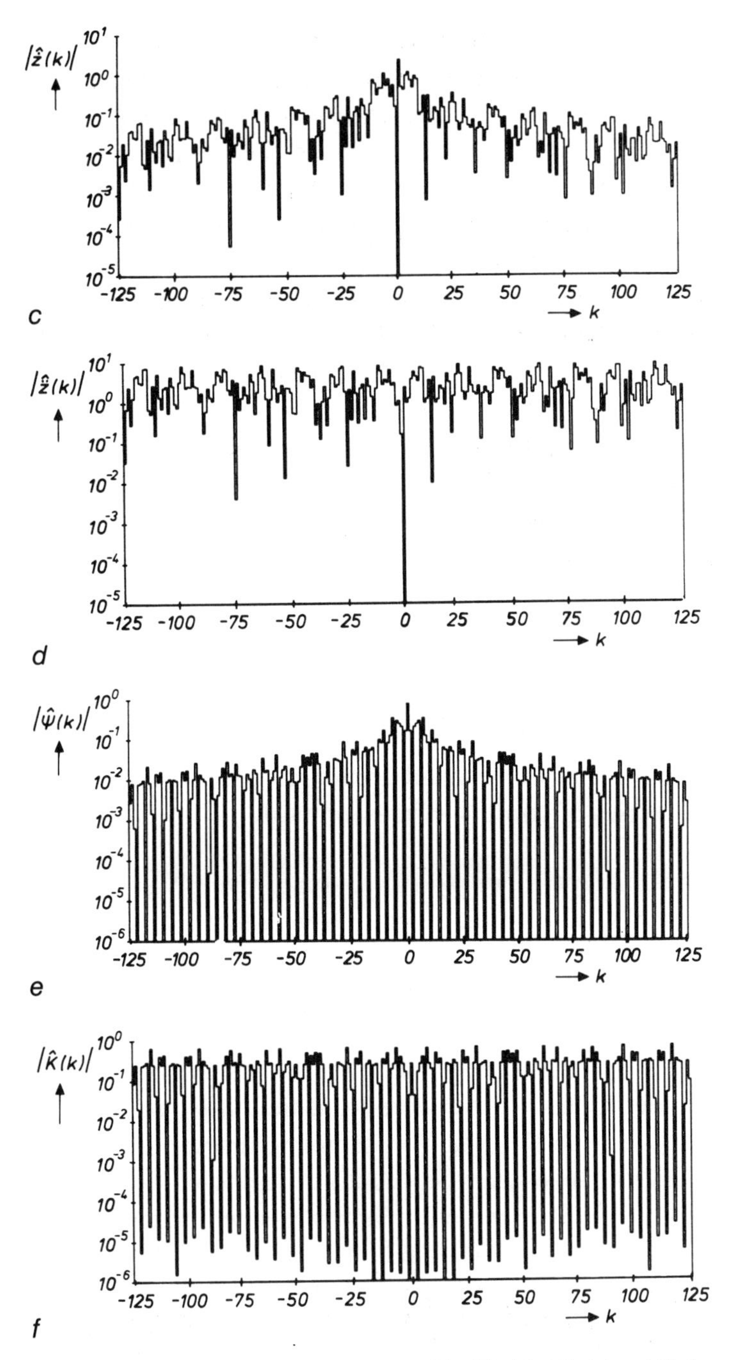


**Figure 3.2.** In (*a*) a polygon of 16 vertices is shown and in (*b*), (*c*), (*d*), (*e*) and (*f*) the magnitudes are displayed of the first 125 Fourier coefficients of the representations $z$, $\dot{z}$, $\ddot{z}$, $\psi$ and $K$ of the polygon. Note the rates of convergence of the Fourier coefficients in (*b*)-(*f*) and compare them with the formulas for $\hat{z}(k)$, $\hat{\dot{z}}(k)$, $\hat{\ddot{z}}(k)$, $\hat{\psi}(k)$ and $\hat{K}(k)$, given in Table 3.1.

contour representations $z$, $\dot{z}$, $\ddot{z}$, $\psi$ and $K$ of contours belonging to the contour classes $\boldsymbol{\Gamma}_{\mathrm{r}}$, $\boldsymbol{\Gamma}_{\mathrm{wr}}$, $\boldsymbol{\Gamma}_{\mathrm{s}}$, $\boldsymbol{\Gamma}_{\mathrm{pr}}$, $\boldsymbol{\Gamma}_{\mathrm{pwr}}$, and $\boldsymbol{\Gamma}_{\mathrm{ps}}$. In Section 2.2 we have shown that these contour representations all preserve the shape information of the contours from these classes. The convergence properties that we found for the Fourier series of these contour representations are sufficient to ensure, in view of the uniqueness theorems for Fourier series (cf. Edwards [1979], pp. 40-41, Edwards [1982], pp. 69-70), that there exists a one-to-one correspondence, in the sense of these convergence properties, between a contour representation and the limit of its Fourier series expansion. Therefore we draw the conclusion that the Fourier series, generated by the aforementioned contour representations, also preserve shape information. For this reason we call $\hat{f} \equiv (\hat{f}(k))_{k \in \mathbb{Z}}$, the sequence of Fourier coefficients generated by a contour representation $f$ of a contour $\gamma$, a *Fourier representation* of $\gamma$. The shape information-preserving properties of Fourier representations will allow us, in Sections 3.5 and 3.6, to formulate necessary and sufficient conditions on sequences of Fourier coefficients, such that a pair of contours is (mirror-)similar or such that a contour has certain symmetry properties.

## 3.3 Consequences of normalized arc length parametrization upon the Fourier series of contour representations

In the previous section we paid attention to the convergence properties of Fourier series of periodic contour representations and to the rate at which the corresponding Fourier coefficients approach to zero. An interesting question, with which we shall be concerned in this section, is whether a periodic contour representation, with a normalized arc length parameter, can be expanded into a Fourier series that has a finite number of nonzero coefficients. If this were true, then it would be theoretically possible, according to the Shannon sampling theorem (Shannon [1949], Jerri [1977]), to reconstruct the contour representation from a finite number of samples. If $M$ were the largest absolute index of the nonzero Fourier coefficients, then $2M$ equidistant samples of the contour representation in one complete period would be needed to allow for an exact reconstruction of the representation.

We first investigate the question just raised for the position function $z$. For the main theorem we need the following lemma, the validity of which is obvious from the uniqueness of Fourier representations.

**Lemma 3.5.**

Let $M, N \in \mathbb{Z}$, $M < N$. Then

$$\sum_{k=M}^{N} a_k \mathrm{e}^{ikt} \equiv 0 \tag{3.3.1}$$

implies

$$a_M = \cdots = a_N = 0. \tag{3.3.2}$$

□

**Theorem 3.1.**
With the exception of a circle, no piecewise differentiable position function $z$, with (normalized arc length parameter, of a simple closed contour $\gamma$ can be expanded into a Fourier series with a finite number of nonzero Fourier coefficients.

*Proof*

Let $M, N \in \mathbb{Z}$, $M < N$, be two finite integers. We assume that the position function $z$, with normalized arc length parameter $t$, of a simple closed contour $\gamma$ can be represented by the finite Fourier series

$$z(t) = \sum_{k=M}^{N} \hat{z}(k)\mathrm{e}^{ikt}. \tag{3.3.3}$$

Our task is now to show that $\gamma$ can only be a circle.
The tangent function $\dot{z}$ is given by

$$\dot{z}(t) = \sum_{k=M}^{N} \hat{\dot{z}}(k)\mathrm{e}^{ikt} \tag{3.3.4}$$

where (cf. Eq. 3.2.15)

$$\hat{\dot{z}}(k) = \mathrm{i}k\hat{z}(k), \qquad M \leqslant k \leqslant N. \tag{3.3.5}$$

The requirement that $t$ is a normalized arc length parameter amounts to

$$|\dot{z}(t)| = \left| \sum_{k=M}^{N} \hat{\dot{z}}(k)e^{ikt} \right| \equiv \frac{L}{2\pi} \tag{3.3.6}$$

where $L$ is the perimeter of $\gamma$. This requirement can be rewritten as

$$|\dot{z}(t)|^2 = \left( \sum_{k=M}^{N} \hat{\dot{z}}(k)e^{ikt} \right)\left( \sum_{l=M}^{N} \overline{\hat{\dot{z}}}(l)e^{-ilt} \right) \equiv \left(\frac{L}{2\pi}\right)^2. \tag{3.3.7}$$

Rewriting the product of sums in Eq. 3.3.7 yields

$$\sum_{n=-(N-M)}^{N-M} \sigma(n)e^{int} \equiv 0 \tag{3.3.8}$$

where

$$\sigma(n) = \begin{cases} \displaystyle\sum_{m=0}^{N-M-n} \hat{\dot{z}}(N-m)\overline{\hat{\dot{z}}}(N-m-n), & 0 < n \leqslant N-M, \quad (3.3.9a) \\ \displaystyle -\left(\frac{L}{2\pi}\right)^2 + \sum_{m=0}^{N-M} |\hat{\dot{z}}(N-m)|^2, & n = 0, \quad (3.3.9b) \\ \displaystyle\sum_{m=0}^{N-M+n} \hat{\dot{z}}(N-m+n)\overline{\hat{\dot{z}}}(N-m), & -(N-M) \leqslant n < 0. \quad (3.3.9c) \end{cases}$$

From Lemma 3.5 it follows that the condition in Eq. 3.3.8 can only be satisfied if

$$\sigma(n) = 0, \qquad -(N-M) \leqslant n \leqslant N-M. \tag{3.3.10}$$

We now assume, without loss of generality, that $\hat{\dot{z}}(N) \neq 0$. Since for any contour $\hat{\dot{z}}(0) = 0$, this assumption implies $N \neq 0$. Through Eqs. 3.3.9a and 3.3.10 we find

$$\sigma(N-M) = \hat{\dot{z}}(N)\overline{\hat{\dot{z}}}(M) = 0 \tag{3.3.11a}$$

which implies, with $\hat{z}(N) \neq 0$

$$\hat{z}(M) = 0 \tag{3.3.11b}$$

With this result we continue

$$\begin{aligned}\sigma(N-M-1) &= \hat{z}(N)\overline{\hat{z}}(M+1) + \hat{z}(N-1)\overline{\hat{z}}(M) \\ &= \hat{z}(N)\overline{\hat{z}}(M+1) = 0 \end{aligned} \tag{3.3.12a}$$

which implies

$$\hat{z}(M+1) = 0. \tag{3.3.12b}$$

We continue this process till $\sigma(1)$

$$\sigma(1) = \sum_{m=0}^{N-m-1} \hat{z}(N-m)\overline{\hat{z}}(N-m-1) \tag{3.3.13a}$$

which implies

$$\hat{z}(N-1) = 0. \tag{3.3.13b}$$

Thus we have found from Eq. 3.3.6 that

$$\hat{z}(n) = 0, \qquad M \leqslant n \leqslant N-1. \tag{3.3.14}$$

Through Eqs. 3.3.5 and 3.3.14 and the fact that $N \neq 0$, the finite Fourier series expansion of $z$ in Eq. 3.3.3 now reduces to

$$z(t) = \hat{z}(0) + \hat{z}(N)e^{iNt}. \tag{3.3.15}$$

We recall from Section 2.1 two conditions that $z$ must satisfy:

- simplicity (cf. Eq. 2.1.5),
- counterclockwise tracing as $t$ increases.

These conditions are satisfied in Eq. 3.3.15 iff $N = 1$.

Thus we find

$$z(t) = \hat{z}(0) + \hat{z}(1)e^{it} \tag{3.3.16}$$

which is the equation of a circle with center at $\hat{z}(0)$ and with radius of size $|\hat{z}(1)|$.

The proof of this theorem is now complete.

□

It is clear from the foregoing that the condition of a (normalized) arc length parametrization plays a crucial role in the proof of Theorem 3.1.

If we discard the condition of (normalized) arc length parametrization for $z$, then a finite Fourier series can indeed represent a simple contour, which will always be closed. The closure of the contour is a result of the periodicity of the complex exponentials that constitute the basis functions for the Fourier series. Figure 3.3a shows an example of such a representation. The speed along the contour, as a function of the parameter $t$, is also shown (Figure 3.3b).

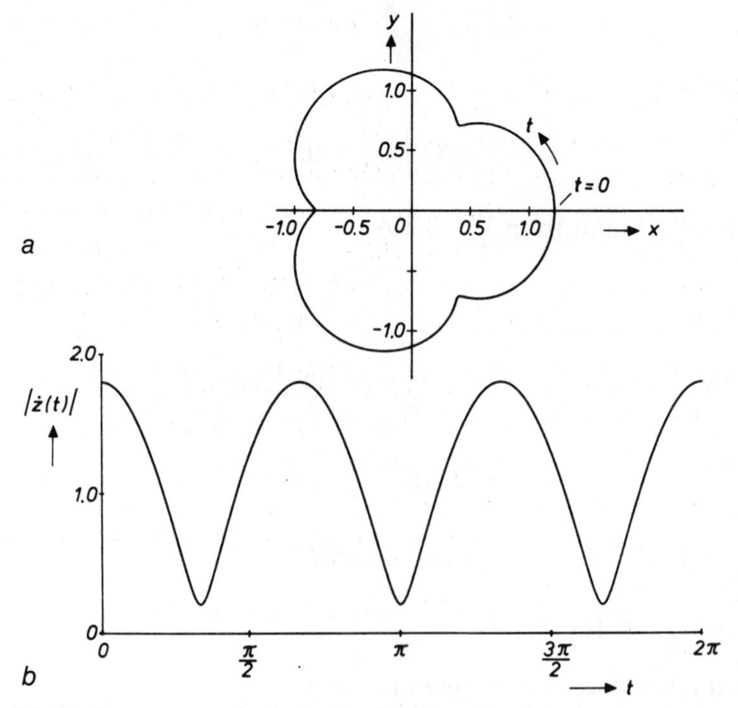


**Figure 3.3.** (a) shows an example of a position function with a finite number of nonzero Fourier coefficients: $z(t) = e^{it} + 0.2e^{4it}$. The speed of motion along the contour: $|\dot{z}(t)| = \{1.64 + 1.6\cos(3t)\}^{1/2}$, is shown in ($b$). Clearly $|\dot{z}(t)|$ varies with $t$.

The following corollaries are an immediate result of Theorem 3.1 and its proof.

**Corollary 3.1.**
If the position function $z(t)$ of an arbitrary simple closed contour is expanded into a finite Fourier series, then the parameter $t$ in the finite expansion is not a (normalized) arc length parameter, unless the contour is a circle.

□

**Corollary 3.2.**
If the parameter $t$ of a position function $z(t)$ of an arbitrary simple closed contour is a (normalized) arc length parameter, then $z(t)$ cannot be reconstructed exactly from a finite number of samples of $z$ in a single period, unless the contour is a circle.

□

If a contour is a circle, then, according to Shannon's sampling theorem (cf. Shannon [1949], Jerri [1977]), two equidistant samples of $z$ in a single period are sufficient to allow for an exact reconstruction of $z(t)$.

The facts, stated in Theorem 3.1 and in Corollary 3.1, were observed previously by Persoon and Fu [1977], but no complete proof was given. There are a number of examples in the literature where these facts have been overlooked. For example, a direct consequence of Theorem 3.1 is that the 'complete set of Fourier descriptors' to characterize the shape of a simple closed contour, as described by Crimmins [1982], contains infinitely many elements, unless the contour is a circle.

The contents of Corollaries 3.1 and 3.2 have important consequences in practice since truncation of a Fourier series expansion of $z$ leads in general to the loss of the linear relation between $t$ and arc length and to an inexact representation of $z$. In the next section we derive upper bounds for the truncation errors resulting from finite Fourier series expansions.

Another corollary to Theorem 3.1, which has implications in practice, is the following.

**Corollary 3.3.**
If $t$ is the (normalized) arc length parameter of the position function $z$

of an arbitrary simple closed contour, then it is not possible to compute the Fourier coefficients $\hat{z}(k)$ of $z(t)$ exactly from a finite number of samples of $z$ in a single period (through a discrete Fourier transform, cf. e.g. Oppenheim and Schafer [1975]), unless the contour is a circle.

□

In signal theory, the phenomenon that causes errors to occur in Fourier transforms and Fourier coefficients, when computed from fewer samples than the Nyquist criterion prescribes, is generally known as *aliasing* (cf. e.g. Oppenheim and Shafer [1975], Hamming [1977]).

The question whether the tangent function $\dot{z}$ or the acceleration function $\ddot{z}$ of a simple closed contour, with a (normalized) arc length parameter, can be expanded into a finite Fourier series can now be answered in a straightforward manner. Since $\hat{\dot{z}}(k) = ik\hat{z}(k)$ and $\hat{\ddot{z}}(k) = -k^2\hat{z}(k)$ (cf. Section 3.2), analogues to Theorem 3.1 and Corollaries 3.1-3.3 can be formulated for $\dot{z}$ and $\ddot{z}$.

For the periodic cumulative angular function $\psi$ and for the curvature function $K$ the situation is different. There exist simple closed contours, with a normalized arc length parameter, for which $\psi$ and $K$ can be expanded into a finite Fourier series. An example of such a contour is given by Zahn and Roskies [1972]

$$\psi(t) = \tfrac{3}{4}e^{-3it} + \tfrac{3}{4}e^{3it} = \tfrac{3}{2}\cos(3t). \tag{3.3.17}$$

Choosing the perimeter of the contour to be $2\pi$, i.e. $t$ corresponds to arc length, we find through Eqs. 2.2.32 and 2.2.34 for $K$ the expression

$$K(t) = 1 - \tfrac{9}{4}ie^{-3it} + \tfrac{9}{4}ie^{3it} = 1 - \tfrac{9}{2}\sin(3t). \tag{3.3.18}$$

A reconstruction of the contour, having Eqs. 3.3.17 and 3.3.18 as periodic cumulative angular function and as curvature function, respectively, is shown in Figure 3.4.

In contrast with the finite or infinite Fourier series expansions of the contour representations $z$, $\dot{z}$ and $\ddot{z}$, those of $\psi$ and $K$ in general do not correspond to a closed contour. Special conditions must be satisfied by the Fourier coefficients $\hat{\psi}(k)$ in order to correspond to a closed contour. In Zahn and Roskies [1972] some sufficient conditions are presented for the (potentially finite) Fourier series expansion to represent a closed contour. However, the necessity of these conditions is not shown. Using the relation $\hat{K}(k) = (2\pi/L)\mathrm{i}k\hat{\psi}(k)$, $\forall k \neq 0$, corresponding sufficient conditions for closure can be derived for $\hat{K}(k)$.

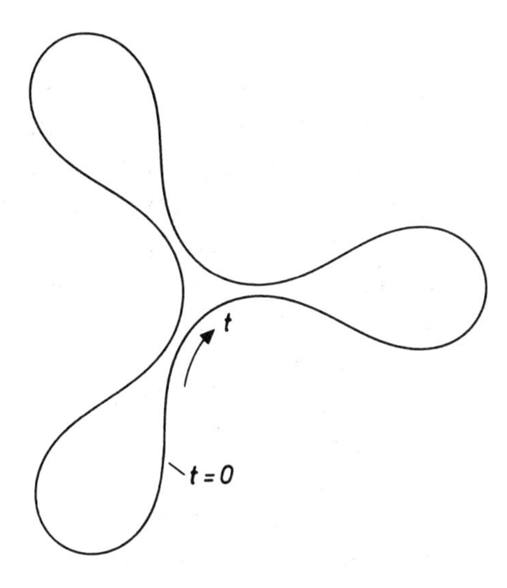


**Figure 3.4.** The contour that has the representations $\psi$ and $K$ as given in Eqs. 3.3.17 and 3.3.18, respectively.

In view of the foregoing, it is clear that the reconstruction of a contour from a truncated Fourier series expansion of $\psi$ or $K$ does in general not lead to a closed contour. Strackee and Nagelkerke [1983] presented an approximation technique to ensure the closure of contours reconstructed from a finite Fourier series expansion of $\psi$. In the context of reconstructing a contour from a finite Walsh sequency expansion of $\psi$ (cf. Section 3.7), another method to obtain a closed contour is described by Sarvarayudu and Sethi [1983]. Their method can also be applied to the reconstruction of closed contours from finite Fourier series expansions of $\psi$ or $K$.

## 3.4 Upper bounds on the truncation errors in finite Fourier series expansions of contour representations

In the previous section we found that the Fourier series of the contour representations $z$, $\dot{z}$ and $\ddot{z}$, with (normalized) arc length parameter, contain an infinite number of nonzero Fourier coefficients, unless the contour is a circle. Apart from the fact that, as a result of a truncation of the Fourier series expansion, the linear relation between the parameter and arc length is lost (Corollary 3.2), the truncated Fourier series will always exhibit a truncation error with respect to the contour representation (again with the exception of a circle). In this section we derive various upper bounds for the truncation errors caused by finite Fourier series expansions. We derive upper bounds for representations of contours that belong to a number of smoothness classes and provide an experimental comparison of the perfomance of some of these upper bounds. Sharp upper bounds may be helpful in deciding whether it is useful, in terms of data reduction, to work with Fourier representations if we wish to maintain a certain approximation accuracy.

We indicate the set of all trigonometric polynomials of degree at most $n$ as $\mathbf{T}_n$ (cf. Appendix A). It is a well-known fact (cf. e.g. Edwards [1979], p. 131) that, of all trigonometric polynomials in $\mathbf{T}_n$, the finite Fourier series expansion of degree $n$ of a periodic function yields the minimum truncation error in mean square sense. In some applications however, the *Chebychev norm* or *sup-norm* (cf. Appendix A) may provide a more appropriate error criterion to judge the truncation error. The finite Fourier series expansion of degree $n$ of a periodic function does in general not minimize this error criterion over $\mathbf{T}_n$.

We define for $p = 1, 2, \ldots$ and for $p = \infty$

$$E_n^{(p)}(f) = \inf \{\|f - P_n\|_p \colon P_n \in \mathbf{T}_n\}. \tag{3.4.1}$$

The optimality of the partial Fourier sum of degree $n$. $S_n f$, in the mean square sense, leads to

$$E_n^{(2)}(f) \equiv \|f - S_n f\|_2. \tag{3.4.2}$$

From the definition of $E_n^{(p)}(f)$ and through Hölder's inequality (cf. e.g. Appendix A or Hardy, Littlewood and Pólya [1952]) we find the inequality

$$E_n^{(p)}(f) \geqslant E_n^{(q)}(f), \quad \text{if } p > q. \tag{3.4.3}$$

From Eqs. 3.4.1-3.4.3 we derive the following chain of inequalities

$$\|f - S_n f\|_\infty \geq E_n^{(\infty)}(f) \geq E_n^{(2)}(f) \equiv \|f - S_n f\|_2. \qquad (3.4.4)$$

From now on we denote $E_n^{(\infty)}(f)$ simply as $E_n(f)$.

In the remainder of this section we derive some upper bounds on $\|f - S_n f\|_\infty$ for the contour representations $z$, $\dot{z}$ and $\psi$ for various contour classes. Part of this discussion can also be found in Dekking and van Otterloo [1986].

First we define a subset of the class of simple closed polygons, the simple closed chains.

**Definition 3.5.** *Simple closed chain.*
A simple closed polygon with $N$ sides of equal length is a simple closed chain, where successive chain links may be collinear.

□

Please note that the fact that successive chain links may be collinear constitutes a slight broadening of the traditional concept of a polygon.

**Theorem 3.2.** *Bound* $V(z)$ (Giardina and Kuhl [1977]).
If $\gamma \in \boldsymbol{\Gamma}_{ps}$, then

$$\|z - S_n z\|_\infty \leq \frac{Var\ (\dot{z})}{\pi n}. \qquad (3.4.5)$$

□

For an arbitrary $N$-sided polygon $\gamma$, $Var\ (\dot{z})$ is given by

$$Var\ (\dot{z}) = \sum_{j=0}^{N-1} \left| \frac{z(t_{j+1}) - z(t_j)}{t_{j+1} - t_j} - \frac{z(t_j) - z(t_{j-1})}{t_j - t_{j-1}} \right|. \qquad (3.4.6)$$

If $\gamma$ is a chain, then Eq. 3.4.5 becomes

$$\|z - S_n z\|_\infty \leq \frac{NL}{2\pi^2 n}. \qquad (3.4.7)$$

**Theorem 3.3.** *Bound* $C(z)$ (Dekking and van Otterloo [1986]).
Let $\gamma$ be a chain with $N$ links. Put $n = Nq + s$, with $q$ and $s$ non-negative integers and $0 \leqslant s < N$. Then

$$\|z - S_n z\|_\infty \leqslant \sum_{r=1}^{N-1} \frac{r^2(|\hat{z}(r)| + |\hat{z}(-r)|)}{(q + 1/2)N^2 + rN} + \sum_{r=s+1}^{N-1} \frac{r^2(|\hat{z}(r)| + |\hat{z}(-r)|)}{(Nq + r)^2} \tag{3.4.8}$$

where the second summation is absent if $s = N - 1$.

*Proof*

Since $\gamma$ is a chain with $N$ links, we derive from Table 3.1 for all integers $p$ the relations

$$(pN + r)^2 \hat{z}(pN + r) = r^2 \hat{z}(r), \qquad \forall r \neq 0 \tag{3.4.9}$$

and

$$\hat{z}(pN) = 0. \tag{3.4.10}$$

Hence

$$\begin{aligned} \|z - S_n z\|_\infty &\leqslant \sum_{|k| > n} |\hat{z}(k)| \\ &= \sum_{r=s+1}^{N-1} (|\hat{z}(Nq + r)| + |\hat{z}(-Nq - r)|) \\ &\quad + \sum_{p=q+1}^{\infty} \sum_{r=1}^{N-1} (|\hat{z}(Np + r)| + |\hat{z}(-Np - r)|) \\ &= \sum_{r=s+1}^{N-1} \frac{r^2(|\hat{z}(r)| + |\hat{z}(-r)|)}{(Nq + r)^2} \\ &\quad + \sum_{r=1}^{N-1} [r^2(|\hat{z}(r)| + |\hat{z}(-r)|)] \sum_{p=q+1}^{\infty} \frac{1}{(Np + r)^2}. \end{aligned} \tag{3.4.11}$$

Since the function $(Nx + r)^{-2}$ is convex, we have

$$\sum_{p=q+1}^{\infty} \frac{1}{(Np + r)^2} \leqslant \int_{q+1/2}^{\infty} \frac{1}{(Nx + r)^2}\, dx. \tag{3.4.12}$$

The integral equals

$$\int_{q+1/2}^{\infty} \frac{1}{(Nx + r)^2}\, dx = \frac{1}{N^2(q + 1/2) + Nr}. \tag{3.4.13}$$

Combining Eqs. 3.4.12 and 3.4.13 and substituting the result into Eq. 3.4.11 yields Eq. 3.4.8, as we required.

□

In practical situations the computation of bound $C(z)$ (Eq. 3.4.8) has fairly low computational complexity if the $2N$ Fourier coefficients that are needed are already known. In that case the complexity of computation of bound $C(z)$ is comparable to that of bound $V(z)$.

The asymptotic behavior of bound $V(z)$ is $O(n^{-1})$ as $n \to \infty$. We recall from Section 3.2 that, if $\gamma \in \boldsymbol{\Gamma}_{\mathrm{pwr}}$, then $\hat{z}(k) = O(k^{-2})$ as $|k| \to \infty$. Then some simple calculations reveal that bound $C(z)$ is also $O(n^{-1})$ as $n \to \infty$. This shows that the asymptotic behavior of $V(z)$ and $C(z)$, in terms of their rate of decay, is equivalent.

In Dekking and van Otterloo [1986] it was shown that the claim of Giardina and Kuhl [1977], that bound $V(z)$ is asymptotically the best possible for a square, is not correct.

Experiments have revealed that, the more significant detail present in a chain, the poorer bound $V(z)$ performs. Bound $C(z)$ performs from three times (for chains with little significant detail) to over ten times (for chains with much significant detail) better than bound $V(z)$ (cf. Figures 3.5, 3.6[1] and 3.7).

[1] Note the open curve in Figure 3.6. Curves need not necessarily be closed in order to allow Fourier analysis of their representations. For that purpose the curve is parametrized such that $z(t) = z(2\pi - t)$, where $t$ is still a normalized arc length parameter, and such that the parametric starting point $t = 0$ coincides with one of the two end points of the curve. In that case the total arc length traversed in one period is twice the length of the curve. For obvious reasons, such a parametrization is called a *retracing*

Bound $V(z)$ performs poorly with complicated contours and bound $C(z)$ has the drawbacks that its computational complexity increases with the complexity of contours and that it applies only to chains. However, a reasonably sharp bound of low computational complexity, that applies to any $\gamma \in \boldsymbol{\Gamma}_{ps}$, can be derived. First we define the Lebesgue constants.

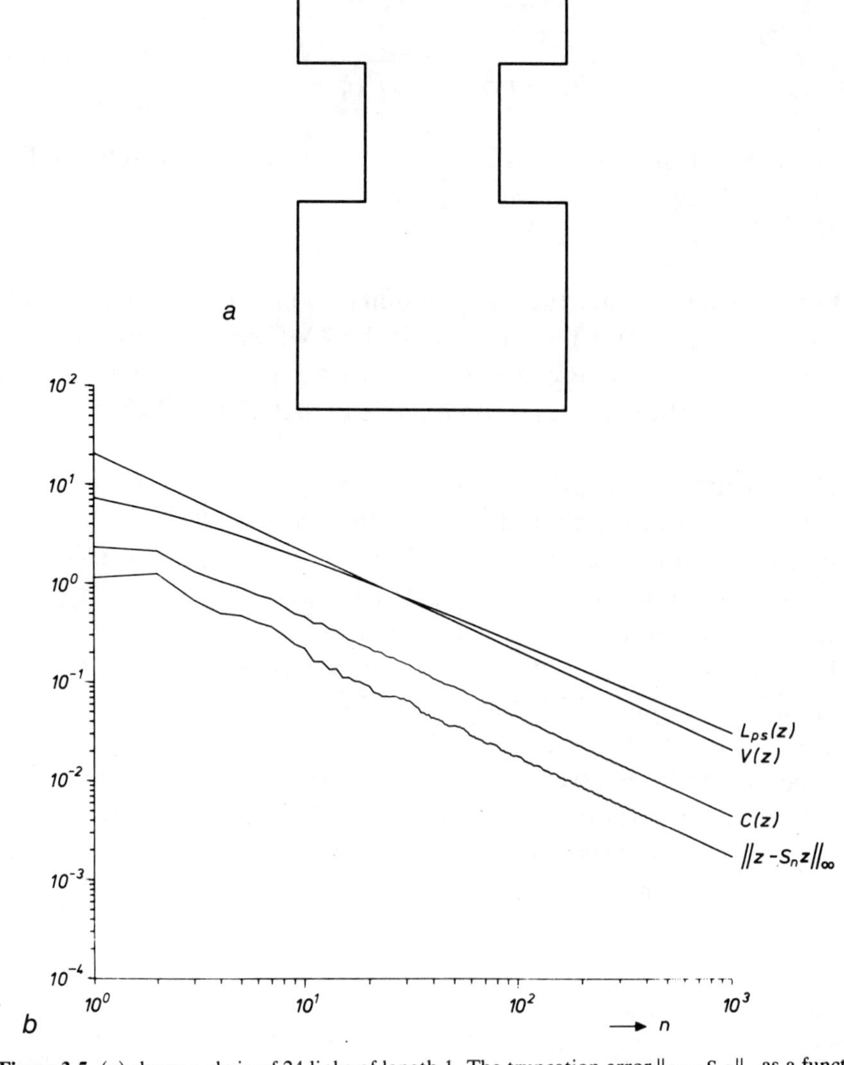


**Figure 3.5.** (*a*) shows a chain of 24 links of length 1. The truncation error $\|z - S_n z\|_\infty$ as a function of $n$, the number of Fourier coefficients used in the finite Fourier series approximation, is displayed in (*b*). In (*b*) we also show a comparison of the performance of the error bounds $V(z)$, $C(z)$ and $L_{ps}(z)$ for this chain.

**Definition 3.6.** *Lebesgue constant $\lambda_n$.*
The $n$-th Lebesgue constant $\lambda_n$ is defined as

$$\lambda_n = \frac{1}{\pi}\int_0^{\pi} \frac{|\sin(n + 1/2)t|}{\sin 1/2 t} \mathrm{d}t. \tag{3.4.14}$$

□

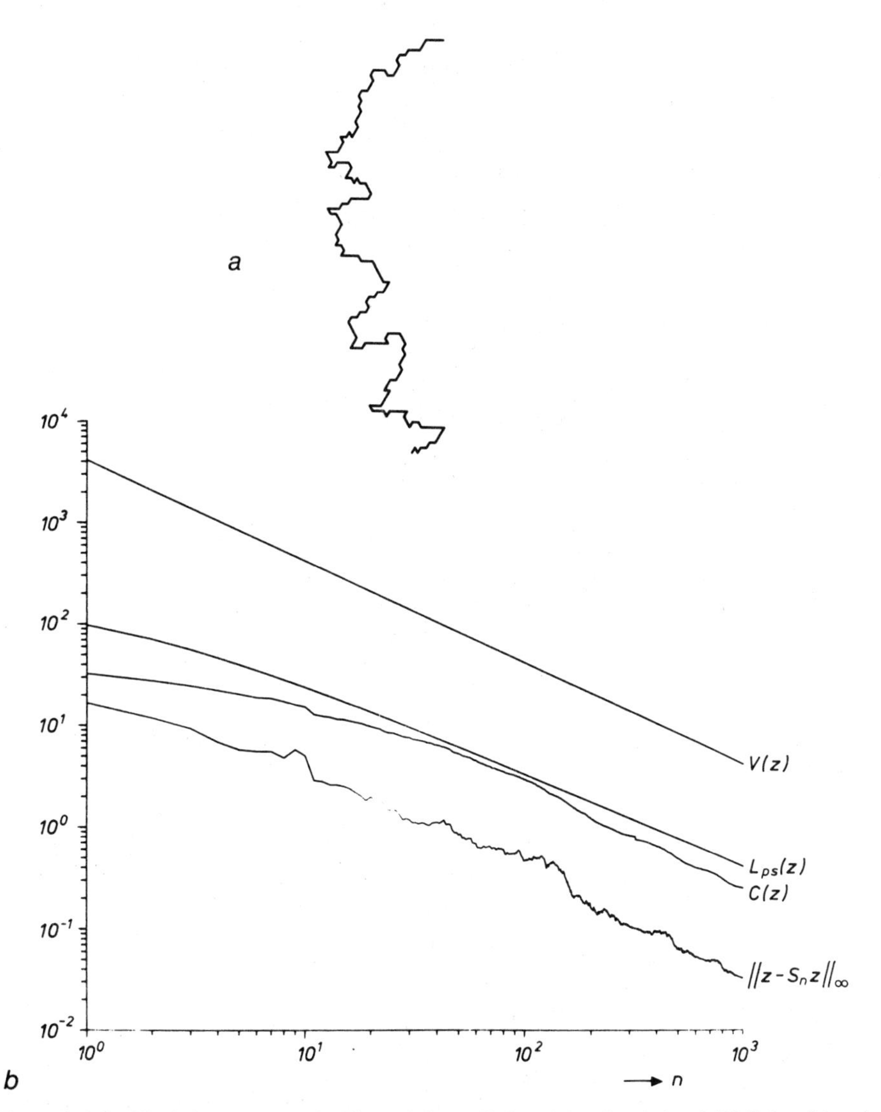


**Figure 3.6.** In ($b$) similar curves as in Figure 3.5 are displayed for the chain of 160 links of length 1 in ($a$).

**Theorem 3.4.** *Bound $L_{ps}(z)$.*
If $\gamma \in \boldsymbol{\Gamma}_{ps}$, then for all integers $n \geqslant 1$

$$\|z - S_n z\|_\infty \leqslant \frac{(\lambda_n + 1)}{4(n+1)} L \tag{3.4.15}$$

where $L$ is the perimeter of $\gamma$.

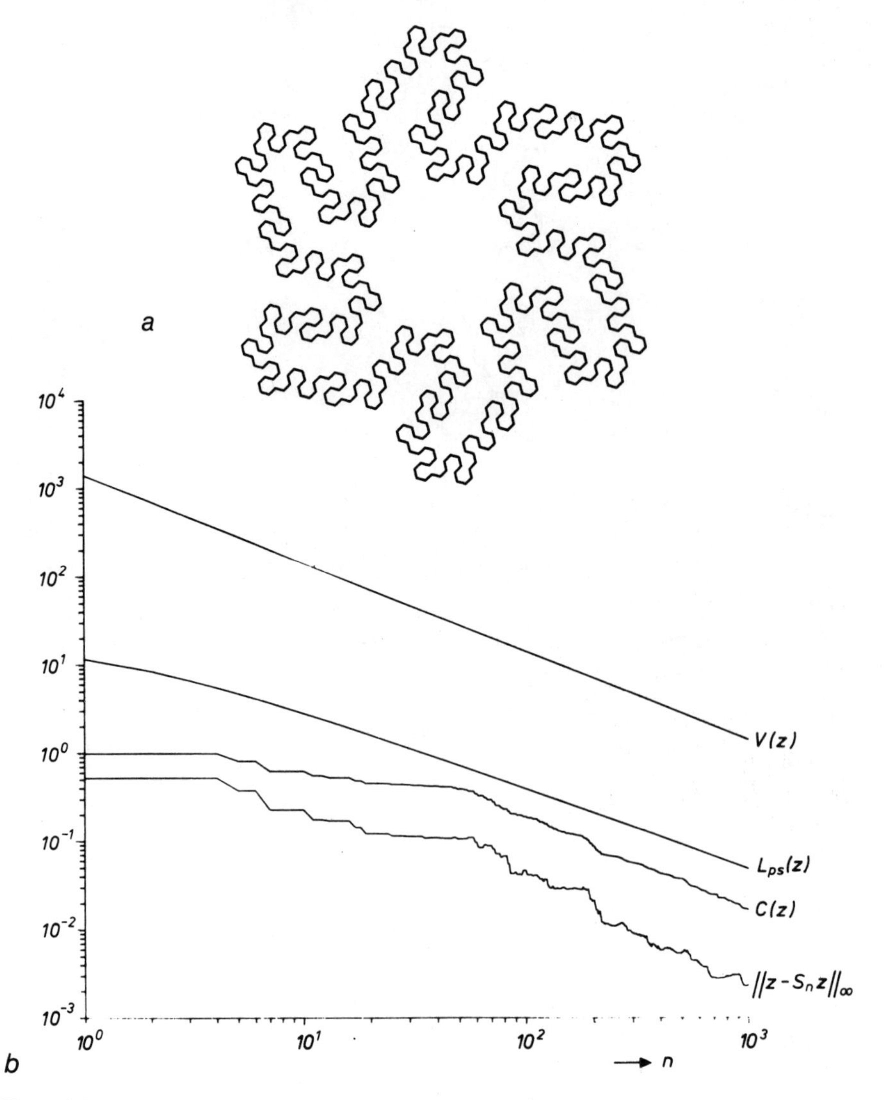


**Figure 3.7.** In (*b*) similar curves as in Figure 3.5 are displayed for the chain of 726 links of length $1/19 \approx 0.05$ in (*a*).

*Proof*

From Lemma 3.1 in Section 3.2 we know that $z \in \boldsymbol{\Lambda}$, with Lipschitz constant $\lambda = L/2\pi$, if $\gamma \in \boldsymbol{\Gamma}_{ps}$.

A classical result of Lebesgue [1910], p. 201, gives the asymptotically best bound on $\|f - S_n f\|_\infty$ for functions $f \in \boldsymbol{\Lambda}$. As Lebesgue's paper does not pay attention to the best possible constant, we refer to Cheney [1966], p. 147, where it is shown that for integers $n \geqslant 1$

$$\|f - S_n f\|_\infty \leqslant (\lambda_n + 1)E_n(f). \tag{3.4.16}$$

(Compare this expression with Eq. 3.4.4.)

Jackson's Theorem II (Cheney [1966], pp. 132-144) says that if $f \in \boldsymbol{\Lambda}$, with Lipschitz constant $\lambda$, then

$$E_n(f) \leqslant \frac{\pi\lambda}{2(n+1)} \tag{3.4.17}$$

and the constant $\pi/2(n+1)$ is the best constant possible.

Substituting $\lambda = L/2\pi$ and applying Eqs. 3.4.16 and 3.4.17 to $z$ yields Eq. 3.4.15.

□

**Remark.**

The paper by Etesami and Uicker [1985] claims the bound

$$\|z - S_n z\|_\infty \leqslant \frac{L}{4(n+1)} \tag{3.4.18}$$

which is not correct (cf. Eq. 3.4.15).

□

It is well known (cf. e.g. Zygmund [1959a]) that

$$\lambda_n \approx \frac{4}{\pi^2} \log n \qquad \text{as } n \to \infty. \tag{3.4.19}$$

In the Figures 3.4, 3.5 and 3.6 we have employed Fejér's expression (cf. e.g. Lebesgue [1910], p. 197)

$$\lambda_n = \frac{1}{2n+1} + \frac{2}{\pi} \sum_{k=1}^{n} \frac{1}{k} \tan\left(\frac{\pi k}{2n+1}\right) \tag{3.4.20}$$

for the computation of the bound $L_{ps}(z)$. In these three examples we see that the cheaply computed bound $L_{ps}(z)$ performs reasonably well in comparison with bound $V(z)$.

Through Eqs. 3.4.15 and 3.4.19 we immediately see that the asymptotic behavior of bound $L_{ps}(z)$ is $O(n^{-1} \log n)$ as $n \to \infty$. Though this is worse than the asymptotic behavior of both bound $V(z)$ and bound $C(z)$, in many situations bound $L_{ps}(z)$ is still sharper than bound $V(z)$ for practical values of $n$.

If $\gamma \in \boldsymbol{\Gamma}_{pr}$ we cannot improve on Theorem 3.4. However, sharper upperbounds can be derived for $\|z - S_n z\|_\infty$ as the contours considered satisfy more severe smoothness conditions. This is a direct consequence of the faster rates of decay of the $|\hat{z}(k)|$ for smoother contours, which in turn lead to sharper bounds in the Jackson Theorems (cf. Cheney [1966], pp. 139-149). This will soon become clear in a theorem that specifies an upper bound for $\|z - S_n z\|_\infty$ if $\gamma \in \boldsymbol{\Gamma}_{wr}$.

The set of all trigonometric polynomials of degree at most $n$, free of a constant term, $\mathbf{t}_n$, have been defined in Appendix A as

$$\mathbf{t}_n = \{p_n\colon\ p_n(t) = \sum_{\substack{|k| \leq n \\ k \neq 0}} a_k e^{ikt},\ a_k \in \mathbb{C}\}. \tag{3.4.21}$$

Furthermore we define

$$e_n^{(p)}(f) = \inf\{\|f - p_n\|_p\colon\ p_n \in \mathbf{t}_n\} \tag{3.4.22}$$

(compare this expression with Eq. 3.4.1). In analogy with $E_n(f)$ we denote $e_n^{(\infty)}(f)$ as $e_n(f)$.

In the proof of the next theorem we need a variant of Jackson's Theorem II (Cheney [1966], p. 143, and Eq. 3.4.17) such that it can be applied to functions that are free of a constant term. We formulate this variant in the following lemma.

**Lemma 3.6.** *Upper bound on* $e_n(f)$.
If $f \in \boldsymbol{\Lambda}$, with Lipschitz constant $\lambda$, is free of a constant term, then

$$e_n(f) \leqslant \frac{\pi\lambda}{2(n+1)} \tag{3.4.23}$$

and the constant $\pi/2(n+1)$ is the best constant possible.

*Proof*

We follow the proof of Jackson's Theorem II (Cheney [1966], pp. 143-144).

Fixing $\delta > 0$, define the auxiliary function

$$\Phi_\delta(t) = \frac{1}{2\delta}\int_{t-\delta}^{t+\delta} f(\tau)\mathrm{d}\tau. \tag{3.4.24}$$

Since $f \in \mathbf{\Lambda}$ we obtain

$$|\dot{\Phi}_\delta(t)| = \frac{1}{2\delta}\,|f(t+\delta) - f(t-\delta)| \leqslant \lambda. \tag{3.4.25}$$

Consequently, by Jackson's Theorem I (Cheney [1966], p. 142) it follows that

$$e_n(\Phi_\delta) \leqslant \frac{\pi}{2(n+1)}\,\|\dot{\Phi}_\delta\|_\infty \leqslant \frac{\pi\lambda}{2(n+1)} \tag{3.4.26}$$

if we show that $\Phi_\delta$ is free of a constant term. This is proved as follows

$$\frac{1}{2\pi}\int_0^{2\pi} \Phi_\delta(t)\mathrm{d}t = \frac{1}{4\pi\delta}\int_0^{2\pi}\int_{t-\delta}^{t+\delta} f(\tau)\mathrm{d}\tau\,\mathrm{d}t$$

$$= \frac{1}{4\pi\delta}\int_0^{2\pi}\int_{-\delta}^{\delta} f(\sigma + t)\mathrm{d}\sigma\,\mathrm{d}t. \tag{3.4.27}$$

Through Fubini's theorem (cf. e.g. Royden [1963], pp. 233-234) we may invert the order of integration in Eq. 3.4.27. This and the periodicity of $f$ lead to

$$\frac{1}{2\pi}\int_0^{2\pi}\Phi_\delta(t)\mathrm{d}t = \frac{1}{4\pi\delta}\int_{-\delta}^{\delta}\int_0^{2\pi} f(t+\sigma)\mathrm{d}t\,\mathrm{d}\sigma$$

$$= \frac{1}{4\pi\delta}\int_{-\delta}^{\delta}\int_0^{2\pi} f(t)\mathrm{d}t\,\mathrm{d}\sigma$$

$$= \frac{1}{2\pi}\int_0^{2\pi} f(t)\mathrm{d}t. \tag{3.4.28}$$

Since $f$ is free of a constant term, Eq. 3.4.28 immediately shows that $\Phi_\delta$ is also free of a constant term. Thus we have shown the validity of Eq. 3.4.26. Furthermore

$$|\Phi_\delta(t) - f(t)| \leqslant \frac{1}{2\delta}\int_{t-\delta}^{t+\delta} |f(\tau) - f(t)|\mathrm{d}\tau$$

$$\leqslant \frac{\lambda}{2\delta}\int_{t-\delta}^{t+\delta} |\tau - t|\mathrm{d}\tau = \frac{\lambda}{2}\delta. \tag{3.4.29}$$

Let $p_n^*$ denote the trigonometric polynomial, free of a constant term, that optimally approximates $\Phi_\delta$. Then, with Eqs. 3.4.26 and 3.4.29

$$e_n(f) \leqslant \|f - p_n^*\|_\infty$$

$$\leqslant \|f - \Phi_\delta\|_\infty + \|\Phi_\delta - p_n^*\|_\infty$$

$$\leqslant \frac{\lambda}{2}\delta + \frac{\pi\lambda}{2(n+1)}. \tag{3.4.30}$$

Since this inequality holds for any $\delta > 0$, it also holds for $\delta = 0$ giving

$$e_n(f) \leqslant \frac{\pi\lambda}{2(n+1)}. \tag{3.4.31}$$

For the proof that the constant $\pi/2(n+1)$ is the best constant possible, we refer again to Cheney [1966], pp. 142-144. The proof of this lemma is now complete. □

We are now in a position to specify an upper bound for $\|z - S_n z\|_\infty$ if $\gamma \in \boldsymbol{\Gamma}_{\mathrm{wr}}$.

**Theorem 3.5.** *Bound* $L_{\mathrm{wr}}(z)$.
If $\gamma \in \boldsymbol{\Gamma}_{\mathrm{wr}}$, then for all integers $n \geq 1$

$$\|z - S_n z\|_\infty \leqslant \frac{(\lambda_n + 1)}{16(n+1)^2} L^2 K_{\max} \tag{3.4.32}$$

where $L$ denotes the perimeter of $\gamma$.

*Proof*

First we note a property of $E_n(f)$ (cf. Eq. 3.4.1)

$$E_n(f - P_n) = E_n(f), \qquad \forall P_n \in \mathbf{T}_n. \tag{3.4.33}$$

Since $\gamma \in \boldsymbol{\Gamma}_{\mathrm{wr}}$ we have $z \in \mathbf{C}^1$ (cf. Appendix A). With these two properties we find from Cheney [1966], p. 146, the inequality

$$E_n(z) \leqslant \frac{\pi}{2(n+1)} e_n(\dot{z}). \tag{3.4.34}$$

Next we derive an upper bound for $e_n(\dot{z})$. According to Lemma 3.2, $\dot{z} \in \boldsymbol{\Lambda}$ if $\gamma \in \boldsymbol{\Gamma}_{\mathrm{wr}}$ with Lipschitz constant $\lambda = (L/2\pi)^2 K_{\max}$, where $L$ denotes the perimeter of $\gamma$. It is well-known that $\dot{z}$ is free of a constant term. As a result of these observations, Lemma 3.6 applies to $\dot{z}$, which yields (cf. Eq. 3.4.23)

$$e_n(\dot{z}) \leqslant \frac{L^2 K_{\max}}{8\pi(n+1)}. \tag{3.4.35}$$

Combining Eqs. 3.4.34 and 3.4.35 leads to

$$E_n(z) \leqslant \frac{L^2 K_{\max}}{16(n+1)^2}. \tag{3.4.36}$$

With the substitution of this result into Eq. 3.4.16 the proof of this theorem is complete. □

If $\gamma \in \boldsymbol{\Gamma}_{\mathrm{r}}$, we can improve slightly on the results in Theorem 3.5, obtained for $\gamma \in \boldsymbol{\Gamma}_{\mathrm{wr}}$. Furthermore, the upper bound on $\|z - S_n z\|_\infty$ can be obtained in a much simpler way, as we will see in the following theorem.

**Theorem 3.6.** *Bound* $L_{\mathrm{r}}(z)$.
If

- $\gamma \in \boldsymbol{\Gamma}_{\mathrm{r}}$,
- $|K(t)| \leqslant K_{\max} < \infty, \quad \forall t \in [0, 2\pi]$,

then for all integers $n \geqslant 3$

$$\|z - S_n z\|_\infty \leqslant \frac{(\lambda_n + 1)}{8\pi(n+1)^2} L^2 K_{\max} \tag{3.4.37}$$

where $L$ denotes the perimeter of $\gamma$.

*Proof*

Jackson's Theorem IV (Cheney [1966], pp. 145-146) states that for $n > k$

$$E_n(f) \leqslant \frac{\pi}{2}\left(\frac{1}{n+1}\right)^k \|f^{(k)}\|_\infty \tag{3.4.38}$$

if the $2\pi$-periodic function $f \in \mathbf{C}^k$, and that the coefficient $\pi/2$ is the best coefficient possible, independent of $f$, $k$ and $n$.

If $\gamma \in \boldsymbol{\Gamma}_{\mathrm{r}}$, then $z \in \mathbf{C}^2$. Thus Eq. 3.4.38 yields

$$E_n(z) \leqslant \frac{\pi}{2}\left(\frac{1}{n+1}\right)^2 \|\ddot{z}\|_\infty. \tag{3.4.39}$$

We recall from Eq. 2.2.25 that

$$\ddot{z}(t) = \left(\frac{L}{2\pi}\right)^2 K(t) n(t)$$

where $n(t)$ is the complex-valued function that corresponds to the unit normal $\boldsymbol{n}(t)$ at $z(t)$ and $K(t)$ is the curvature at $z(t)$. From this equation we derive

$$\|\ddot{z}\|_\infty = \left(\frac{L}{2\pi}\right)^2 \|K\|_\infty \leqslant \left(\frac{L}{2\pi}\right)^2 K_{\max}. \tag{3.4.40}$$

Since $z \in \mathbf{C}^2$ the second step in Eq. 3.4.40 constitutes an equality in this case.

Combining Eqs. 3.4.16, 3.4.39 and 3.4.40 immediately leads to the required result.

□

Through Eqs. 3.4.19, 3.4.32 and 3.4.37 we can see immediately that the asymptotic behavior of both bound $L_{wr}(z)$ and bound $L_r(z)$ is $O(n^{-2}\log n)$ as $n \to \infty$. We recall from Section 3.2 that we found for both $\gamma \in \boldsymbol{\Gamma}_{wr}$ and $\gamma \in \boldsymbol{\Gamma}_r$ that $\hat{z}(k) = O(k^{-2})$ as $|k| \to \infty$.

In Section 3.2 we showed in the Lemmas 3.2 and 3.3 that both $\dot{z}$ and $\psi$ satisfy a Lipschitz condition if $\gamma \in \boldsymbol{\Gamma}_{wr}$. These facts can be used to derive upper bounds on $\|\dot{z} - S_n\dot{z}\|_\infty$ and on $\|\psi - S_n\psi\|_\infty$.

**Theorem 3.7.** *Bound* $L_{wr}(\dot{z})$.
If $\gamma \in \boldsymbol{\Gamma}_{wr}$, then for all integers $n \geqslant 1$

$$\|\dot{z} - S_n\dot{z}\|_\infty \leqslant \frac{(\lambda_n + 1)}{8\pi(n+1)} L^2 K_{\max}. \tag{3.4.41}$$

*Proof*

In Lemma 3.2 it was shown that, if $\gamma \in \boldsymbol{\Gamma}_{wr}$, then $\dot{z} \in \boldsymbol{\Lambda}$, with Lipschitz constant $\lambda = (L/2\pi)^2 K_{\max}$. Consequently, Jackson's Theorem II, Eq. 3.4.17, applies to $\dot{z}$. Through substitution of the value of $\lambda$ into this equation and through Eq. 3.4.16, the required result is obtained immediately.

□

**Theorem 3.8.** *Bound* $L_{wr}(\psi)$.
If $\gamma \in \boldsymbol{\Gamma}_{wr}$, then for all integers $n \geqslant 1$

$$\|\psi - S_n\psi\|_\infty \leqslant \frac{(\lambda_n + 1)}{4(n+1)} (LK_{\max} + 2\pi). \tag{3.4.42}$$

*Proof*

In Lemma 3.3 it was shown that, if $\gamma \in \boldsymbol{\Gamma}_{\mathrm{wr}}$, then $\psi \in \mathbf{\Lambda}$, with Lipschitz constant $\lambda = (LK_{\max}/2\pi) + 1$. Consequentially, Jackson's Theorem II, Eq. 3.4.17, applies to $\psi$ and through a substitution of the value of $\lambda$ into this equation and through Eq. 3.4.16, the required result is obtained immediately.

□

If $\gamma \in \boldsymbol{\Gamma}_{\mathrm{r}}$, then both $\dot{z}$ and $\psi$ belong to $\mathbf{C}^1$. This enables us to derive upperbounds $L_{\mathrm{r}}(\dot{z})$ and $L_{\mathrm{r}}(\psi)$ on $\|\dot{z} - S_n\dot{z}\|_\infty$ and $\|\psi - S_n\psi\|_\infty$, respectively, through Jackson's Theorem I (Cheney [1966], pp. 142-143). However, these bounds are not sharper than the bounds $L_{\mathrm{wr}}(\dot{z})$ and $L_{\mathrm{wr}}(\psi)$ (cf. Theorems 3.7 and 3.8).

Through Eq. 3.4.19 and Eqs. 3.4.41 and 3.4.42 we observe that the asymptotic behavior of both bound $L_{\mathrm{wr}}(\dot{z})$ and bound $L_{\mathrm{wr}}(\psi)$ is $O(n^{-1}\log n)$ as $n \to \infty$. We recall from Section 3.2 that $\hat{\dot{z}}(k) = O(|k|^{-1})$ and $\hat{\psi}(k) = O(|k|^{-1})$ as $|k| \to \infty$ if $\gamma \in \boldsymbol{\Gamma}_{\mathrm{wr}}$.

Similar to the upper bounds on the finite Fourier series truncation error, no improvements on the rates of decay of Fourier coefficients were found if $\gamma \in \boldsymbol{\Gamma}_{\mathrm{r}}$.

The truncation error bounds, derived in this section for various contour classes and for various representations of these contours, enable us to determine the number of Fourier coefficients needed to guarantee a certain approximation accuracy of a Fourier representation. Thus we are also able to decide whether it is useful at all to use a Fourier representation instead of the corresponding contour representation. We found that especially for contours with much significant detail this may not be the case (cf. also Dekking and van Otterloo [1986]).

## 3.5 Geometric similarity and geometric mirror-similarity in terms of Fourier representations of contours

When no confusion can arise, we use $f$ in the following paragraphs again as a generic symbol to indicate any of the $2\pi$-periodic contour representations $z$, $\dot{z}$, $\ddot{z}$, $\psi$ and $K$.

In Definition 2.4 we have given a formal definition of similarity between contours. In Eq. 2.3.1 we have translated this concept into a necessary and sufficient condition on representations of pairs of contours that preserve shape information. We did the same for geometric mirror-similarity, in Definition 2.5 and in Eq. 2.3.2, respectively.

In Section 3.2 we have examined the characteristics of the convergence of the Fourier series and the Fourier coefficients of the periodic contour representations $z$, $\dot{z}$, $\ddot{z}$, $\psi$ and $K$ for contours from the contour classes $\boldsymbol{\Gamma}_{\mathrm{r}}$, $\boldsymbol{\Gamma}_{\mathrm{wr}}$, $\boldsymbol{\Gamma}_{\mathrm{s}}$, $\boldsymbol{\Gamma}_{\mathrm{pr}}$, $\boldsymbol{\Gamma}_{\mathrm{pwr}}$ and $\boldsymbol{\Gamma}_{\mathrm{ps}}$. This examination has shown that Fourier series, generated by these periodic contour representations, preserve shape information. Thereby we established that the sequence of Fourier coefficients $\hat{f} \equiv (\hat{f}(k))_{k \in \mathbb{Z}}$, generated by the shape information-preserving periodic contour representation $f$, also constitutes a shape information-preserving contour representation. This allows us to formulate necessary and sufficient conditions for geometric similarity and for geometric mirror-similarity in terms of the sequences of Fourier coefficients, generated by the representations of a pair of contours. This is in analogy with such conditions on the representations of the pair of contours themselves.

**Theorem 3.9.**
Two contours $\gamma_1$ and $\gamma_2$, with Fourier representations $\hat{f}_1$ and $\hat{f}_2$ respectively, are geometrically similar iff there exist scalars $\zeta \in \mathbb{C}$, $\beta \in \mathbb{R}^+$, $\alpha$, $\tau \in \mathbb{R}$ such that

$$\hat{f}_2 = \mathcal{T}_\tau \mathcal{R}_\alpha \mathcal{S}_\beta \mathcal{D}_\zeta \hat{f}_1, \tag{3.5.1}$$

where $\mathcal{T}_\tau$, $\mathcal{R}_\alpha$, $\mathcal{S}_\beta$ and $\mathcal{D}_\zeta$ are similarity operators which are defined in Table 3.2.

*Proof*

In Section 2.2 we have shown that there exists a one-to-one correspondence, in the sense of the relevant convergence properties, between a contour representation and its Fourier series expansion. This means that Eq. 2.3.1 can be obtained from Eq. 3.5.1, and therefore Theorem 2.4 applies, thus showing the validity of this theorem.

□

In Table 3.2 also the mirror-similarity operator $\mathcal{M}_x$ is defined. These (mirror-)similarity operators constitute analogues to the corresponding operators that were previously defined in Table 2.1. In Table 2.1 the

**Table 3.2.** Representations of the (mirror-)similarity operators in the sequence spaces of the Fourier representations.

| Fourier representations | Operator | | | | |
|---|---|---|---|---|---|
| | $\mathcal{S}_\beta$ | $\mathcal{D}_\zeta$ | $\mathcal{R}_\alpha$ | $\mathcal{T}_\tau$ | $\mathcal{M}_x$ |
| $\hat{z}$ | $\beta\hat{z}(k)$ | $\hat{z}(0)+\zeta$<br>$\hat{z}(k),\quad k \neq 0$ | $e^{i\alpha}\hat{z}(k)$ | $e^{-ik\tau}\hat{z}(k)$ | $\overline{\hat{z}}(k)$ |
| $\hat{\dot{z}}$ | $\beta\hat{\dot{z}}(k)$ | $\hat{\dot{z}}(k)$ | $e^{i\alpha}\hat{\dot{z}}(k)$ | $e^{-ik\tau}\hat{\dot{z}}(k)$ | $-\overline{\hat{\dot{z}}}(k)$ |
| $\hat{\ddot{z}}$ | $\beta\hat{\ddot{z}}(k)$ | $\hat{\ddot{z}}(k)$ | $e^{i\alpha}\hat{\ddot{z}}(k)$ | $e^{-ik\tau}\hat{\ddot{z}}(k)$ | $\overline{\hat{\ddot{z}}}(k)$ |
| $\hat{\psi}$ | $\hat{\psi}(k)$ | $\hat{\psi}(k)$ | $\hat{\psi}(k)$ | $\hat{\psi}(0)-\psi(-\tau)$<br>$e^{-ik\tau}\hat{\psi}(k),\quad k \neq 0$ | $-\overline{\hat{\psi}}(k)$ |
| $\hat{K}$ | $\beta^{-1}\hat{K}(k)$ | $\hat{K}(k)$ | $\hat{K}(k)$ | $e^{-ik\tau}\hat{K}(k)$ | $\overline{\hat{K}}(k)$ |

mirror-similarity operators operate on parametric contour representations whereas here they operate on the Fourier sequences, generated by these contour representations. The transformation of operators is obtained through straightforward calculation, using the expressions in Table 2.1 and the definition of Fourier coefficients (cf. Eq. 3.2.2).

**Remark.**
Though the (mirror-)similarity operators, when applied to Fourier representations, are formally different from the corresponding operators that are applied to contour representations, we have chosen not to express these differences by differences in notation, since both the meaning of these operators and the domain on which they operate is always clear.

□

Using Table 3.2 we can derive a formulation of the condition for geometric similarity in Eq. 3.5.1 for each of the individual Fourier representations. A survey of these conditions is presented in Table 3.3.

The formulation of the conditions on $\hat{\psi}$ in Table 3.3 still requires some explanation since no condition on $\hat{\psi}(0)$ is mentioned. From Table 2.3 and from the definition of Fourier coefficients in Eq. 3.2.2 we

**Table 3.3.** Necessary and sufficient conditions that the Fourier coefficients of the individual contour representations must satisfy, for some $\zeta \in \mathbb{C}, \beta \in \mathbb{R}^+, \alpha, \tau \in \mathbb{R}$ and $\forall k \in \mathbb{Z}$ (unless stated otherwise), in order to render a pair of geometrically similar contours.

| Fourier representations | Condition for geometric similarity |
|---|---|
| $\hat{z}$ | $\hat{z}_2(0) = \beta e^{i\alpha}\{\hat{z}_1(0) + \zeta\}$<br>$\hat{z}_2(k) = \beta e^{i(\alpha - k\tau)}\hat{z}_1(k), \quad k \neq 0$ |
| $\hat{\dot{z}}$ | $\hat{\dot{z}}_2(k) = \beta e^{i(\alpha - k\tau)}\hat{\dot{z}}_1(k)$ |
| $\hat{\ddot{z}}$ | $\hat{\ddot{z}}_2(k) = \beta e^{i(\alpha - k\tau)}\hat{\ddot{z}}_1(k)$ |
| $\hat{\psi}$ | $\hat{\psi}_2(k) = e^{-ik\tau}\hat{\psi}_1(k), \qquad k \neq 0$ |
| $\hat{K}$ | $\hat{K}_2(k) = \beta^{-1}e^{-ik\tau}\hat{K}_1(k)$ |

obtain the condition on $\hat{\psi}(0)$ for geometric similarity of a pair of contours as

$$\hat{\psi}_2(0) = \hat{\psi}_1(0) - \psi_1(-\tau). \tag{3.5.2}$$

It can be shown, however, that the conditions stated in Table 3.3 are sufficient and that Eq. 3.5.2 follows from these conditions. The reason for this is the fact that the property

$$\psi(0) = \sum_{k \in \mathbb{Z}} \hat{\psi}(k) = 0 \tag{3.5.3}$$

reduces the number of degrees of freedom in the Fourier representation $\hat{\psi}$ by one.

Next we formulate the condition on pairs of Fourier representations for geometric mirror-similarity.

**Theorem 3.10.**
Two contours $\gamma_1$ and $\gamma_2$, with Fourier representations $\hat{f}_1$ and $\hat{f}_2$ respectively, are geometrically mirror-similar iff there exist scalars $\zeta \in \mathbb{C}$, $\beta \in \mathbb{R}^+$, $\alpha, \tau \in \mathbb{R}$ such that

$$\hat{f}_2 = \mathcal{M}_x \mathcal{T}_\tau \mathcal{R}_\alpha \mathcal{S}_\beta \mathcal{D}_\zeta \hat{f}_1. \tag{3.5.4}$$

□

The proof of this theorem is similar to that of Theorem 3.9.

A survey of the formulations of this condition for geometric mirror-similarity, in terms of the individual Fourier representations, can be found in Table 3.4.

From Table 2.4 and from the definition of Fourier coefficients in Eq. 3.2.2 we obtain the condition on $\hat{\psi}(0)$ for geometric mirror-similarity of a pair of contours as

$$\hat{\psi}_2(0) = -\hat{\psi}_1(0) + \psi(\tau). \tag{3.5.5}$$

However, it can be shown along the same lines as we did for the corresponding condition on $\hat{\psi}(0)$ for geometrically similar contours, Eq. 3.5.2, that the condition in Eq. 3.5.5 is automatically satisfied if the conditions in Table 3.4 are satisfied.

**Table 3.4.** Necessary and sufficient conditions that the Fourier coefficients of the individual contour representations must satisfy, for some $\zeta \in \mathbb{C}$, $\beta \in \mathbb{R}^+$, $\alpha$, $\tau \in \mathbb{R}$ and $\forall k \in \mathbb{Z}$ (unless stated otherwise), in order to render a pair of geometrically mirror-similar contours.

| Fourier representations | Condition for geometric mirror-similarity |
|---|---|
| $\hat{z}$ | $\hat{z}_2(0) = \beta e^{-i\alpha}\{\overline{\hat{z}}_1(0) + \overline{\zeta}\}$<br>$\hat{z}_2(k) = \beta e^{-i(\alpha - k\tau)}\overline{\hat{z}}_1(k), \quad k \neq 0$ |
| $\hat{\dot{z}}$ | $\hat{\dot{z}}_2(k) = -\beta e^{-i(\alpha - k\tau)}\overline{\hat{\dot{z}}}_1(k)$ |
| $\hat{\ddot{z}}$ | $\hat{\ddot{z}}_2(k) = \beta e^{-i(\alpha - k\tau)}\overline{\hat{\ddot{z}}}_1(k)$ |
| $\hat{\psi}$ | $\hat{\psi}_2(k) = -e^{ik\tau}\overline{\hat{\psi}}_1(k), \quad k \neq 0$ |
| $\hat{K}$ | $\hat{K}_2(k) = \beta^{-1}e^{ik\tau}\overline{\hat{K}}_1(k)$ |

## 3.6 Symmetry in terms of Fourier representations of contours

In Definitions 2.6 and 2.7 we have formally defined geometric symmetry $\boldsymbol{m}$ and geometric symmetry $\boldsymbol{n}$, respectively. The argumentation in Section 3.2 concerning the shape information-preserving properties of a Fourier representation $\hat{f}$, generated by a shape information-preserving periodic contour representation $f$, enables us to formulate necessary and sufficient conditions on $\hat{f}$ such that this sequence of Fourier coefficients represents a contour that has geometric symmetry $\boldsymbol{m}$ or geometric symmetry $\boldsymbol{n}$. Such conditions for geometric symmetry in terms of $\hat{f}$ constitute the Fourier analogues of the corresponding conditions in terms of the contour representation $f$ itself, which were formulated in Eqs. 2.4.1 and 2.4.2.

**Theorem 3.11.**
A contour $\gamma$, with Fourier representation $\hat{f}$, has geometric symmetry $\boldsymbol{m}$, or geometric mirror-symmetry, iff there exist scalars $\zeta \in \mathbb{C}$ and $\alpha, \tau \in \mathbb{R}$ such that

$$\mathcal{D}_\zeta \hat{f} = \mathcal{M}_x \mathcal{T}_\tau \mathcal{R}_\alpha \mathcal{D}_\zeta \hat{f}. \tag{3.6.1}$$

□

The proof of this theorem is similar to that of Theorem 3.9.

A survey of the formulations of this condition for geometric symmetry $\boldsymbol{m}$, in terms of the individual Fourier representations, can be found in Table 3.5. These formulations merit a closer examination.

First we consider the conditions on $\hat{z}$ in order to represent a contour that has geometric symmetry $\boldsymbol{m}$. A suitable $\zeta \in \mathbb{C}$ always exists. This can be seen by choosing for example $\zeta = -\hat{z}(0)$.

Another way of formulating the condition on $\hat{z}$, in order to represent a contour $\gamma$ with symmetry $\boldsymbol{m}$, is the following: a contour $\gamma$, with Fourier representation $\hat{z}$, has geometric symmetry $\boldsymbol{m}$ iff there exist constants $\alpha, \tau \in \mathbb{R}$ such that $\mathcal{T}_\tau \mathcal{R}_\alpha \hat{z}(k)$ is real-valued, $\forall k \in \mathbb{Z} - \{0\}$. This formulation corresponds exactly to the necessary and sufficient conditions on $\hat{z}$ for geometric symmetry $\boldsymbol{m}$ in Wallace and Wintz [1980].

Along the same lines we derive from Table 3.5 similar necessary and sufficient conditions for geometric symmetry $\boldsymbol{m}$ in terms of the remaining Fourier representations:

- $\mathcal{T}_\tau \mathcal{R}_\alpha \hat{\dot{z}}(k)$ is imaginary,
- $\mathcal{T}_\tau \mathcal{R}_\alpha \hat{\ddot{z}}(k)$ is real-valued,
- $\mathcal{T}_\tau \mathcal{R}_\alpha \hat{\psi}(k)$ is imaginary,
- $\mathcal{T}_\tau \mathcal{R}_\alpha \hat{K}(k)$ is real-valued,

for some $\alpha, \tau \in \mathbb{R}$ and $\forall k \in \mathbb{Z} - \{0\}$.

From Table 2.5 and from the definition of Fourier coefficients in Eq. 3.2.2 we also obtain a condition on $\hat{\psi}(0)$ for a contour with geometric symmetry ***m***

$$\hat{\psi}(0) = -\hat{\psi}(0) + \psi(\tau) \tag{3.6.2}$$

which results from the special property of $\psi$ that $\mathcal{T}_\tau \psi(0) = 0$, independent of the value of $\tau$ (cf. Table 2.1). However, it can be shown, along the same lines as we did for a similar condition on $\hat{\psi}(0)$ for geometrically similar contours in Eq. 3.5.2, that the condition in Eq. 3.6.2 is automatically satisfied if the conditions in Table 3.5 are satisfied.

**Table 3.5.** Necessary and sufficient conditions that the Fourier coefficients of the individual representations must satisfy, for some $\zeta \in \mathbb{C}$, $\alpha, \tau \in \mathbb{R}$ and $\forall k \in \mathbb{Z}$ (unless stated otherwise), in order to render a contour with symmetry ***m***.

| Fourier representations | Condition for geometric symmetry ***m*** |
|---|---|
| $\hat{z}$ | $\hat{z}(0) + \zeta = e^{-i\alpha}\{\overline{\hat{z}}(0) + \overline{\zeta}\}$<br>$\hat{z}(k) = e^{-i(\alpha - k\tau)}\overline{\hat{z}}(k), \qquad k \neq 0$ |
| $\hat{\dot{z}}$ | $\hat{\dot{z}}(k) = -e^{-i(\alpha - k\tau)}\overline{\hat{\dot{z}}}(k)$ |
| $\hat{\ddot{z}}$ | $\hat{\ddot{z}}(k) = e^{i(\alpha - k\tau)}\overline{\hat{\ddot{z}}}(k)$ |
| $\hat{\psi}$ | $\hat{\psi}(k) = -e^{ik\tau}\overline{\hat{\psi}}(k), \qquad k \neq 0$ |
| $\hat{K}$ | $\hat{K}(k) = e^{ik\tau}\overline{\hat{K}}(k)$ |

We now turn to a discussion of geometric symmetry ***n*** or ***n***-fold geometric rotational symmetry.

**Theorem 3.12.**
A contour $\gamma$, with Fourier representation $\hat{f}$, has geometric symmetry ***n*** or **n**-fold geometric rotational symmetry, iff there exists a scalar $\zeta \in \mathbb{C}$

$$\mathcal{D}_\zeta \hat{f} = \mathcal{T}_{2\pi/n} \mathcal{R}_{2\pi/n} \mathcal{D}_\zeta \hat{f}. \tag{3.6.3}$$

□

The proof of this theorem is similar to that of Theorem 3.9.

A survey of the formulations of this condition for geometric symmetry $\boldsymbol{n}$, in terms of the individual Fourier representations, can be found in Table 3.6.

**Table 3.6.** Necessary and sufficient conditions that the Fourier coefficients of the individual contour representations must satisfy, for some $\zeta \in \mathbb{C}$ and $\forall k \in \mathbb{Z}$ (unless stated otherwise), in order to render a contour with symmetry $\boldsymbol{n}$.

| Fourier representation | Condition for geometric symmetry $\boldsymbol{n}$ |
|---|---|
| $\hat{z}$ | $\hat{z}(0) = -\zeta$<br>$\hat{z}(k) = 0, \quad \forall k\colon\ k \neq 1 \bmod \boldsymbol{n}$ |
| $\hat{\dot{z}}$ | $\hat{\dot{z}}(k) = 0, \quad \forall k\colon\ k \neq 1 \bmod \boldsymbol{n}$ |
| $\hat{\ddot{z}}$ | $\hat{\ddot{z}}(k) = 0, \quad \forall k\colon\ k \neq 1 \bmod \boldsymbol{n}$ |
| $\hat{\psi}$ | $\hat{\psi}(k) = 0, \quad \forall k\colon\ k \neq 0 \bmod \boldsymbol{n}$ |
| $\hat{K}$ | $\hat{K}(k) = 0, \quad \forall k\colon\ k \neq 0 \bmod \boldsymbol{n}$ |

Granlund [1972] was the first to mention conditions for geometric symmetry $\boldsymbol{n}$ in terms of $\hat{z}$, while Zahn and Roskies [1972] were the first to mention such conditions on $\hat{\psi}$, be it in a slightly different form. Crimmins [1982] finds Granlund's proof of the validity of the conditions for geometric symmetry $\boldsymbol{n}$ not logically conclusive. He shows that, for a contour that has geometric symmetry $\boldsymbol{n}$, there exists a $k_0$ such that, $\forall k \in \mathbb{Z} - \{0\}$, $\hat{z}(k) \neq 0$ implies $k = k_0 \bmod \boldsymbol{n}$. To this observation he adds the conjecture that $k_0 = 1$ if $\hat{z}(1) \neq 0$. In Dekking and van Otterloo [1986] a short proof is given for the statement that a contour has geometric symmetry $\boldsymbol{n}$ iff, $\forall k \in \mathbb{Z} - \{0\}$, $\hat{z}(k) \neq 0$ implies $k = 1 \bmod \boldsymbol{n}$, thereby confirming the correctness of Granlund's conclusions. On the other hand we emphasize here that the validity of the conditions for geometric symmetry $\boldsymbol{n}$ is subject to two conventions, which we introduced in Section 2.1, that both need to be satisfied:

- simplicity of contours,
- counterclockwise positive sense of the parametrization.

## 3.7 Concluding remarks

In Section 3.1 we reviewed the literature on the application of Fourier series theory to shape analysis.

In Section 3.2 we analyzed the convergence properties of Fourier series and Fourier coefficients, generated by the parametric contour representations $z$, $\dot{z}$, $\ddot{z}$, $\psi$ and $K$ for contours belonging to the contour classes $\boldsymbol{\Gamma}_{\mathrm{r}}$, $\boldsymbol{\Gamma}_{\mathrm{wr}}$, $\boldsymbol{\Gamma}_{\mathrm{s}}$, $\boldsymbol{\Gamma}_{\mathrm{pr}}$, $\boldsymbol{\Gamma}_{\mathrm{pwr}}$ and $\boldsymbol{\Gamma}_{\mathrm{ps}}$. It was established that the sequence of Fourier coefficients, generated by any of these contour representations, preserves shape information. We also found that the type and rate of convergence of the Fourier series heavily depends upon the smoothness properties of a contour, and thereby upon the differentiability properties of its representations. These results may provide guidelines in practice to determine whether it is appropriate to use a particular Fourier representation or not.

In Section 3.3 we have shown that, as a result of (normalized) arc length parametrization, with the exception of a circle, no position function $z$ can be expanded into a Fourier series with a finite number of nonzero Fourier coefficients. Consequently, the same holds for $\dot{z}$ and for $\ddot{z}$. Through an example we have shown that in some cases $\psi$ and $K$ may be expanded into a finite Fourier series, without affecting the linear relation between the parameter and arc length.

In practice we always use a finite number of Fourier coefficients. In Section 3.4 we have derived, for various contour representations and for various contour classes, upper bounds on the truncation error that is caused by a finite Fourier series expansion.

Conditions for geometric similarity and for geometric mirror-similarity, in terms of pairs of Fourier representations, have been presented in Section 3.5. In Chapter 4 these conditions will provide boundary conditions for similarity measures based on Fourier representations.

Similarly, the conditions for geometric symmetry $\boldsymbol{m}$ and for geometric symmetry $\boldsymbol{n}$, in terms of Fourier representations, which were formulated in Section 3.6, will provide boundary conditions for symmetry measures on the basis of Fourier representations in Chapter 4.

Also the application of Fourier coefficients for contour representation normalization will be studied in Chapter 4.

Apart from Fourier expansions, Walsh expansions of parametric contour representations have also been proposed for shape representation.

Before we turn to Chapter 4, we discuss these expansions briefly.

Searle [1970] proposes the use of Walsh expansions of the radial distance function for shape analysis purposes.

Shapiro [1976] compares, in the context of cell analysis, the performance of various orthogonal expansions through the results of reconstructing contours from a finite number of expansion coefficients. He studies, amongst others, the performance of Walsh expansions of the polar representation $R(\xi)$. He found that, apart from potential problems of multiple-valuedness of $R(\xi)$ (cf. Section 2.1), a reconstruction from Walsh coefficients gives a reasonable approximation if a sufficient number of coefficients is used. Shapiro also observes that, for a given accuracy of approximation, more Walsh coefficients than Fourier coefficients of $R(\xi)$ are needed.

In the context of the discrimination of handwritten numerals, Dinstein and Silberberg [1980] propose the average Walsh power spectrum of the periodic cumulative angular function $\psi$, defined as the Walsh expansion of the autocorrelation function of $\psi$, for shape representation. This representation clearly does not preserve all shape information.

Sethi and Sarvarayudu [1980] expand $\psi$ itself into a Walsh sequence. They used the magnitudes of the Walsh coefficients for the classification of handwritten numerals. A slightly lower error rate in the classification was achieved with the magnitudes of Walsh coefficients than with the same number of magnitudes of Fourier coefficients. Unlike the magnitudes of Fourier coefficients, the magnitudes of Walsh coefficients are sensitive to the location of the parametric starting point on the contour. Therefore they proposed two starting point normalization methods. In Sarvarayudu [1982] and in Sarvarayudu and Sethi [1983] this work is extended further. Geometrical properties are linked with properties of Walsh coefficients of $\psi$. Their methods are somewhat biased towards dealing with polygonal contours. This can be explained from the fact that a finite Walsh sequency expansion consists of a linear combination of step functions and the periodic cumulative angular function $\psi$ of a polygon is a step function. A method for the reconstruction and closing of a contour from a finite number of Walsh coefficients of $\psi$ is also presented. Finally they report on experiments with the classification of hand-printed numerals and characters using a two-stage classifier. The first stage uses the magnitudes of Walsh coefficients of $\psi$ as features, while the second stage uses their phases as features. However, the classification results they obtained with Walsh coefficients are no better than those known for classifiers based on Fourier expansions.

Summarizing this exposé on Walsh coefficients, we observe from the literature that Walsh coefficients have a computational advantage over Fourier coefficients. The power of Walsh coefficients to represent shape information in few coefficients may vary somewhat with the particular contour representation that generates them, but it seems in general to be no better or worse than that of the Fourier coefficients generated by that contour representation. Finite Walsh sequency expansions have more problems with the approximation of a smooth contour representation than finite Fourier series expansions. Furthermore, both magnitude and phase of Walsh coefficients are sensitive to the location of the parametric starting point. In conclusion we state that the advantages of Walsh sequency expansions are overshadowed by their disadvantages. Therefore we will not discuss the use of Walsh sequency expansions of parametric contour representations any further in this book.

Both Fourier series expansions and Walsh sequency expansions, being global orthogonal transformations, suffer from the inherent drawback that they are unable to deal properly with local perturbations on a contour. This problem limits their usefulness in applications where such phenomena are likely to occur. On the other hand, in inherently global contour operations, such as for example contour normalization, Fourier coefficients seem to be particularly useful, as we will see in Section 4.3. As for shape similarity measurement, our attention will be somewhat biased towards the contour representations themselves.

## References

Apostol, T.M. [1974]
*Mathematical Analysis*, Second Edition, Reading, MA: Addison-Wesley.

Beddow, J.K. and T.P. Meloy (Eds.) [1980]
*Advanced Particulate Morphology*, Boca Raton, FL: C.R.C. Press.

Beddow, J.K., G.C. Philip, A.F. Vetter and M.D. Nasta [1977]
'On Relating Some Particle Profile Characteristics to the Profile Fourier Coefficients', Powd. Technol. **18**: 19-25.

Bertrand, O., R. Queval and H. Maitre [1982]
'Shape Interpolation Using Fourier Descriptors with Application to Animation Graphics', Sign. Proc. **4**: 53-58.

Burkhardt, H. [1979]
*Transformationen zur Lageinvarianten Merkmalgewinnung* (Habilitationsschrift), Düsseldorf, Germany: VDI-Verlag, Fortschritt-Berichte der VDI-Zeitschriften, Reihe 10 (Angewandte Informatik), Nr. 7.

Carleson, L. [1966]
'On Convergence and Growth of Partial Sums of Fourier Series', Acta Math. **116**: 135-157.

Chen, C.-J. and Shi Q.-Y. [1980]
'Shape Features for Cancer Cell Recognition', Proc. Fifth Intl. Conf. on Patt. Recogn., Miami Beach, FL: 579-581.

Cheney, E.W. [1966]
*Introduction to Approximation Theory*, New York: McGraw-Hill Book Co., Inc.

Crimmins, T.R. [1982]
'A Complete Set of Fourier Descriptors for Two-Dimensional Shapes', IEEE Trans. Syst., Man and Cybern. **SMC-12**: 848-855.

Dekking, F.M. and P.J. van Otterloo [1986]
'Fourier Coding and Reconstruction of Complicated Contours', IEEE Trans. Syst., Man and Cybern. **SMC-16**: 395-404.

Dinstein, I. and T. Silberberg [1980]
'Shape Discrimination with Walsh Descriptors', Proc. Fifth Intl. Conf. on Patt. Recogn., Miami Beach, FL: 1055-1061.

Edwards, R.E. [1979]
*Fourier Series, a Modern Introduction, Volume 1*, New York: Springer-Verlag.

Edwards, R.E. [1982]
*Fourier Series, a Modern Introduction, Volume 2*, New York: Springer-Verlag.

Ehrlich, R. and B. Weinberg [1970]
'An Exact Method for Characterization of Grain Shape', Journ. of Sedim. Petrol. **40**: 205-212.

Etesami, F. and J.J. Uicker, Jr. [1985]
'Automatic Dimensional Inspection of Machine Part Cross-Sections Using Fourier Analysis', Comp. Vis., Graph. and Im. Proc. **29**: 216-247.

Giardina, C.R. and F.P. Kuhl [1977]
'Accuracy of Curve Approximation by Harmonically Related Vectors with Elliptical Loci', Comp. Graph. and Im. Proc. **6**: 277-285.

Granlund, G.H. [1972]
'Fourier Preprocessing for Hand Print Character Recognition', IEEE Trans. Comp. **C-21**: 195-201.

Hamming, R.W. [1977]
*Digital Filters*, Englewood Cliffs, NJ: Prentice-Hall, Inc.

Hardy, G.H., J.E. Littlewood and G. Pólya [1952]
*Inequalities*, Second Edition, Cambridge, England: Cambridge University Press.

Hewitt, E. and R.E. Hewitt [1979]
'The Gibbs-Wilbraham Phenomenon: An Episode in Fourier Analysis', Arch. Hist. Exact Sci. **21**: 129-160.

Hunt, R.A. [1967]
'On the Convergence of Fourier Series'. In: *Orthogonal Expansions and Their Continuous Analogues*, D. Tepper Haimo (Ed.), Proc. Conf. Southern Illinois University, Edwardsville, IL, Southern Illinois University Press, Feffer and Simons [1968]: 235-255.

Jerri, A.J. [1977]
'The Shannon Sampling Theorem – Its Various Extensions and Applications: A Tutorial Review', Proc. IEEE **65**: 1565-1596.

Katznelson, Y. [1968]
*An Introduction to Harmonic Analysis*, New York: John Wiley and Sons, Inc.

Kreyszig, E. [1968]
*Introduction to Differential Geometry and Riemannian Geometry*, Toronto: University of Toronto Press.

Kuhl, F.P. and C.R. Giardina [1982]
'Elliptic Fourier Features of a Closed Contour', Comp. Graph. and Im. Proc. **18**: 236-258.

Lebesgue, H. [1910]
'Sur la Représentation Trigonométrique Approchée des Fonctions Satisfaisant à une Condition de Lipschitz', Bull. Soc. Math. France **38**: 184-210.

Lighthill, M.J. [1962]
*Introduction to Fourier Analysis and Generalised Functions*, Cambridge, England: Cambridge University Press.

Luerkens, D.W., J.K. Beddow and A.F. Vetter [1982a]
'Morphological Fourier Descriptors', Powd. Technol. **31**: 209-215.

Luerkens, D.W., J.K. Beddow and A.F. Vetter [1982b]
'A Generalized Method of Morphological Analysis (the (R,S) Method)', Powd. Technol. **31**: 217-220.

Meloy, T.P. [1977a]
'A Hypothesis for Morphological Characterization of Particle Shape and Physiochemical Properties', Powd. Technol. **16**: 233-253.

Meloy, T.P. [1977b]
'Fast Fourier Transforms Applied to Shape Analysis of Particle Silhouettes to Obtain Morphological Data', Powd. Technol. **17**: 27-35.

Mitchell, O.R. and T.A. Grogan [1984]
'Shape Descriptors of Object Boundaries for Computer Vision'. In: *Intelligent Robots: Third International Conference on Robot Vision and Sensory Controls RoViSeC3*, D.P. Casasent and E.L. Hall (Eds.), Proc. SPIE **449**: 685-692.

Nguyen, N.G., R.S. Poulsen and C. Louis [1983]
'Some New Color Features and Their Application to Cervical Cell Classification', Patt. Recogn. **16**: 401-411.

Oppenheim, A.V. and R.W. Schafer [1975]
*Digital Signal Processing*, Englewood Cliffs, NJ: Prentice-Hall, Inc.

Persoon, E. and K.-S. Fu [1977]
'Shape Discrimination Using Fourier Descriptors', IEEE Trans. Syst., Man and Cybern. **SMC-7**: 170-179.

Proffitt, D. [1982]
'Normalization of Discrete Planar Objects', Patt. Recogn. **15**: 137-143.

Richard, Jr., C.W. and H. Hemami [1974]
'Identification of Three-Dimensional Objects Using Fourier Descriptors of the Boundary Curve', IEEE Trans. Syst., Man and Cybern. **SMC-4**: 371-378.

Royden, H.L. [1963]
*Real Analysis*, New York: The Macmillan Company.

Sarvarayudu, G.P.R. [1982]
*Shape Analysis Using Walsh Functions* (Ph.D. Thesis, Dept. of Electron. and Electr. Commun. Engin., Indian Institute of Technology, Kharagpur, India).

Sarvarayudu, G.P.R. and I.K. Sethi [1983]
'Walsh Descriptors for Polygonal Curves', Patt. Recogn. **16**: 327-336.

Schwarcz, H.P. and K.C. Shane [1969]
'Measurement of Particle Shape by Fourier Analysis', Sedimentology **13**: 213-231.

Searle, N.H. [1970]
'Shape Analysis by Use of Walsh Functions'. In: *Machine Intelligence 5*, B. Meltzer and D. Michie (Eds.): 395-410, Edinburgh: Edinburgh University Press.

Sethi, I.K. and G.P.R. Sarvarayudu [1980]
'Boundary Approximation Using Walsh Series Expansion for Numeral Recognition', Proc. Fifth Intl. Conf. on Patt. Recogn., Miami Beach, FL: 879-881.

Shannon, C.E. [1949]
'Communication in the Presence of Noise', Proc. IRE **37**: 10-21.

Shapiro, B. [1976]
'The Use of Orthogonal Expansions for Biological Shape Description', Computer Science Technical Report Series TR-472, University of Maryland, College Park, MD.

Strackee, J. and N.J.D. Nagelkerke [1983]
'On Closing the Fourier Descriptor Presentation', IEEE Trans. Patt. Anal. and Mach. Intell. **PAMI-4**: 660-661.

Sychra, J.J., P.H. Bartels, J. Taylor, M. Bibbo and G.L. Wied [1976]
'Cytoplasmic and Nuclear Shape Analysis for Computerized Cell Recognition', Acta Cytol. **20**: 68-78.

Tai, H.T., C.C. Li and S.H. Chiang [1982]
'Application of Fourier Shape Descriptors to Classification of Fine Particles', Proc. Sixth Intl. Conf. on Patt. Recogn., Munich. Germany: 748-751.

Titchmarsh, E.C. [1939]
*The Theory of Functions*, Second Edition, Oxford: Oxford University Press.

Wallace, T.P. [1981]
'Comments on *Algorithms for Shape Analysis of Contours and Waveforms*', IEEE Trans. Patt. Anal. and Mach. Intell. **PAMI-3**: 593.

Wallace, T.P. and O.R. Mitchell [1979]
'Local and Global Shape Description of Two- and Three-Dimensional Objects', Techn. Report TR-EE 79-43, School of Electrical Engineering, Purdue University, West-Lafayette, IN.

Wallace, T.P. and O.R. Mitchell [1980]
'Analysis of Three-Dimensional Movement Using Fourier Descriptors', IEEE Trans. Patt. Anal. and Mach. Intell. **PAMI-2**: 583-588.

Wallace, T.P. and P.A. Wintz [1980]
'An Efficient Three-Dimensional Aircraft Recognition Algorithm Using Normalized Fourier Descriptors', Comp. Graph. and Im. Proc. **13**: 99-126.

Watson, L.T. and L.G. Shapiro [1982]
'Identification of Space Curves from Two-Dimensional Perspective Views', IEEE Trans. Patt. Anal. and Mach. Intell. **PAMI-4**: 469-475.

Young, I.T., J.E. Walker and J.E. Bowie [1974]
'An Analysis Technique for Biological Shape – I', Inform. and Contr. **25**: 357-370.

Zahn, C.T. and R.Z. Roskies [1972]
'Fourier Descriptors for Plane Closed Curves', IEEE Trans. Comp. **C-21**: 269-281.

Zygmund, A. [1959a]
*Trigonometric Series, Volume I*, Second Edition, Cambridge, England: Cambridge University Press.

Zygmund, A. [1959b]
*Trigonometric Series, Volume II*, Second Edition, Cambridge, England: Cambridge University Press.

# Chapter 4

# Measurement of similarity, mirror-similarity and symmetry

## 4.1 Introductory considerations

In Chapter 4 we present a detailed discussion on the measurement of similarity, mirror-similarity and symmetry, based on the contour representations and Fourier representations introduced in the previous chapters.

This section gives some introductory considerations on similarity measurement and dissimilarity measurement.

In Section 4.2 various measures of dissimilarity and mirror-dissimilarity are defined and some of their properties are evaluated. For practical purposes, sampled-data formulations of these measures are given as well as an analysis of their computational complexity.

In Section 4.3 we study the trade-off between normalization of contours and optimization in dissimilarity measurement. The fundamental requirements that normalization procedures must satisfy are given in each case and a number of proposals for such procedures are made.

Section 4.4 contains a further theoretical analysis of the dissimilarity measures, defined in Section 4.2. Through a number of experiments we evaluate the relative behavior of the dissimilarity measures. By analyzing the experimental results of individual dissimilarity measures we obtain an insight into which aspects of geometric dissimilarity they measure.

In Section 4.5 we define measures for mirror-dissymmetry and for ***n***-fold rotational dissymmetry. The proposed measures for dissymmetry are closely related to the (mirror-)dissimilarity measures, defined in Section 4.2. A comparison with earlier proposals in the literature is also made. By an experiment we evaluate the performance of the dissymmetry measures defined in this section.

Finally, in Section 4.6 we review the results of this chapter.

In Section 2.5 we remarked that the concepts of geometric similarity and geometric mirror-similarity are mathematical abstractions. In that section we made the same remark concerning the three types of geometric symmetry, introduced in Section 2.4. In reality we will not encounter pairs of objects that have geometrically similar or geometrically mirror-similar contours or objects that are geometrically symmetric. Also the finite precision with which we can perform measurements would make the establishment of such facts virtually impossible. Therefore there is a need to dispose of quantitative methods by means of which the extent of similarity and symmetry can be measured.

In everyday life we use subjective criteria in our assessment of how similar figures are and this assessment will in general have a rather qualitative character. Our perception of a figure is not only determined by its geometric properties but also by its semantic content, which in turn is a result of our cultural and social background. This, and the context in which figures appear, also influence our notion of similarity between them. The parametric contour representations, introduced in Chapter 2, only describe the geometry of a figure. Consequently, if we measure similarity on the basis of these representations, then such a measurement can only express some geometric characteristics of similarity between figures.

In a number of pattern recognition and image analysis applications it is feasible to perform clustering and classification solely on the basis of geometric information. In many problems, however, geometric information alone will not suffice. For example, the analysis of decorations on objects of primitive art by computer, as studied in van Otterloo [1978], is bound to be of little use from an anthropological point of view if such an analysis is merely based on the geometric properties of these decorations and if it refrains from considering their semantic connotations. On the other hand, there is no reason to neglect the usefulness of geometric information for such applications. Apart from semantic and contextual information, geometric information plays an important role in our perception of the world that surrounds us. If required by a particular pattern recognition or image analysis application, the information, obtained by measuring one or more geometric aspects of similarity between figures, may be passed on to a higher level of processing where it may be combined with topological, contextual, semantic and other types of information.

The measurement of similarity between figures in a geometrical sense is also by no means a trivial problem. Here too our notion of similarity is affected by subjective considerations. Intuitively, if two figures are similar by subjective standards, a similarity measure should give a high value and if two figures are very dissimilar according to the same standards, a similarity measure should give a low value. Many measures for geometrical similarity can be formulated which satisfy the boundary conditions that geometrically similar figures give the maximum value of the measure, while figures that are not geometrically similar give a value below the maximum. We will soon see that in our approach these boundary conditions influence to a large extent the mathematical form of a similarity measure. However, even if a similarity measure satisfies these conditions, then this guarantees in no way that the measure has the aforementioned intuitive properties. It remains a very difficult and open problem to determine which measures are in reasonable correspondence with certain subjective notions of shape similarity. Obviously there exists no unique 'best' or 'optimal' measure that will give satisfactory results in all circumstances. The choice of a particular similarity measure will mainly be governed by the nature of the problem at hand, though also the robustness of a measure for noise and distortion, its computational requirements and the computational means available will influence such a choice. The quality of a similarity measure can be judged, for example, by the clustering or classification results obtained. On the basis of such evaluations we can get insight into how well a similarity measure performs with regard to that particular problem. On the other hand, it is usually not possible to make general statements about the quality of a similarity measure on the basis of results in a particular application: a measure that performs well in character recognition does not necessarily perform well in industrial inspection.

Though in practice we usually group objects on the basis of our subjective notions of similarity, in the context of pattern recognition and image analysis we will use the concept of dissimilarity for that purpose. Dissimilarity is usually measured by means of a distance measure (Sneath and Sokal [1973], Anderberg [1973]) and should, for ease of interpretation, preferably satisfy the conditions of a metric. For the properties of a metric, we refer to Appendix A.

Similarity and dissimilarity are complementary concepts, i.e. given a similarity measure we can always define a dissimilarity measure as a function of the similarity measure and vice versa (Späth [1980]).

It may happen that we need a similarity measure instead of a dissimilarity measure. Späth [1980] defines the concept of a *metric similarity function* and gives a number of examples of mappings of a metric dissimilarity function into a metric similarity function and vice versa. Since all such mappings set up a one-to-one correspondence between a metric dissimilarity function and a metric similarity function, which is necessary to preserve the metric properties after the mapping, the ordering of pairs of elements by either the metric dissimilarity function or by the metric similarity function is exactly reversed by the mapping. Therefore the information provided by either of the measures is exactly the same.

It is understood that the dissimilarity measures, that will be defined in Section 4.2, only pretend to measure geometric aspects of dissimilarity, and thereby geometric aspects of similarity. Whenever possible, a geometric or physical interpretation of the measures will be given. Such interpretations are important in judging what aspects of dissimilarity are measured and can be of help in predicting the usefulness of a measure in a given application. However, also the computational complexity of a measure has a definite influence upon its usefulness in practice. Therefore attention will be given to this aspect with the actual definition of the dissimilarity measures.

We already mentioned that there does not exist a unique 'best' or 'optimal' measure to quantify geometric dissimilarity between contours. Each dissimilarity measure will emphasize a different aspect of geometric dissimilarity. Therefore we may consider each dissimilarity measure as a *feature* of dissimilarity and combine a number of dissimilarity measures into a new one that possibly reflects the dissimilarity between contours more appropriately in a given application. Some possibilities to combine metrics to form a new metric will now be reviewed (cf. Anderberg [1973], Späth [1980]):

- metrics are closed under addition, i.e given two metrics $d_1$ and $d_2$ then

$$d = d_1 + d_2 \tag{4.1.1}$$

is also a metric,

- metrics are closed under scaling by a real-valued positive constant, i.e. given a metric $d_1$ and a constant $\beta \in \mathbb{R}^+$ then

$$d = \beta d_1 \tag{4.1.2}$$

is also a metric,

- if $d_1$ is a metric and $\gamma \in \mathbb{R}^+$, then

$$d = \frac{d_1}{\gamma + d_1} \tag{4.1.3}$$

is also a metric.

The operations in Eqs. 4.1.1-4.1.3 may of course be combined. For example, if $\{d_n: n = 1, \ldots, N\}$ is a set of metrics, then $\forall \beta_n, \gamma_n \in \mathbb{R}^+$

$$d = \sum_{n=1}^{N} \frac{\beta_n d_n}{\gamma_n + d_n} \tag{4.1.4}$$

is also a metric.

The possibilities just mentioned to map metrics into new metrics are certainly not the only possibilities. We refer to Anderberg [1973] and Späth [1980] for further information. We still note that metrics are not closed under multiplication, i.e. the product of two metrics is not necessarily a metric. This is because the triangle inequality, Eq. A.2, may not be satisfied by the product.

## 4.2 Measures of dissimilarity and mirror-dissimilarity

In Section 2.5 we marked a number of information-preserving contour representations as candidates for use in the analysis of similarity and symmetry of contours of two-dimensional objects. In Section 2.3 the conditions that these contour representations must satisfy in order to render geometrically similar or geometrically mirror-similar contours have been formulated.

In Section 3.2 we verified that the sequences of Fourier coefficients generated by the information-preserving contour representations also preserve shape information, for which reason we called them Fourier representations. In the previous section we mentioned the need for dissimilarity measures and mirror-dissimilarity measures in pattern recognition and image analysis applications. In this section we define measures of dissimilarity and mirror-dissimilarity, based on the contour representations and the Fourier representations, defined in Chapters 2 and 3 respectively. We show that the dissimilarity measures are metrics on equivalence classes of geometrically similar contours.

The (mirror-)dissimilarity measures contain an index $p$, by which we can control whether local or global differences in the contour representations or the Fourier representations are emphasized. The index value $p = 2$ constitutes a special case, because it leads to a greater mathematical tractability of the (mirror-)dissimilarity measures. We will find that for $p = 2$, a (mirror-)dissimilarity measure based on a contour representation is equivalent to the measure based on the corresponding Fourier representation.

In practice dissimilarity measurement is performed on the basis of a finite number of contour representation samples or a finite number of Fourier coefficients. Therefore we also present sampled-data formulations of the (mirror-)dissimilarity measures and analyze their computational complexity. For $p = 2$ we will find that a substantial reduction in computational complexity can be achieved.

### *4.2.1 Measures of dissimilarity and mirror-dissimilarity based on parametric contour representations*

Since we consider the shape of an object not to depend upon its position, size and orientation, or upon the choice of a starting point of a parametric representation of its contour, we require (mirror-)dissimilarity measures to be invariant for the application of equiform transformations on the contours involved. In general this invariance can be achieved in two ways:

- normalization of the position, size, orientation and parametric starting point of a contour such that we obtain for each equivalence class of geometrically similar contours a unique normalized representant,

- pairwise optimization of the position, size, orientation and parametric starting point of the contours, so as to yield a minimal dissimilarity value.

Combinations of these two methods are also possible. For the time being we normalize the position and size of the contours and we optimize their orientation and their parametric starting point in the (mirror-)dissimilarity measures. At first we exclude the periodic cumulative angular function $\psi$ from the discussion and deal with it separately later in this section.

We denote an *appropriate translation normalization parameter* by $\zeta^*$, $\zeta^* \in \mathbb{C}$, and an *appropriate scale normalization parameter* by $\beta^*$, $\beta^* \in \mathbb{R}^+$. (What we mean by 'appropriate' translation and scale normalization parameters will be discussed in detail in Section 4.3.) Then the relation between a contour representation $f$, where $f$ stands for any of the representations $z$, $\dot{z}$, $\ddot{z}$ and $K$, and its translation- and scale-normalized version $f^*$ is given by

$$f^* = \mathcal{S}_{\beta^*}\mathcal{D}_{\zeta^*}f. \tag{4.2.1}$$

For a survey of the formulation of the effects of translation and scaling upon the individual contour representations we refer to Table 2.1. We present a detailed discussion on optimization versus normalization in Section 4.3. In that section we will also indicate appropriate translation and scale normalization parameters.

In Appendix A the Lebesgue spaces $\mathbf{L}^p(2\pi)$, $1 \leqslant p \leqslant \infty$, are defined, as well as the usual norm $\|.\|_p$ on $\mathbf{L}^p(2\pi)$. It is mentioned there that if $f_1, f_2 \in \mathbf{L}^p(2\pi)$, $1 \leqslant p \leqslant \infty$, then $\|f_1 - f_2\|_p$ defines a metric on $\mathbf{L}^p(2\pi)$, the Minkowski-metrics or $\mathbf{L}^p$-metrics (cf. Appendix A). The family of dissimilarity measures that we now define is directly based upon the $\mathbf{L}^p$-metrics.

**Definition 4.1.** *Dissimilarity measure of index p.*
Let $f$ act as a generic symbol for any of the contour representations $z$, $\dot{z}$, $\ddot{z}$ and $K$. Then a measure of dissimilarity of index $p$ between a pair of contours $\gamma_1$ and $\gamma_2$, with contour representations $f_1$ and $f_2$ respectively, $f_1, f_2 \in \mathbf{L}^p(2\pi)$, is defined as

$$d^{(p)}(f_1, f_2) = \min_{\alpha,\tau} \|f_1^* - \mathcal{T}_\tau \mathcal{R}_\alpha f_2^*\|_p, \qquad 1 \leqslant p \leqslant \infty. \tag{4.2.2}$$

□

We now consider the periodic cumulative angular function $\psi$ in some detail. Since $\psi$ is invariant for translation, scaling and rotation of a contour, the righthand side of Eq. 4.2.2, upon substitution of $\psi$ for $f^*$, becomes

$$\min_{\tau} \| \psi_1 - \mathcal{T}_\tau \psi_2 \|_p. \tag{4.2.3}$$

We require of dissimilarity measures that they are invariant for shifts in the starting points of the contours involved. In particular, if Eq. 4.2.3 would be a valid dissimilarity measure in this respect, then it is required that

$$\min_{\tau} \| \mathcal{T}_\sigma \psi_1 - \mathcal{T}_\tau \psi_2 \|_p = \min_{\tau} \| \psi_1 - \mathcal{T}_\tau \psi_2 \|_p, \qquad \forall \sigma \in \mathbb{R}. \tag{4.2.4}$$

Unfortunately, Eq. 4.2.4 is in general not satisfied. The reason for this is the effect of a starting point shift upon $\psi$ (cf. Table 2.1)

$$\mathcal{T}_\tau \psi(t) = \psi(t - \tau) - \psi(-\tau). \tag{4.2.5}$$

The formula in Eq. 4.2.3 does not even define a symmetrical measure, as we will show in the following example.

**Example 4.1.**
Consider two contours $\gamma_1$ and $\gamma_2$ as displayed in Figures 4.1a and 4.2a, respectively. In the contour $\gamma_1$ the straight line segments have the same length as the circular arcs. The representation $\psi_1$ of $\gamma_1$ is displayed in Figure 4.1b and $\psi_2$ of $\gamma_2$ in Figure 4.2b. Note that $\psi_2$ is identically zero because $\gamma_2$ is a circle.

We are interested in the behavior of $\min_\tau \| \mathcal{T}_\sigma \psi_1 - \mathcal{T}_\tau \psi_2 \|_p$ as a function of $\sigma$. Since $\psi_2$ is identically zero we find

$$\min_{\tau} \| \mathcal{T}_\sigma \psi_1 - \mathcal{T}_\tau \psi_2 \|_p = \| \mathcal{T}_\sigma \psi_1 \|_p. \tag{4.2.6}$$

To illustrate the effect of a starting shift upon $\psi$ (cf. Eq. 4.2.5) we have displayed $\psi_1'(t) = \mathcal{T}_{\pi/16} \psi_1(t)$ and $\psi_1''(t) = \mathcal{T}_{\pi/8} \psi_1(t)$ in Figure 4.3 (compare with Figure 4.1).

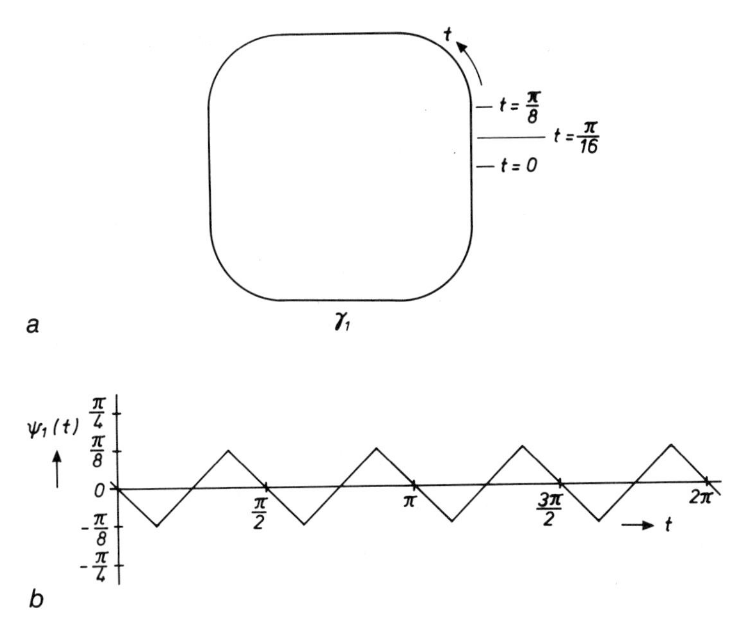


**Figure 4.1.** Contour $\gamma_1$ is shown in ($a$), consisting of four circular arcs and four straight line segments. The length of each straight line segment is the same as that of each circular arc. The periodic cumulative angular function $\psi_1$ of $\gamma_1$ is displayed in ($b$).

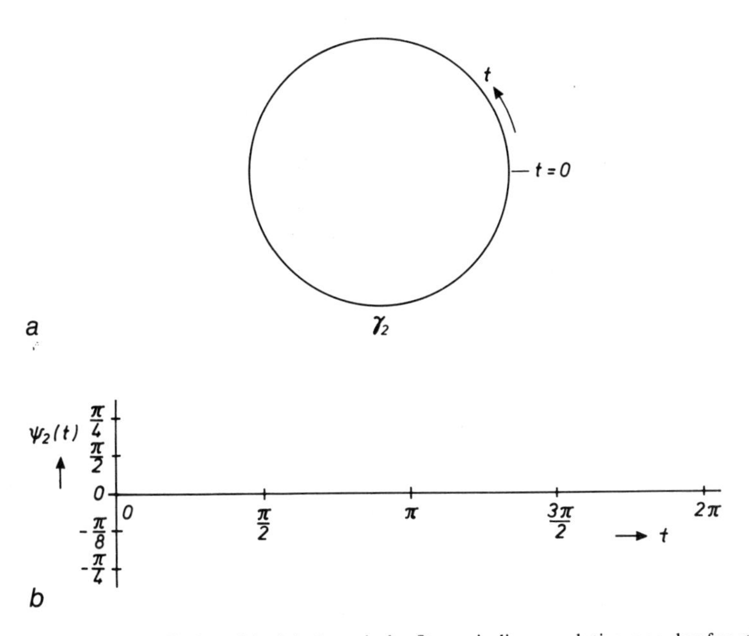


**Figure 4.2.** Contour $\gamma_2$, displayed in ($a$), is a circle. Its periodic cumulative angular function $\psi_2$, shown in ($b$), is identically zero.

Through straightforward integration we find for $\|\mathcal{T}_\sigma\psi_1\|_p$

$$\|\mathcal{T}_\sigma\psi_1\|_p = \begin{cases} \left[\dfrac{4}{\pi(p+1)}\left\{\left(\dfrac{\pi}{8}+k\dfrac{\pi}{4}-\sigma\right)^{p+1} + \left(\dfrac{\pi}{8}-k\dfrac{\pi}{4}+\sigma\right)^{p+1}\right\}\right]^{1/p} \\ \quad \text{for } k\dfrac{\pi}{4} \leqslant \sigma \leqslant \dfrac{\pi}{8}+k\dfrac{\pi}{4} & (4.2.7\text{a}) \\ \left[\dfrac{4}{\pi(p+1)}\left\{\left(-\dfrac{\pi}{8}-k\dfrac{\pi}{4}+\sigma\right)^{p+1} + \left(\dfrac{3\pi}{8}+k\dfrac{\pi}{4}-\sigma\right)^{p+1}\right\}\right]^{1/p} \\ \quad \text{for } \dfrac{\pi}{8}+k\dfrac{\pi}{4} \leqslant \sigma \leqslant \dfrac{\pi}{4}+k\dfrac{\pi}{4} & (4.2.7\text{b}) \end{cases}$$

for all $k \in \mathbb{Z}$.

These expressions show that $\min_\tau \|\mathcal{T}_\sigma\psi_1 - \mathcal{T}_\tau\psi_2\|_p$ is not independent of the starting point shift $\mathcal{T}_\sigma$ in $\psi_1$. A graph of $\|\mathcal{T}_\sigma\psi_1\|_p$, as a function of $\sigma$, is displayed in Figure 4.4 for $p = 1$, $p = 2$, $p = 5$ and, in the limit, for $p = \infty$.

We will also show through this example that $\min_\tau \|\psi_1 - \mathcal{T}_\tau\psi_2\|_p$ is not a symmetric measure, i.e. in general

$$\min_\tau \|\psi_1 - \mathcal{T}_\tau\psi_2\|_p \neq \min_\tau \|\psi_2 - \mathcal{T}_\tau\psi_1\|_p. \tag{4.2.8}$$

This can be seen as follows.

We found the results of the lefthand side of Eq. 4.2.8, as a function of the starting point shift in $\psi_1$ in Eq. 4.2.7a, b. For the righthand side of Eq. 4.2.8 we find

$$\min_\tau \|\psi_2 - \mathcal{T}_\tau\psi_1\|_p = \min_\tau \|\mathcal{T}_\tau\psi_1\|_p = \frac{1}{p+1}\left(\frac{\pi}{8}\right)^p \tag{4.2.9}$$

for $\tau = k \cdot \pi/4$, $k \in \mathbb{Z}$. So only for starting point shifts of $k \cdot \pi/4$, $k \in \mathbb{Z}$, $\min_\tau \|\psi_1 - \mathcal{T}_\tau\psi_2\|_p$ and $\min_\tau \|\psi_2 - \mathcal{T}_\tau\psi_1\|_p$ yield the same result in this example.

□

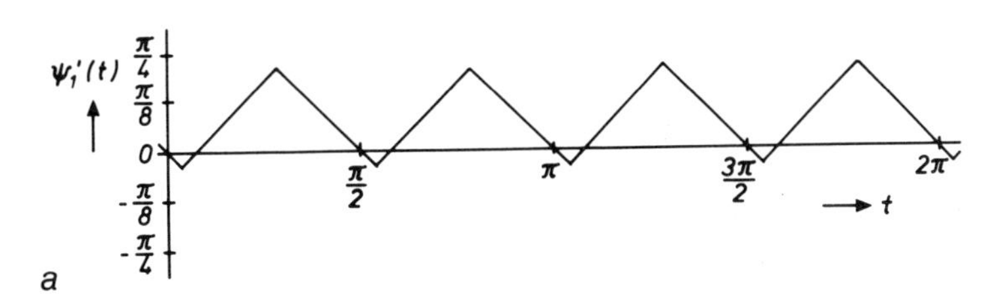


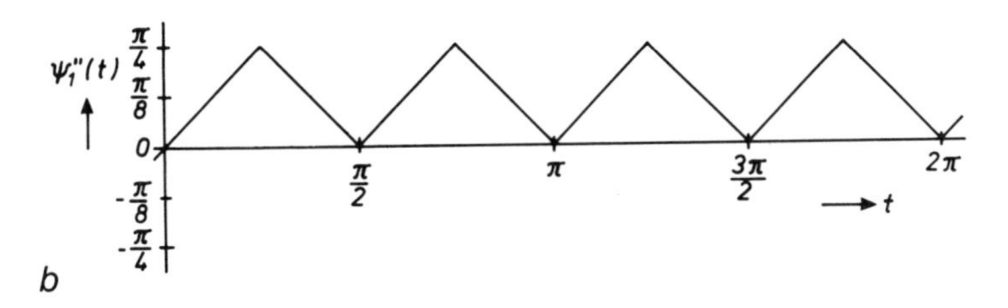


**Figure 4.3.** Periodic cumulative angular functions of contour $\gamma_1$ in Figure 4.1a after starting point shifts. In $(a)$ $\psi_1' = \mathcal{T}_{\pi/16}\psi_1$ is shown and in $(b)$ $\psi_1'' = \mathcal{T}_{\pi/8}\psi_1$.

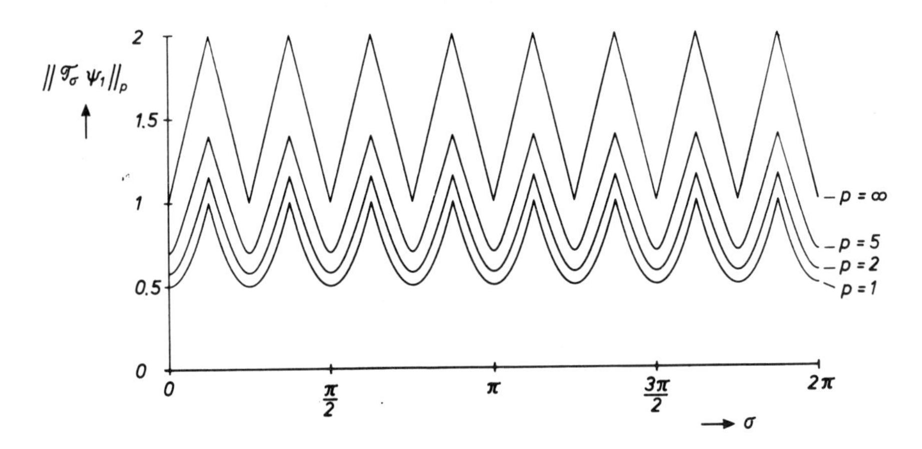


**Figure 4.4.** Graph of $\|\mathcal{T}_\sigma\psi_1\|_p$ as a function of the starting point shift $\sigma$ for various values of the index $p$.

One way to cure the problems described in the foregoing example is to propose the formula

$$\min_{\sigma,\tau} \| \mathcal{T}_\sigma \psi_1 - \mathcal{T}_\tau \psi_2 \|_p \tag{4.2.10}$$

as a dissimilarity measure on the basis of $\psi$. Though this formula would yield a valid dissimilarity measure, it has the drawback that it leads to a considerably increased computational effort.

Another way to get rid of the peculiar starting point dependence of $\psi$ is to base a dissimilarity measure on a *normalized periodic cumulative angular function* $\psi^*$ that preserves the information in $\psi$ and that has the property

$$\mathcal{T}_\tau \psi^*(t) = \psi^*(t - \tau). \tag{4.2.11}$$

One way to find such a function $\psi^*$ is as follows. Let

$$\psi^*(t) = \psi(t) + \lambda(\psi) \tag{4.2.12}$$

where $\lambda(\psi)$ is a, not yet specified, real-valued and single-valued function of $\psi$. Since $\psi(0) = 0$ it is clear that $\psi(t)$ can always be obtained from $\psi^*(t)$ as

$$\psi(t) = \psi^*(t) - \psi^*(0) \tag{4.2.13}$$

which shows that $\psi^*$ preserves the information in $\psi$.

From Eq. 4.2.12 we derive the equations

$$\psi^*(t - \tau) = \psi(t - \tau) + \lambda(\psi) \tag{4.2.14}$$

and

$$\mathcal{T}_\tau \psi^*(t) = \mathcal{T}_\tau \psi(t) + \lambda(\mathcal{T}_\tau \psi). \tag{4.2.15}$$

Through substitution of Eqs. 4.2.14 and 4.2.15 into the required property, expressed in Eq. 4.2.11, we find with the aid of Eq. 4.2.5 that the function $\lambda(\psi)$ must satisfy

$$\lambda(\mathcal{T}_\tau \psi) = \lambda(\psi) + \psi(-\tau). \tag{4.2.16}$$

Many functions $\lambda$, that have this property, can be defined.

**Definition 4.2.** *Dissimilarity measure of index p on the basis of* $\psi^*$.
A measure of dissimilarity of index $p$ between a pair of contours $\gamma_1$ and $\gamma_2$, with contour representations $\psi_1$ and $\psi_2$ respectively, $\psi_1, \psi_2 \in \mathbf{L}^p(2\pi)$, is defined as

$$d^{(p)}(\psi_1, \psi_2) = \min_{\tau} \| \psi_1^* - \mathcal{T}_\tau \psi_2^* \|_p, \qquad 1 \leq p \leq \infty \tag{4.2.17}$$

where $\psi^*$ is a normalized version of $\psi$, according to Eqs. 4.2.12 and 4.2.16.

□

Note that the form of Eq. 4.2.17 conforms with the definition of the dissimilarity measures in Eq. 4.2.2, since $\mathcal{R}_\alpha \psi^* = \psi^*$ (cf. Table 2.1).

In order to give an example of a normalization of $\psi$ we define the contour average of a periodic contour representation.

**Definition 4.3.** *Contour average.*
Let $f$ be a periodic contour representation or another contour-related function, with a normalized arc length parameter $t$. The contour average $\langle f \rangle$ of $f$ is defined as

$$\langle f \rangle = \frac{1}{2\pi} \int_{2\pi} f(t) \mathrm{d}t. \tag{4.2.18}$$

□

We note from the definition of the complex Fourier coefficients in Eq. 3.2.2 that

$$\langle f \rangle \equiv \hat{f}(0). \tag{4.2.19}$$

We recall from Table 3.3 the sufficient conditions that the Fourier coefficients of $\psi$ must satisfy in order to render a pair of geometrically similar contours

$$\hat{\psi}_2(k) = \mathrm{e}^{-ik\tau} \hat{\psi}_1(k), \qquad \forall k \in \mathbb{Z} - \{0\}$$

for some fixed value of $\tau \in \mathbb{R}$.

Using

$$\psi(t) = \sum_{k \in \mathbb{Z}} \hat{\psi}(k) e^{ikt}$$

we find as a sufficient condition for geometric similarity in terms of $\psi$

$$\begin{aligned}\psi_2(t) - \langle \psi_2 \rangle &= \psi_1(t - \tau) - \langle \psi_1 \rangle \\ &= \mathcal{T}_\tau \psi_1(t) - \langle \mathcal{T}_\tau \psi_1 \rangle, \qquad \forall t \in [0, 2\pi] \qquad (4.2.20)\end{aligned}$$

for some fixed value of $\tau$.

Compare the sufficient conditions in Eq. 4.2.20 with those in Table 2.3. From Eq. 4.2.20 we draw the conclusion that an appropriate choice for $\lambda(\psi)$ is

$$\lambda(\psi) = -\langle \psi \rangle. \qquad (4.2.21)$$

It is easily verified that $\lambda(\psi)$ in Eq. 4.2.21 satisfies the condition in Eq. 4.2.16. This choice results in $\langle \psi^* \rangle = 0$ (cf. Eq. 4.2.12).

In the following theorem we show that the normalization of $\psi$, as defined in Eq. 4.2.21, is optimal if the index $p$ in $d^{(p)}(\psi_1, \psi_2)$ equals 2.

**Theorem 4.1.** *Optimal normalization in $d^{(p)}(\psi_1, \psi_2)$ for $p = 2$.*
If $p = 2$ in the dissimilarity measure $d^{(p)}(\psi_1, \psi_2)$, defined in Eq. 4.2.17, then the normalization of $\psi$ according to Eq. 4.2.21, i.e. by choosing $\lambda(\psi) = -\langle \psi \rangle$ in Eq. 4.2.12, is optimal.

*Proof*

Consider

$$\begin{aligned}d^{(2)}(\psi_1, \psi_2) &= \min_\tau \| \psi_1^* - \mathcal{T}_\tau \psi_2^* \|_2 \\ &= \min_\tau \left[ \frac{1}{2\pi} \int_{2\pi} |\psi_1^*(t) - \psi_2^*(t - \tau)|^2 dt \right]^{1/2}. \qquad (4.2.22)\end{aligned}$$

Applying Eq. 4.2.12 to Eq. 4.2.22 and expanding the integrand yields

$$d^{(2)}(\psi_1, \psi_2) = \min_{\tau}\left[\frac{1}{2\pi}\int_{2\pi}\{\psi_1(t) - \psi_2(t-\tau)\}^2 \mathrm{d}t + 2\{\langle\psi_1\rangle - \langle\psi_2\rangle\}\{\lambda(\psi_1) - \lambda(\psi_2)\} + \{\lambda(\psi_1) - \lambda(\psi_2)\}^2\right]^{1/2} \tag{4.2.23}$$

which can be rewritten as

$$d^{(2)}(\psi_1, \psi_2) = \min_{\tau}\left[\frac{1}{2\pi}\int_{2\pi}\{\psi_1(t) - \psi_2(t-\tau)\}^2 \mathrm{d}t + \{\langle\psi_1\rangle + \lambda(\psi_1) - \langle\psi_2\rangle - \lambda(\psi_2)\}^2 - \{\langle\psi_1\rangle - \langle\psi_2\rangle\}^2\right]^{1/2}. \tag{4.2.24}$$

Only the middle quadratic term in Eq. 4.2.24 depends upon the choice of a normalization function $\lambda(\psi)$. It is clear from the expression in Eq. 4.2.24 that $d^{(2)}(\psi_1, \psi_2)$ is always minimized if we choose $\lambda(\psi) = -\langle\psi\rangle$.

□

A survey of the dissimilarity measures of index $p$, defined in Definitions 4.1 and 4.2, in terms of the individual contour representations, is given in Table 4.1.

**Remark.**
In the tables in this section, dealing with (mirror-)dissimilarity measures for general values of the index $p$, the value $p = \infty$ has been excluded, although these measures are also defined for this value of $p$. The reason for this exclusion is the fact that the formula for $\|\cdot\|_\infty$ is somewhat different from $\|\cdot\|_p$, $1 \leqslant p < \infty$ (cf. Eqs. A.12 and A.13). However, given the formula of a (mirror-)dissimilarity measure for $1 \leqslant p < \infty$, the derivation of the formula for $p = \infty$ is straightforward. □

**Table 4.1.** Dissimilarity measures of index $p$, $1 \leqslant p < \infty$, in terms of the individual contour representations.

| Contour representation | Dissimilarity measures of index $p$ |
|---|---|
| $z$ | $d^{(p)}(z_1, z_2) = \min_{\alpha,\tau} \left[ \frac{1}{2\pi} \int_{2\pi} \left\| z_1^*(t) - e^{i\alpha} z_2^*(t-\tau) \right\|^p dt \right]^{1/p}$ |
| $\dot{z}$ | $d^{(p)}(\dot{z}_1, \dot{z}_2) = \min_{\alpha,\tau} \left[ \frac{1}{2\pi} \int_{2\pi} \left\| \dot{z}_1^*(t) - e^{i\alpha} \dot{z}_2^*(t-\tau) \right\|^p dt \right]^{1/p}$ |
| $\ddot{z}$ | $d^{(p)}(\ddot{z}_1, \ddot{z}_2) = \min_{\alpha,\tau} \left[ \frac{1}{2\pi} \int_{2\pi} \left\| \ddot{z}_1^*(t) - e^{i\alpha} \ddot{z}_2^*(t-\tau) \right\|^p dt \right]^{1/p}$ |
| $\psi$ | $d^{(p)}(\psi_1, \psi_2) = \min_{\tau} \left[ \frac{1}{2\pi} \int_{2\pi} \left\| \psi_1^*(t) - \psi_2^*(t-\tau) \right\|^p dt \right]^{1/p}$ |
| $K$ | $d^{(p)}(K_1, K_2) = \min_{\tau} \left[ \frac{1}{2\pi} \int_{2\pi} \left\| K_1^*(t) - K_2^*(t-\tau) \right\|^p dt \right]^{1/p}$ |

An important property of the dissimilarity measures is established in the following theorem.

**Theorem 4.2.** *Metric properties of dissimilarity measures.*
The families of dissimilarity measures defined in Definitions 4.1 and 4.2 constitute metrics on the equivalence classes of representations of geometrically similar contours.

*Proof*

Let $f$ stand for any of the contour representations $z$, $\dot{z}$, $\ddot{z}$, $\psi$ and $K$. If $f_1$, $f_2$ and $f_3$ are the representations of the arbitrary contours $\gamma_1$, $\gamma_2$ and $\gamma_3$, with $f_1, f_2, f_3 \in \mathbf{L}^p(2\pi)$, then we have to show (cf. Definition A.1):

(a) $d^{(p)}(f_1, f_2) = 0$ iff $f_1$ and $f_2$ belong to the same equivalence class of representations of geometrically similar contours.

(b) $d^{(p)}(f_1, f_2) \leqslant d^{(p)}(f_3, f_1) + d^{(p)}(f_3, f_2)$, the triangle inequality.

Re. (a) The families of dissimilarity measures $d^{(p)}(f_1, f_2)$ are directly based upon the necessary and sufficient conditions that $f_1$ and $f_2$ must satisfy in order to render geometrically similar contours (cf. Eq. 2.3.1). This ensures that, if $d^{(p)}(f_1, f_2) = 0$, then the conditions for geometric similarity in Eq. 2.3.1 are satisfied almost everywhere for some $\alpha$ and $\tau$, which allows us to conclude that $\gamma_1$ and $\gamma_2$ are indeed geometrically similar. On the other hand, if $\gamma_1$ and $\gamma_2$ are geometrically similar we find $d^{(p)}(f_1, f_2) = 0$, whereas if $\gamma_1$ and $\gamma_2$ are not geometrically similar we find $d^{(p)}(f_1, f_2) > 0$. In this discussion we have assumed that the normalized representation $f^*$ can always be determined uniquely from $f$ (cf. Eqs. 4.2.1 and 4.2.12).

Re. (b) Through Eqs. 4.2.2 and 4.2.17 we have

$$d^{(p)}(f_3, f_1) = \min_{\alpha,\tau} \|f_3^* - \mathcal{T}_\tau \mathcal{R}_\alpha f_1^*\|_p. \tag{4.2.25}$$

Let $\alpha_{13}$ and $\tau_{13}$ be the solutions of the minimizations over $\alpha$ and $\tau$ in $d^{(p)}(f_3, f_1)$.

Similarly, let $\alpha_{23}$ and $\tau_{23}$ be the corresponding solutions in $d^{(p)}(f_3, f_2)$. Then we obtain

$$d^{(p)}(f_3, f_1) + d^{(p)}(f_3, f_2) = \|f_3^* - \mathcal{T}_{\tau_{13}} \mathcal{R}_{\alpha_{13}} f_1^*\|_p + \|f_3^* - \mathcal{T}_{\tau_{23}} \mathcal{R}_{\alpha_{23}} f_2^*\|_p. \tag{4.2.26}$$

Since $\|f_1 - f_2\|_p$, with $f_1, f_2 \in \mathbf{L}^p(2\pi)$, constitutes a metric on $\mathbf{L}^p(2\pi)$ (cf. Theorems A.1 and A.2), the triangle inequality leads to

$$\begin{aligned} d^{(p)}(f_3, f_1) + d^{(p)}(f_3, f_2) &\geqslant \|\mathcal{T}_{\tau_{13}} \mathcal{R}_{\alpha_{13}} f_1^* - \mathcal{T}_{\tau_{23}} \mathcal{R}_{\alpha_{23}} f_2^*\|_p \\ &= \|f_1^* - \mathcal{T}_{(\tau_{23} - \tau_{13})} \mathcal{R}_{(\alpha_{23} - \alpha_{13})} f_2^*\|_p \\ &\geqslant d^{(p)}(f_1, f_2) \end{aligned} \tag{4.2.27}$$

which is the required inequality.

The properties (a) and (b) in this proof correspond to the requirements of a metric in Eqs. A.1 and A.2, respectively. Since these properties are sufficient for a metric (cf. Appendix A), the proof of the theorem is now complete. □

Along the same lines as we defined families of dissimilarity measures in Definitions 4.1 and 4.2 we now define mirror-dissimilarity measures. The mirror-dissimilarity measures are directly based on the necessary and sufficient conditions on contour representations of geometrically mirror-similar contours, as defined in Eq. 2.3.2.

**Definition 4.4.** *Mirror-dissimilarity measure of index p.*
Let $f$ act as a generic symbol for any of the contour representations $z$, $\dot{z}$, $\ddot{z}$, $\psi$ and $K$. Then a measure of mirror-dissimilarity of index $p$ between a pair of contours $\gamma_1$ and $\gamma_2$, with contour representations $f_1$ and $f_2$ respectively, $f_1, f_2 \in \mathbf{L}^p(2\pi)$, is defined as

$$\tilde{d}^{(p)}(f_1, f_2) = \min_{\alpha,\tau} \| f_1^* - \mathcal{M}_x \mathcal{T}_\tau \mathcal{R}_\alpha f_2^* \|_p, \qquad 1 \leqslant p \leqslant \infty. \tag{4.2.28}$$

□

It is easily verified that $\tilde{d}^{(p)}(f_1, f_2)$ is in fact a dissimilarity measure between a contour and another contour, that is mirror-reflected about the $x$-axis, i.e.

$$\tilde{d}^{(p)}(f_1, f_2) = d^{(p)}(f_1, \mathcal{M}_x f_2). \tag{4.2.29}$$

In analogy with the dissimilarity measures defined in Definitions 4.1 and 4.2, the families of mirror-dissimilarity measures $\tilde{d}^{(p)}$ act as measures between equivalence classes of geometrically similar contours, i.e. apart from the type of contour representation and the value of the index $p$, the value of $\tilde{d}^{(p)}$ will only depend upon the pair of equivalence classes to which the contours belong and not upon the particular specimens from these equivalence classes.

The families of mirror-dissimilarity measures $\tilde{d}^{(p)}$ possess the property of symmetry: $\tilde{d}^{(p)}(f_1, f_2) = \tilde{d}^{(p)}(f_2, f_1)$. The property of reflexivity, i.e. $\tilde{d}^{(p)}(f_1, f_1) = 0$, is only satisfied by a special subset of contours, namely contours that are mirror-symmetric. We will use this fact later in Section 4.5 to define families of measures of mirror-dissymmetry.

The triangle inequality does not hold for $\tilde{d}^{(p)}$ and is in fact meaningless in this case. The same applies to the concept of a metric in relation to $\tilde{d}^{(p)}$.

A survey of the mirror-dissimilarity measures of index $p$, defined in Definition 4.4, in terms of the individual contour representations, is given in Table 4.2.

**Table 4.2.** Mirror-dissimilarity measures of index $p$, $1 \leqslant p < \infty$, in terms of the individual contour representations.

| Contour representation | Mirror-dissimilarity measures of index $p$ |
|---|---|
| $z$ | $\tilde{d}^{(p)}(z_1, z_2) = \min_{\alpha,\tau} \left[ \frac{1}{2\pi} \int_{2\pi} \lvert z_1^*(t) - e^{-i\alpha} \bar{z}_2^*(-t+\tau) \rvert^p dt \right]^{1/p}$ |
| $\dot{z}$ | $\tilde{d}^{(p)}(\dot{z}_1, \dot{z}_2) = \min_{\alpha,\tau} \left[ \frac{1}{2\pi} \int_{2\pi} \lvert \dot{z}_1^*(t) + e^{-i\alpha} \dot{\bar{z}}_2^*(-t+\tau) \rvert^p dt \right]^{1/p}$ |
| $\ddot{z}$ | $\tilde{d}^{(p)}(\ddot{z}_1, \ddot{z}_2) = \min_{\alpha,\tau} \left[ \frac{1}{2\pi} \int_{2\pi} \lvert \ddot{z}_1^*(t) - e^{-i\alpha} \ddot{\bar{z}}_2^*(-t+\tau) \rvert^p dt \right]^{1/p}$ |
| $\psi$ | $\tilde{d}^{(p)}(\psi_1, \psi_2) = \min_{\tau} \left[ \frac{1}{2\pi} \int_{2\pi} \lvert \psi_1^*(t) + \psi_2^*(-t+\tau) \rvert^p dt \right]^{1/p}$ |
| $K$ | $\tilde{d}^{(p)}(K_1, K_2) = \min_{\tau} \left[ \frac{1}{2\pi} \int_{2\pi} \lvert K_1^*(t) - K_2^*(-t+\tau) \rvert^p dt \right]^{1/p}$ |

An important special case of the (mirror-)dissimilarity measures of index $p$ is formed by the measures of index $p = 2$. These are discussed in the next subsection.

### *4.2.2 Measures of dissimilarity and mirror-dissimilarity of index p = 2*

In this subsection a special case of the measures of dissimilarity and mirror-dissimilarity of index $p$ is discussed: the measures of index $p = 2$. These measures are directly based upon the $\mathbf{L}^2$-metric or Euclidean metric on $\mathbf{L}^2(2\pi)$. The main reasons for the importance of this case are its greater mathematical tractability, leading to computationally efficient implementations, and, as we will see later on, its isometric relation with corresponding Fourier representation-based (mirror-)dissimilarity measures. For $p = 2$, the concepts of correlation and convolution appear in the expressions of the dissimilarity and mirror-dissimilarity measures.

**Definition 4.5.** *Cyclic cross-correlation function.*
Let $f_1$ and $f_2$ be a pair of complex-valued $2\pi$-periodic functions, with $f_1, f_2 \in \mathbf{L}^2(2\pi)$. Then the cross-correlation function between $f_1$ and $f_2$ is defined as

$$\varrho_{12}(\tau; f) = \frac{1}{2\pi} \int_{2\pi} f_1(t)\overline{f_2}(t - \tau)\mathrm{d}t. \tag{4.2.30}$$

□

**Definition 4.6.** *Cyclic convolution function.*
Let $f_1$ and $f_2$ be a pair of complex-valued $2\pi$-periodic functions, with $f_1, f_2 \in \mathbf{L}^2(2\pi)$. Then the cyclic convolution function of $f_1$ and $f_2$ is defined as

$$h_{12}(\tau; f) = \frac{1}{2\pi} \int_{2\pi} f_1(t)f_2(-t + \tau)\mathrm{d}t. \tag{4.2.31}$$

□

It is readily understood that, as a special case, the functions $f_1$ and $f_2$ in Definitions 4.5 and 4.6 may also be real-valued.

We will treat the (mirror-)dissimilarity measures based on complex-valued contour representations and those based on real-valued contour representations separately, because there is a slight difference in their analysis.

Let $f$ act as a generic symbol for any of the complex-valued contour representations $z$, $\dot{z}$ and $\ddot{z}$. Substitution of $p = 2$ into $d^{(p)}(f_1, f_2)$, Eq. 4.2.2, and expanding the norm gives

$$\begin{aligned} d^{(2)}(f_1, f_2) &= \min_{\alpha,\tau} \| f_1^* - \mathcal{T}_\tau \mathcal{R}_\alpha f_2^* \|_2 \\ &= \min_{\alpha,\tau} \left[ \frac{1}{2\pi} \int_{2\pi} \left| f_1^*(t) - \mathrm{e}^{\mathrm{i}\alpha} f_2^*(t - \tau) \right|^2 \mathrm{d}t \right]^{1/2} \\ &= \left[ \|f_1^*\|_2^2 + \|f_2^*\|_2^2 - \max_{\alpha,\tau} \{ \mathrm{e}^{-\mathrm{i}\alpha} \varrho_{12}(\tau; f^*) + \mathrm{e}^{\mathrm{i}\alpha} \overline{\varrho_{12}}(\tau; f^*) \} \right]^{1/2} \end{aligned} \tag{4.2.32}$$

where $f^*$ in $\varrho_{12}(\tau; f^*)$ indicates that the translation- and scale-normalized versions of $f_1$ and $f_2$ are used in the cyclic cross-correlation function (cf. Eq. 4.2.1). Analyzing the optimization over $\alpha$ and $\tau$ in Eq. 4.2.32 we observe that

$$\max_{\tau}\Big[\max_{\alpha}\big\{e^{-i\alpha}\varrho_{12}(\tau;f^*)+e^{i\alpha}\overline{\varrho_{12}}(\tau;f^*)\big\}\Big]=2\max_{\tau}\big|\varrho_{12}(\tau;f^*)\big| \tag{4.2.33}$$

where the solution for $\alpha$, as a function of $\tau$, is given by

$$\alpha=\arg\big\{\varrho_{12}(\tau;f^*)\big\}\equiv\arctan\left(\frac{\mathrm{Im}\,\{\varrho_{12}(\tau;f^*)\}}{\mathrm{Re}\,\{\varrho_{12}(\tau;f^*)\}}\right). \tag{4.2.34}$$

Through Eq. 4.2.33 we find for $d^{(2)}(f_1, f_2)$ the end result

$$d^{(2)}(f_1,f_2)=\Big[\,\|f_1^*\|_2^2+\|f_2^*\|_2^2-2\max_{\tau}\big|\varrho_{12}(\tau;f^*)\big|\,\Big]^{1/2}. \tag{4.2.35}$$

Let $g$ act as a generic symbol for any of the two real-valued contour representations $\psi$ and $K$. Substitution of $p = 2$ into the dissimilarity measure of index $p$, Eqs. 4.2.2 and 4.2.17, and expanding the norm gives

$$\begin{aligned}
d^{(2)}(g_1,g_2)&=\min_{\alpha,\tau}\|g_1^*-\mathcal{T}_\tau\mathcal{R}_\alpha g_2^*\|_2\\
&=\min_{\tau}\left[\frac{1}{2\pi}\int_{2\pi}\big|g_1^*(t)-g_2^*(t-\tau)\big|^2\mathrm{d}t\right]^{1/2}\\
&=\Big[\,\|g_1^*\|_2^2+\|g_2^*\|_2^2-2\max_{\tau}\varrho_{12}(\tau;g^*)\Big]^{1/2}.
\end{aligned} \tag{4.2.36}$$

A survey of the results in Eqs. 4.2.35 and 4.2.36 for the dissimilarity measures of index $p = 2$, in terms of the individual contour representations, is given in Table 4.3.

Note that in the expressions for the dissimilarity measures of index $p = 2$ on the basis of the real-valued contour representations $\psi$ and $K$ an optimization over the cyclic correlation function appears, whereas in the corresponding measures on the basis of the complex-valued contour representations $z$, $\dot{z}$ and $\ddot{z}$ the optimization is over the modulus of the cyclic correlation function.

**Table 4.3.** Dissimilarity measures of index $p = 2$ in terms of the individual contour representations.

| Contour representation | Dissimilarity measures of index $p = 2$ |
|---|---|
| $z$ | $d^{(2)}(z_1, z_2) = \left[\|z_1^*\|_2^2 + \|z_2^*\|_2^2 - 2 \max_\tau \|\varrho_{12}(\tau; z^*)\|\right]^{1/2}$ |
| $\dot{z}$ | $d^{(2)}(\dot{z}_1, \dot{z}_2) = \left[\|\dot{z}_1^*\|_2^2 + \|\dot{z}_2^*\|_2^2 - 2 \max_\tau \|\varrho_{12}(\tau; \dot{z}^*)\|\right]^{1/2}$ |
| $\ddot{z}$ | $d^{(2)}(\ddot{z}_1, \ddot{z}_2) = \left[\|\ddot{z}_1^*\|_2^2 + \|\ddot{z}_2^*\|_2^2 - 2 \max_\tau \|\varrho_{12}(\tau; \ddot{z}^*)\|\right]^{1/2}$ |
| $\psi$ | $d^{(2)}(\psi_1, \psi_2) = \left[\|\psi_1^*\|_2^2 + \|\psi_2^*\|_2^2 - 2 \max_\tau \varrho_{12}(\tau; \psi^*)\right]^{1/2}$ |
| $K$ | $d^{(2)}(K_1, K_2) = \left[\|K_1^*\|_2^2 + \|K_2^*\|_2^2 - 2 \max_\tau \varrho_{12}(\tau; K^*)\right]^{1/2}$ |

**Table 4.4.** Mirror-dissimilarity measures of index $p = 2$ in terms of the individual contour representations.

| Contour representation | Mirror-dissimilarity measures of index $p = 2$ |
|---|---|
| $z$ | $\tilde{d}^{(2)}(z_1, z_2) = \left[\|z_1^*\|_2^2 + \|z_2^*\|_2^2 - 2 \max_\tau \|h_{12}(\tau; z^*)\|\right]^{1/2}$ |
| $\dot{z}$ | $\tilde{d}^{(2)}(\dot{z}_1, \dot{z}_2) = \left[\|\dot{z}_1^*\|_2^2 + \|\dot{z}_2^*\|_2^2 - 2 \max_\tau \|h_{12}(\tau; \dot{z}^*)\|\right]^{1/2}$ |
| $\ddot{z}$ | $\tilde{d}^{(2)}(\ddot{z}_1, \ddot{z}_2) = \left[\|\ddot{z}_1^*\|_2^2 + \|\ddot{z}_2^*\|_2^2 - 2 \max_\tau \|h_{12}(\tau; \ddot{z}^*)\|\right]^{1/2}$ |
| $\psi$ | $\tilde{d}^{(2)}(\psi_1, \psi_2) = \left[\|\psi_1^*\|_2^2 + \|\psi_2^*\|_2^2 + 2 \min_\tau h_{12}(\tau; \psi^*)\right]^{1/2}$ |
| $K$ | $\tilde{d}^{(2)}(K_1, K_2) = \left[\|K_1^*\|_2^2 + \|K_2^*\|_2^2 - 2 \max_\tau h_{12}(\tau; K^*)\right]^{1/2}$ |

The analysis of the mirror-dissimilarity measures of index $p = 2$ evolves along the same lines as that of the dissimilarity measures of index $p = 2$. Instead of the cyclic correlation function the cyclic convolution function appears in the expressions for the mirror-dissimilarity measures of index $p = 2$. A survey of these expressions, in terms of the individual contour representations, is given in Table 4.4.

Note in Table 4.4 the somewhat differing expression for $\tilde{d}^{(2)}(\psi_1, \psi_2)$ which is a consequence of the expression for $\tilde{d}^{(p)}(\psi_1, \psi_2)$ in Table 4.2.

### *4.2.3 Measures of dissimilarity and mirror-dissimilarity based on Fourier representations of contours*

In analogy with the (mirror-)dissimilarity measures defined in Definitions 4.1, 4.2 and 4.4, which are based on the information-preserving contour representations introduced in Chapter 2, we can also define such measures on the basis of the corresponding Fourier representations of contours. In Section 3.2 we have shown that these Fourier representations are also information-preserving.

In Section 3.5 we have formulated the necessary and sufficient conditions that Fourier representations must satisfy in order to render geometrically (mirror-)similar contours. In analogy with (mirror-)dissimilarity measures based on parametric contour representations (cf. Section 4.2.1), these conditions will form the basis for the definition of (mirror-)dissimilarity measures based on Fourier representations of contours.

We consider Fourier representations as elements in the sequence space $\ell^p(\mathbb{Z})$, $1 \leqslant p \leqslant \infty$. In the definition of (mirror-)dissimilarity measures based on Fourier representations we employ the fact that, if $\hat{f}_1, \hat{f}_2 \in \ell^p(\mathbb{Z})$, then $\|\hat{f}_1 - \hat{f}_2\|_p$ defines a metric on $\ell^p(\mathbb{Z})$, where $\|\cdot\|_p$ denotes the usual norm on $\ell^p(\mathbb{Z})$ (cf. Appendix A).

This section constitutes a direct parallel with Sections 4.2.1 and 4.2.2. The required invariance of the (mirror-)dissimilarity measures for the position, size and orientation of the contours and for the position of the parametric starting point on the contours will be realized in the same manner. Therefore, the conversion of the formulations and derivations in the Sections 4.2.1 and 4.2.2 to Fourier analogues is straightforward, replacing contour representations by Fourier representations and integrals by appropriate sums. For this reason we limit ourselves to the major formulations and leave the details to the reader.

**Definition 4.7.** *Dissimilarity measure of index p based on Fourier representations.*

Let $\hat{f}$ act as a generic symbol for any of the Fourier representations $\hat{z}$, $\hat{\dot{z}}$, $\hat{\ddot{z}}$, $\hat{\psi}$ and $K$. Then a measure of dissimilarity of index $p$ between a pair of contours $\gamma_1$ and $\gamma_2$, with Fourier representations $\hat{f}_1$ and $\hat{f}_2$ respectively, $\hat{f}_1, \hat{f}_2 \in \ell^p(\mathbb{Z})$, is defined as

$$d^{(p)}(\hat{f}_1, \hat{f}_2) = \min_{\alpha,\tau} \|\hat{f}_1^* - \mathcal{T}_\tau \mathcal{R}_\alpha \hat{f}_2^*\|_p, \qquad 1 \leqslant p \leqslant \infty. \qquad (4.2.37)$$

□

Compare Definition 4.7 with Definition 4.1. In Eq. 4.2.37, $\hat{f}^*$ stands for the Fourier representation, generated by the normalized contour representation $f^*$, as defined in Eqs. 4.2.1, 4.2.12 and 4.2.21. It is easily verified that

$$\hat{f}^* = \mathcal{S}_{\beta^*} \mathcal{D}_{\zeta^*} \hat{f} \qquad (4.2.38)$$

when $f$ stands for $z$, $\dot{z}$, $\ddot{z}$ or $K$, and that

$$\hat{\psi}^* = \hat{\psi} + \lambda(\psi) = \hat{\psi} - \hat{\psi}(0). \qquad (4.2.39)$$

A survey of the formulations of the effects of translation and scaling upon the individual Fourier representations can be found in Table 3.2. With the aid of this table it is straightforward to derive expressions for $d^{(p)}(\hat{f}_1, \hat{f}_2)$ in terms of the individual Fourier representations.

The dissimilarity measures $d^{(p)}(\hat{f}_1, \hat{f}_2)$ also possess metric properties.

**Theorem 4.3.** *Metric properties of dissimilarity measures based on Fourier representations.*
The families of dissimilarity measures, $d^{(p)}(\hat{f}_1, \hat{f}_2)$, defined in Definition 4.7, constitute metrics over the equivalence classes of Fourier representations of geometrically similar contours.

□

The proof of this theorem follows exactly the proof of Theorem 4.2, replacing $f \in \mathbf{L}^p(2\pi)$ by $\hat{f} \in \ell^p(\mathbb{Z})$.

**Definition 4.8.** *Mirror-dissimilarity measure of index p based on Fourier representations.*
Let $\hat{f}$ act as a generic symbol for any of the Fourier representations $\hat{z}$, $\hat{\dot{z}}$, $\hat{\ddot{z}}$, $\hat{\psi}$ and $\hat{K}$. Then a measure of mirror-dissimilarity of index $p$ between a pair of contours $\gamma_1$ and $\gamma_2$, with Fourier representations $\hat{f}_1$ and $\hat{f}_2$ respectively, $\hat{f}_1, \hat{f}_2 \in \ell^p(\mathbb{Z})$, is defined as

$$\tilde{d}^{(p)}(\hat{f}_1, \hat{f}_2) = \min_{\alpha,\tau} \|\hat{f}_1^* - \mathcal{M}_x \mathcal{T}_\tau \mathcal{R}_\alpha \hat{f}_2^* \|_p, \qquad 1 \leqslant p \leqslant \infty. \tag{4.2.40}$$

□

Compare Definition 4.8 with Definition 4.4. An analogon to Table 4.2 for $\tilde{d}^{(p)}(\hat{f}_1, \hat{f}_2)$ can be derived by using Table 3.2.

For index $p = 2$ the dissimilarity measures $d^{(2)}(f_1, f_2)$ and $d^{(2)}(\hat{f}_1, \hat{f}_2)$ are isometrics

$$d^{(2)}(f_1, f_2) = d^{(2)}(\hat{f}_1, \hat{f}_2), \qquad \forall f_1, f_2 \in \mathbf{L}^2(2\pi). \tag{4.2.41}$$

This fact is a direct consequence of Parseval's formula (cf. e.g. Edwards [1979], pp. 131-132):

$$\|f\|_2^2 \equiv \frac{1}{2\pi} \int_{2\pi} |f(t)|^2 \mathrm{d}t = \sum_{k \in \mathbb{Z}} |\hat{f}(k)|^2 \equiv \|\hat{f}\|_2^2, \qquad \forall f \in \mathbf{L}^2(2\pi). \tag{4.2.42}$$

For the same reason the mirror-dissimilarity measures of index $p = 2$ satisfy

$$\tilde{d}^{(2)}(f_1, f_2) = \tilde{d}^{(2)}(\hat{f}_1, \hat{f}_2), \qquad \forall f_1, f_2 \in \mathbf{L}^2(2\pi). \tag{4.2.43}$$

In practice the (mirror-)dissimilarity measures of index $p = 2$ can be computed efficiently via the Fourier domain. One of the reasons for this efficiency is the form that the cyclic cross-correlation function and the cyclic convolution function take when they are expressed in the Fourier coefficients of the corresponding contour representations:

$$\varrho_{12}(\tau; f) = \sum_{k \in \mathbb{Z}} \hat{f}_1(k)\overline{\hat{f}_2(k)}e^{ik\tau} \tag{4.2.44}$$

and

$$h_{12}(\tau; f) = \sum_{k \in \mathbb{Z}} \hat{f}_1(k)\hat{f}_2(k)e^{ik\tau}. \tag{4.2.45}$$

Another reason that (mirror-)dissimilarity measures can be computed efficiently for $p = 2$ is the existence of fast algorithms for the (approximate) computation of Fourier coefficients. More attention will be given to these issues in the next subsection.

### *4.2.4 Sampled-data formulations of measures of dissimilarity and mirror-dissimilarity and analysis of computational complexity*

In the previous sections we have defined measures of dissimilarity and mirror-dissimilarity on the basis of various contour representations and Fourier representations. In the formulation of the measures the contour representations are functions of the continuous normalized arc length parameter $t$.

In practice the measures are computed on the basis of a finite number of samples of the contour representations or on the basis of a finite number of Fourier coefficients. In this section we present the discrete formulations of the previously defined measures of dissimilarity and mirror-dissimilarity in terms of sampled contour representations and finite Fourier representations and analyze the computational complexities of the measures.

We assume that in practice we have $N$ samples of a contour representation $f$, taken equidistantly in terms of arc length along the contour. The problem of estimating these contour representation samples from segmented digital images will be dealt with in Section 4.4 and Appendix C.

Since a $2\pi$-normalized arc length parametrization is used for the contour representations the discrete contour sample $f[n]$ can be related to the contour representation as

$$f[n] = f\left(\tau + n\frac{2\pi}{N}\right), \qquad n \in \{0, \ldots, N-1\} \tag{4.2.46}$$

for some $\tau \in [0, 2\pi/N)$. In practice we choose $\tau = 0$, which corresponds to the convention $f[0] = f(0)$.

The periodicity of a contour representation $f$ in terms of its $N$ discrete samples is expressed as

$$f[n] = f[n + N]. \tag{4.2.47}$$

An estimate of $\|f\|_p$, on the basis of discrete samples of $f$, is defined as

$$\|f[\ ]\|_p = \left[\frac{1}{N}\sum_{n=0}^{N-1} |f[n]|^p\right]^{1/p}. \tag{4.2.48}$$

**Remark.**
In a number of situations we denote the sampled version of a contour representation $f$ by $f[\ ]$.

In analogous situations we denote the truncated version of a Fourier representation $\hat{f}$ by $\hat{f}[\ ]$.

□

In the dissimilarity measures of index $p$, $d^{(p)}(f_1, f_2)$, the orientation of the contour $\gamma_2$, represented by $f_2$, is optimized with respect to that of contour $\gamma_1$, represented by $f_1$. In practice this optimization is performed over a finite number of orientations, say $M$. An estimate of $d^{(p)}(f_1, f_2)$ based on $N$ equidistant samples of both $f_1$ and $f_2$, denoted as $d^{(p)}[f_1, f_2]$, is given by

$$d^{(p)}[f_1, f_2] = \min_{m,q} \|f_1^*[\ ] - \mathcal{T}_{[q]}\mathcal{R}_{(2\pi/M)m} f_2^*[\ ]\|_p,$$

$$m \in \{0, \dots, M-1\}, \quad q \in \{0, \dots, N-1\} \tag{4.2.49}$$

where the discrete starting point shift operator $\mathcal{T}_{[q]}$ is defined as

$$\mathcal{T}_{[q]}f[n] = f[n - q]. \tag{4.2.50}$$

Similarly, an estimate of the mirror-dissimilarity measure of index $p$ $\tilde{d}^{(p)}(f_1, f_2)$, denoted as $\tilde{d}^{(p)}[f_1, f_2]$, is given by

$$\tilde{d}^{(p)}[f_1, f_2] = \min_{m,q} \left\| f_1^*[\ ] - \mathcal{M}_x \mathcal{T}_{[q]} \mathcal{R}_{(2\pi/M)m} f_2^*[\ ] \right\|_p,$$

$$m \in \{0, \ldots, M-1\}, \quad q \in \{0, \ldots, N-1\}. \tag{4.2.51}$$

A survey of the expressions for $d^{(p)}[f_1, f_2]$ and $\tilde{d}^{(p)}[f_1, f_2]$ can be found in Tables 4.5 and 4.6 respectively.

In the righthand columns of these tables the *computational complexity* of the (mirror-)dissimilarity measures is listed. By the computational complexity of the measures we mean the order of the number of arithmetic operations that has to be performed in order to compute the measure. We have assumed that the number of arithmetic operations, necessary to compute $|f|^p$, is not a function of $p$. The concept of computational complexity that we use here corresponds with the time com-

**Table 4.5.** Discrete dissimilarity measures of index $p$, $1 \leq p < \infty$, in terms of $N$ equidistant samples of the individual contour representations. The minimizations are over $N$ equidistant starting point shifts and $M$ equally spaced orientations.

| Contour representation | Discrete dissimilarity measures of index $p$ | Computational complexity |
|---|---|---|
| $z[\ ]$ | $d^{(p)}[z_1, z_2] = \min_{m,q} \left[ \frac{1}{N} \sum_{n=0}^{N-1} \left\| z_1^*[n] - e^{i(2\pi/M)m} z_2^*[n-q] \right\|^p \right]^{1/p}$ | $O(M \cdot N^2)$ |
| $\dot{z}[\ ]$ | $d^{(p)}[\dot{z}_1, \dot{z}_2] = \min_{m,q} \left[ \frac{1}{N} \sum_{n=0}^{N-1} \left\| \dot{z}_1^*[n] - e^{i(2\pi/M)m} \dot{z}_2^*[n-q] \right\|^p \right]^{1/p}$ | $O(M \cdot N^2)$ |
| $\ddot{z}[\ ]$ | $d^{(p)}[\ddot{z}_1, \ddot{z}_2] = \min_{m,q} \left[ \frac{1}{N} \sum_{n=0}^{N-1} \left\| \ddot{z}_1^*[n] - e^{i(2\pi/M)m} \ddot{z}_2^*[n-q] \right\|^p \right]^{1/p}$ | $O(M \cdot N^2)$ |
| $\psi[\ ]$ | $d^{(p)}[\psi_1, \psi_2] = \min_{q} \left[ \frac{1}{N} \sum_{n=0}^{N-1} \left\| \psi_1^*[n] - \psi_2^*[n-q] \right\|^p \right]^{1/p}$ | $O(N^2)$ |
| $K[\ ]$ | $d^{(p)}[K_1, K_2] = \min_{q} \left[ \frac{1}{N} \sum_{n=0}^{N-1} \left\| K_1^*[n] - K_2^*[n-q] \right\|^p \right]^{1/p}$ | $O(N^2)$ |

plexity for most traditional computing devices (Aho, Hopcroft and Ullman [1974]). However, if we use multi-processor architectures for the computation of the measures, then the specified order of complexity, in terms of the number of arithmetic operations, may give an overly pessimistic impression of the actual time complexity involved.

In the previous section we have found that for index $p = 2$ the concepts of correlation and convolution appear in the dissimilarity and mirror-dissimilarity measures, respectively. The *discrete cyclic cross-correlation function* $\varrho_{12}[q; f]$, on the basis of $N$ equidistant samples of both $f_1$ and $f_2$ is defined as (cf. Definition 4.5)

$$\varrho_{12}[q;f] = \frac{1}{N} \sum_{n=0}^{N-1} f_1[n]\overline{f_2}[n-q]. \tag{4.2.52}$$

**Table 4.6.** Discrete mirror-dissimilarity measures of index $p, 1 \leqslant p < \infty$, in terms of $N$ equidistant samples of the individual contour representations. The minimizations are over $N$ equidistant starting point shifts and $M$ equally spaced orientations.

| Contour representation | Discrete mirror-dissimilarity measures of index $p$ | Computational complexity |
|---|---|---|
| $z[\ ]$ | $\tilde{d}^{(p)}[z_1, z_2] = \min\limits_{m,q} \left[ \frac{1}{N} \sum\limits_{n=0}^{N-1} \left\| z_1^*[n] - e^{-i(2\pi/M)m} \overline{z}_2^*[-n+q] \right\|^p \right]^{1/p}$ | $O(M \cdot N^2)$ |
| $\dot{z}[\ ]$ | $\tilde{d}^{(p)}[\dot{z}_1, \dot{z}_2] = \min\limits_{m,q} \left[ \frac{1}{N} \sum\limits_{n=0}^{N-1} \left\| \dot{z}_1^*[n] - e^{-i(2\pi/M)m} \dot{\overline{z}}_2^*[-n+q] \right\|^p \right]^{1/p}$ | $O(M \cdot N^2)$ |
| $\ddot{z}[\ ]$ | $\tilde{d}^{(p)}[\ddot{z}_1, \ddot{z}_2] = \min\limits_{m,q} \left[ \frac{1}{N} \sum\limits_{n=0}^{N-1} \left\| \ddot{z}_1^*[n] - e^{-i(2\pi/M)m} \ddot{\overline{z}}_2^*[-n+q] \right\|^p \right]^{1/p}$ | $O(M \cdot N^2)$ |
| $\psi[\ ]$ | $\tilde{d}^{(p)}[\psi_1, \psi_2] = \min\limits_{q} \left[ \frac{1}{N} \sum\limits_{n=0}^{N-1} \left\| \psi_1^*[n] + \psi_2^*[-n+q] \right\|^p \right]^{1/p}$ | $O(N^2)$ |
| $K[\ ]$ | $\tilde{d}^{(p)}[K_1, K_2] = \min\limits_{q} \left[ \frac{1}{N} \sum\limits_{n=0}^{N-1} \left\| K_1^*[n] - K_2^*[-n+q] \right\|^p \right]^{1/p}$ | $O(N^2)$ |

Likewise the *discrete cyclic convolution function* $h_{12}[q; f]$ of $f_1$ and $f_2$ is defined as (cf. Definition 4.6):

$$h_{12}[q;f] = \frac{1}{N} \sum_{n=0}^{N-1} f_1[n] f_2[-n+q]. \tag{4.2.53}$$

As a result of the periodicity of $f_1[\ ]$ and $f_2[\ ]$, both $\varrho_{12}[q; f]$ and $h_{12}[q; f]$ are periodic, with period $N$.

The derivation of the discrete (mirror-)dissimilarity measures of index $p = 2$ from those of general index $p$ is along the same lines as described in Section 4.2.2 for the continuous-parameter representations. A survey of the results is given in Tables 4.7 and 4.8.

The computational complexity of the discrete (mirror-)dissimilarity measures of index $p = 2$ is dominated by the computation of $\varrho_{12}[q; f]$ and $h_{12}[q; f]$. Straightforward computation would lead to a computational complexity of $O(N^2)$ for all measures of index $p = 2$. However, the following analysis shows that often more efficient implementations are possible.

**Table 4.7.** Discrete dissimilarity measures of index $p = 2$ in terms of $N$ equidistant samples of the individual contour representations. The optimizations are over $N$ equidistant starting point shifts.

| Contour representation | Discrete dissimilarity measures of index $p = 2$ | Computational complexity |
|---|---|---|
| $z[\ ]$ | $d^{(2)}[z_1, z_2] = \left[\lVert z_1^*[\ ]\rVert_2^2 + \lVert z_2^*[\ ]\rVert_2^2 - 2 \max_q \lvert \varrho_{12}[q; z^*]\rvert\right]^{1/2}$ | $O(N \log_2 N)$ |
| $\dot{z}[\ ]$ | $d^{(2)}[\dot{z}_1, \dot{z}_2] = \left[\lVert \dot{z}_1^*[\ ]\rVert_2^2 + \lVert \dot{z}_2^*[\ ]\rVert_2^2 - 2 \max_q \lvert \varrho_{12}[q; \dot{z}^*]\rvert\right]^{1/2}$ | $O(N \log_2 N)$ |
| $\ddot{z}[\ ]$ | $d^{(2)}[\ddot{z}_1, \ddot{z}_2] = \left[\lVert \ddot{z}_1^*[\ ]\rVert_2^2 + \lVert \ddot{z}_2^*[\ ]\rVert_2^2 - 2 \max_q \lvert \varrho_{12}[q; \ddot{z}^*]\rvert\right]^{1/2}$ | $O(N \log_2 N)$ |
| $\psi[\ ]$ | $d^{(2)}[\psi_1, \psi_2] = \left[\lVert \psi_1^*[\ ]\rVert_2^2 + \lVert \psi_2^*[\ ]\rVert_2^2 - 2 \max_q \varrho_{12}[q; \psi^*]\right]^{1/2}$ | $O(N \log_2 N)$ |
| $K[\ ]$ | $d^{(2)}[K_1, K_2] = \left[\lVert K_1^*[\ ]\rVert_2^2 + \lVert K_2^*[\ ]\rVert_2^2 - 2 \max_q \varrho_{12}[q; K^*]\right]^{1/2}$ | $O(N \log_2 N)$ |

The *discrete Fourier transform* (DFT) of a periodic sequence $f[n]$, with period $N$, is defined as (cf. Oppenheim and Schafer [1975])

$$\hat{f}[k] = \frac{1}{N} \sum_{n=0}^{N-1} f[n] \mathrm{e}^{-\mathrm{i}k(2\pi/N)n} \tag{4.2.54a}$$

and the *inverse discrete Fourier transform* (IDFT) as

$$f[n] = \sum_{k=0}^{N-1} \hat{f}[k] \mathrm{e}^{\mathrm{i}k(2\pi/N)n}. \tag{4.2.54b}$$

Note from Eq. 4.2.54a that $\hat{f}[k]$ is also a periodic sequence with period $N$.

Through the cyclic correlation theorem (cf. Tretter [1976]) we can express $\varrho_{12}[q; f]$ as the IDFT of the sequence $\hat{f}_1[k]\overline{\hat{f}_2[k]}$

$$\varrho_{12}[q; f] = \sum_{k=0}^{N-1} \hat{f}_1[k]\overline{\hat{f}_2[k]} \mathrm{e}^{\mathrm{i}k(2\pi/N)q}. \tag{4.2.55}$$

**Table 4.8.** Discrete mirror-dissimilarity measures of index $p = 2$ in terms of $N$ equidistant samples of the individual contour representations. The optimizations are over $N$ equidistant starting point shifts.

| Contour representation | Discrete mirror-dissimilarity measures of index $p = 2$ | Computational complexity |
|---|---|---|
| $z[\ ]$ | $\tilde{d}^{(2)}[z_1, z_2] = \left[ \lVert z_1^*[\ ] \rVert_2^2 + \lVert z_2^*[\ ] \rVert_2^2 - 2 \max_q \lvert h_{12}[q; z^*] \rvert \right]^{1/2}$ | $O(N \log_2 N)$ |
| $\dot{z}[\ ]$ | $\tilde{d}^{(2)}[\dot{z}_1, \dot{z}_2] = \left[ \lVert \dot{z}_1^*[\ ] \rVert_2^2 + \lVert \dot{z}_2^*[\ ] \rVert_2^2 - 2 \max_q \lvert h_{12}[q; \dot{z}^*] \rvert \right]^{1/2}$ | $O(N \log_2 N)$ |
| $\ddot{z}[\ ]$ | $\tilde{d}^{(2)}[\ddot{z}_1, \ddot{z}_2] = \left[ \lVert \ddot{z}_1^*[\ ] \rVert_2^2 + \lVert \ddot{z}_2^*[\ ] \rVert_2^2 - 2 \max_q \lvert h_{12}[q; \ddot{z}^*] \rvert \right]^{1/2}$ | $O(N \log_2 N)$ |
| $\psi[\ ]$ | $\tilde{d}^{(2)}[\psi_1, \psi_2] = \left[ \lVert \psi_1^*[\ ] \rVert_2^2 + \lVert \psi_2^*[\ ] \rVert_2^2 + 2 \min_q h_{12}[q; \psi^*] \right]^{1/2}$ | $O(N \log_2 N)$ |
| $K[\ ]$ | $\tilde{d}^{(2)}[K_1, K_2] = \left[ \lVert K_1^*[\ ] \rVert_2^2 + \lVert K_2^*[\ ] \rVert_2^2 - 2 \min_q h_{12}[q; K^*] \right]^{1/2}$ | $O(N \log_2 N)$ |

Similarly, through the cyclic convolution theorem (cf. Tretter [1976]) $h_{12}[q; f]$ can be expressed as

$$h_{12}[q; f] = \sum_{k=0}^{N-1} \hat{f}_1[k]\hat{f}_2[k]e^{ik(2\pi/N)q}. \tag{4.2.56}$$

Compare Eqs. 4.2.55 and 4.2.56 with Eqs. 4.2.44 and 4.2.45.

With the specification of the computational complexities in Tables 4.7 and 4.8 it has been assumed that the number of samples $N$ has been chosen such that $N$ is a power of 2. In that case $\varrho_{12}[q; f]$ and $h_{12}[q; f]$ can be computed efficiently through Eqs. 4.2.55 and 4.2.56, using the Radix-2 Fast Fourier Transform (FFT) algorithm (Cooley and Tukey [1965]). If $N$ is not a power of 2, but another highly composite number, then the Mixed-Radix FFT algorithm can be applied (Singleton [1969]). Number-theoretic transform methods (McClellan and Rader [1979], Nussbaumer [1981]) or special-purpose hardware may lead to even greater computational efficiency.

Another method to compute the Fourier coefficients is to determine a polygonal approximation of the contour and to apply the formulas for the Fourier coefficients of polygonal representations in Table 3.1. In the latter case the number of arithmetic operations per Fourier coefficient is proportional to the number of vertices of the polygon. Efficiency in computation is only achieved if the number of vertices in the polygonal approximation is relatively small and if the number of Fourier coefficients can be kept limited.

In practice, the (mirror-)dissimilarity measures based on Fourier representations, as defined in Section 4.2.3, are computed by means of truncated or windowed Fourier representations, resulting in discrete measures of dissimilarity $d^{(p)}[\hat{f}_1, \hat{f}_2]$ and mirror-dissimilarity $\tilde{d}^{(p)}[\hat{f}_1, \hat{f}_2]$.

We consider truncated Fourier representations and assume that $N$ Fourier coefficients are used, with $N$ even. Adaptations of formulas for $N$ odd are straightforward.

The coefficients in the Fourier representation $\hat{f}$ and those in the truncated Fourier representation $\hat{f}[\ ]$ are related as

$$\hat{f}[k] = \begin{cases} \hat{f}(k), & -\dfrac{N}{2} \leqslant k \leqslant \dfrac{N}{2} - 1, \quad (4.2.57a) \\ 0, & k \text{ otherwise} \quad (4.2.57b) \end{cases}$$

where $k \in \mathbb{Z}$.

In analogy with Eq. 4.2.48 we denote an estimate of $\|\hat{f}\|_p$ on the basis of $N$ Fourier coefficients as $\|\hat{f}[\ ]\|_p$ and define this estimate as

$$\|\hat{f}[\ ]\|_p = \left[ \sum_{k=-N/2}^{N/2-1} |\hat{f}[k]|^p \right]^{1/p}. \tag{4.2.58}$$

If we perform in the discrete (mirror-)dissimilarity measures the optimization of contour orientation over $M$ discrete orientations and the optimization of parametric starting point over $N$ discrete starting points, then $d^{(p)}[\hat{f}_1, \hat{f}_2]$ is given by

$$d^{(p)}[\hat{f}_1, \hat{f}_2] = \min_{m,q} \|\hat{f}_1^*[\ ] - \mathcal{T}_{[q]}\mathcal{R}_{(2\pi/M)m}\hat{f}_2^*[\ ]\|_p,$$

$$m \in \{0, \ldots, M-1\}, \qquad q \in \{0, \ldots, N-1\} \tag{4.2.59}$$

where the discrete starting point shift operator $\mathcal{T}_{[q]}$ is defined in the Fourier domain as (cf. Table 3.2 and Eq. 4.2.50)

$$\mathcal{T}_{[q]}\hat{f}[k] = e^{-ik(2\pi/N)q}\hat{f}[k]. \tag{4.2.60}$$

Compare Eq. 4.2.59 with Eq. 4.2.49.

Similarly, the discrete mirror-dissimilarity measure $\tilde{d}^{(p)}[\hat{f}_1, \hat{f}_2]$ is given by

$$\tilde{d}^{(p)}[\hat{f}_1, \hat{f}_2] = \min_{m,q} \|\hat{f}_1^*[\ ] - \mathcal{M}_x\mathcal{T}_{[q]}\mathcal{R}_{(2\pi/M)m}\hat{f}_2^*[\ ]\|_p,$$

$$m \in \{0, \ldots, M-1\}, \qquad q \in \{0, \ldots, N-1\}. \tag{4.2.61}$$

Compare Eq. 4.2.61 with Eq. 4.2.51.

It is straightforward to derive the expressions for $d^{(p)}[\hat{f}_1, \hat{f}_2]$ and $\tilde{d}^{(p)}[\hat{f}_1, \hat{f}_2]$ for the individual truncated Fourier representations. Under the conditions for orientation and starting point optimization indicated in Eqs. 4.2.59 and 4.2.61, the computational complexities of the (mirror-)dissimilarity measures $d^{(p)}[\hat{f}_1, \hat{f}_2]$ and $\tilde{d}^{(p)}[\hat{f}_1, \hat{f}_2]$ are identical to the computional complexities of the corresponding (mirror-)dissimilarity measures $d^{(p)}[f_1, f_2]$ and $\tilde{d}^{(p)}[f_1, f_2]$, which are listed in Tables 4.5 and 4.6.

Due to the relations observed in Eqs. 4.2.41 and 4.2.43 for index $p = 2$, we will not treat this case separately for $d^{(2)}[\hat{f}_1, \hat{f}_2]$ and $\tilde{d}^{(2)}[\hat{f}_1, \hat{f}_2]$, but merely refer to what has been said about $d^{(2)}[f_1, f_2]$ and $\tilde{d}^{(2)}[f_1, f_2]$ earlier in this section.

## 4.3 Normalization versus optimization in dissimilarity and mirror-dissimilarity measures

In the previous section families of (mirror-)dissimilarity measures have been defined, based on various periodic contour representations or on the sequences of Fourier coefficients generated by these representations. Each of these measures satisfies the required property of invariance for equiform transformations. As a result we were able to show in Theorem 4.2 that the dissimilarity measures constitute metrics on the equivalence classes of representations of geometrically similar contours.

The invariance of the (mirror-)dissimilarity measures for equiform transformations was achieved through an appropriate normalization of the contour representations for the position and the size of the contours and through an optimization of the orientation and the parametric starting point on one contour with respect to those of the other. What constitutes 'appropriate' translation and scale normalization of contour representations was left unspecified. This will be discussed in this section.

In Tables 4.5-4.8 we saw that the optimization of the (mirror-)dissimilarity measures for orientation and starting point of the contours leads to a considerable computational complexity of the (mirror-)dissimilarity measures, which may be prohibitive for some applications. Therefore we will also discuss methods to normalize the orientation and starting point of a contour.

A general requirement that a normalization process must satisfy is that it leads to a unique solution. If this were not the case a dissimilarity measure could even give a nonzero value for geometrically similar contours, just by choosing different solutions of the normalization process in the contours. Consequentially, the metric properties of the dissimilarity measure would be lost. Unfortunately, the requirement that a normalization process must have a unique solution does not lead to a unique definition of such a process, as will become clear in the following sections.

In Sections 4.3.1-4.3.3 we will subsequently discuss the normalization of contour position, of contour size and of contour orientation and parametric starting point. In Section 4.3.4 the results of the previous sections are reviewed and normalized dissimilarity measures are defined. Also a combined optimization/normalization method of dissimilarity measurement is described and its computational complexity is analyzed. The latter method may be very interesting in practice, since it combines a low computational complexity with a limited risk of using an incorrect normalization for dissimilarity measurement.

### *4.3.1 Normalization of contour position*

From Table 2.2 we know that the position function $z$ is the only contour representation, introduced in Chapter 2, that is variant under a translation of the contour it represents. For the other contour representations, $\dot{z}$, $\ddot{z}$, $\psi$ and $K$, the translation operator $\mathcal{D}_\zeta$ is equivalent to the identity operator (cf. Table 2.1).

We denote a translation normalization parameter of a particular position function $z$ as $\zeta^*(z)$. The requirement that a normalization process must have a unique solution means that $\zeta^*$ must be a single-valued function of $z$.

A proper translation normalization process has the property that a position function $z$ and the position function $\mathcal{D}_\zeta z$, resulting from a translation of the contour over $\zeta \in \mathbb{C}$, lead to the same contour representation after translation normalization, i.e.

$$\mathcal{D}_{\zeta^*(z)} z = \mathcal{D}_{\zeta^*(\mathcal{D}_\zeta z)}(\mathcal{D}_\zeta z), \quad \forall \zeta \in \mathbb{C} \tag{4.3.1}$$

which leads to the requirement (cf. Table 2.1)

$$\zeta^*(\mathcal{D}_\zeta z) = \zeta^*(z) - \zeta. \tag{4.3.2}$$

In the literature we find two propositions for $\zeta^*(z)$ that both satisfy the requirements just mentioned.

Many authors (e.g. Granlund [1972], Richard and Hemami [1974], Sychra et al. [1976], Persoon and Fu [1977], Burkhardt [1979], Wallace and Wintz [1980], Wallace and Mitchell [1980], Chen and Shi [1980], Kuhl and Giardina [1982], Proffitt [1982], Parui and Dutta Majumder [1982], Nguyen, Poulsen and Louis [1983], Mitchell and Grogan [1984]) use the average position along the contour to normalize its position, i.e. they define

$$\zeta_c^*(z) = -\frac{1}{2\pi}\int_{2\pi} z(t)\mathrm{d}t. \tag{4.3.3}$$

Note that $\zeta_c^*(z) = -\langle z \rangle = -\hat{z}(0)$ (cf. Eqs. 4.2.18 and 3.2.2). In shape analysis techniques based on the Fourier coefficients generated by $z$, this translation normalization is usually implicit by not considering $\hat{z}(0)$.

In the shape analysis literature, where the shape of a region is represented by the gravitational moments of that region, the translation normalization of that region is also based on its moments.

**Definition 4.9.** *Moment $m_{pq}$.*
The moment $m_{pq}$ of a region $R$ in the plane is defined as

$$m_{pq} = m_{pq}(R) = \iint_R x^p y^q \mathrm{d}x\,\mathrm{d}y, \qquad p, q = 0, 1, \ldots. \tag{4.3.4}$$

The moment $m_{pq}$ is said to be of *order* $(p + q)$.

□

Please note that the definition of $m_{pq}$ in Eq. 4.3.4 is a special case of the general definition of two-dimensional moments of order $(p + q)$ which is given by (cf. e.g. Hu [1962])

$$\int_{-\infty}^{\infty}\int_{-\infty}^{\infty} f(x, y)x^p y^q \mathrm{d}x\,\mathrm{d}y, \qquad p, q = 0, 1, \ldots. \tag{4.3.5}$$

In Eq. 4.3.4 we have chosen $f(x, y)$ to be the characteristic function of the region $R$, i.e. $f(x, y) = 1$ inside $R$ and $f(x, y) = 0$ outside $R$ (cf. Appendix B), which suffices in the context of contour-oriented shape analysis, where we do not consider the internal structure of an object to be part of its shape.

Since $\mathrm{d}x\mathrm{d}y$ is equivalent to the element of area $\mathrm{d}A$, it is clear that $m_{00}$ (i.e. $p = 0$, $q = 0$ in Eq. 4.3.4) represents the area $A$ of the two-dimensional region $R$.

Translation normalization based on moments is accomplished by using the centroid (or center of gravity) of the region enclosed by the contour, i.e. by defining

$$\zeta_{\mathrm{r}}^{*}(z) = -\frac{1}{m_{00}}(m_{10} + \mathrm{i}m_{01}) \tag{4.3.6}$$

(cf. Hu [1962], Alt [1962], Ehrlich and Weinberg [1970], Nagy and Tuong [1970], Casey [1970], Dudani, Breeding and McGhee [1977], Wong and Hall [1978], Zvolanek [1981], Reeves and Rostampour [1981], Reeves and Wittner [1983]). Note that the centroid corresponds to the regional average, i.e.

$$\frac{1}{m_{00}}(m_{10} + \mathrm{i}m_{01}) = \frac{\iint_R z\mathrm{d}A}{\iint_R \mathrm{d}A}. \tag{4.3.7}$$

It is easily verified that both $\zeta_{\mathrm{c}}^{*}(z)$ and $\zeta_{\mathrm{r}}^{*}(z)$ satisfy the requirement of uniqueness and Eq. 4.3.2.

In the literature the contour average $\langle z \rangle$ is frequently called the centroid or the center of gravity. The following example shows that the contour average and the center of gravity are in general not the same.

**Example 4.2.**
In this example we consider a simply-connected region, bounded by the simple closed polygon specified in Figure 4.5.

We recall from Table 3.1 that the average position along a closed polygon is given by

$$\langle z \rangle \equiv \hat{z}(0) = \frac{1}{2L} \sum_{n=0}^{N-1} |\Delta z(t_n)| \{z(t_{n+1}) + z(t_n)\} \tag{4.3.8a}$$

where $L$ is the perimeter of the polygon

$$L = \sum_{n=0}^{N-1} |\Delta z(t_n)|. \tag{4.3.8b}$$

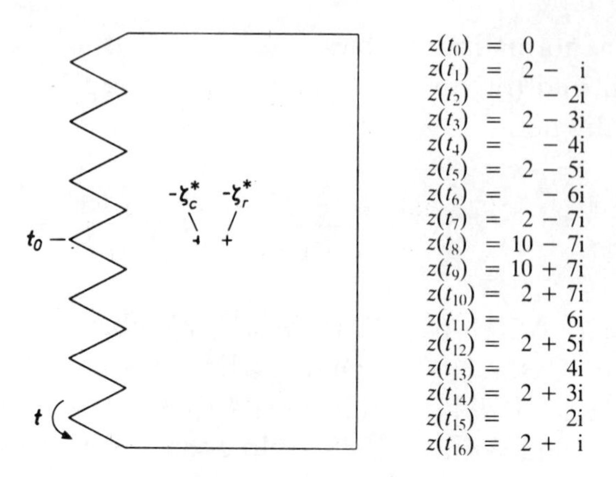


**Figure 4.5.** A region bounded by a simple closed polygon with 17 vertices. The complex-valued vertices are listed on the right. The contour average $\langle z \rangle = \hat{z}(0) = -\zeta_c^*$ and the centroid $(m_{10} + im_{01})/m_{00} = -\zeta_r^*$ are not the same.

It is straightforward to derive the following expression for the centroid of the region enclosed by a polygon

$$\frac{1}{m_{00}}(m_{10} + \mathrm{i}m_{10})$$

$$= \frac{1}{6A}\sum_{n=0}^{N-1}\Big[\{2x(t_n)y(t_n) + x(t_n)y(t_{n+1})$$

$$+ x(t_{n+1})y(t_n) + 2x(t_{n+1})y(t_{n+1})\}\{\bar{z}(t_n) - \bar{z}(t_{n+1})\}\Big] \tag{4.3.9a}$$

where $A$ is the area of the polygonal region given by

$$A = \frac{1}{2}\sum_{n=0}^{N-1}\{x(t_n)y(t_{n+1}) - x(t_{n+1})y(t_n)\}. \tag{4.3.9b}$$

Substitution of the coordinates of the vertices of the polygon specified in Figure 4.5 yields

$$\hat{z}(0) = \frac{1}{60 + 28\sqrt{5}}(472 + 28\sqrt{5} + 0\cdot \mathrm{i}) \approx 4.360 + 0\cdot \mathrm{i}$$

and

$$\frac{1}{m_{00}}(m_{10} + \mathrm{i}m_{01}) = \frac{1}{756}(4144 + 0 \cdot \mathrm{i}) \approx 5.481 + 0 \cdot \mathrm{i}$$

which clearly shows that the average position along a contour and the centroid of the region, bounded by that contour, are in general not the same.

□

In Section 2.4 we remarked that the symmetry point $\boldsymbol{n}$ of an $\boldsymbol{n}$-fold geometrically rotationally symmetric contour coincides with the centroid of the region bounded by that contour. In the following theorem we show that $\boldsymbol{n}$-fold rotational symmetry, $\boldsymbol{n} \geqslant \mathbf{2}$, is a sufficient condition for the centroid and the average position along the contour to coincide.

**Theorem 4.4.**
If a contour is $\boldsymbol{n}$-fold geometrically rotationally symmetric, $\boldsymbol{n} \geqslant \mathbf{2}$, then the average position along the contour and the centroid of the region, bounded by that contour, coincide.

*Proof*

Assume that a contour $\gamma$, with position function $z$, possesses $\boldsymbol{n}$-fold geometrical rotational symmetry. It follows from Table 2.6 and Eq. 2.4.3 that the following relation holds for $z$

$$z(t) + \zeta = \mathrm{e}^{\mathrm{i}m(2\pi/\boldsymbol{n})}\{z(t - m(2\pi/\boldsymbol{n})) + \zeta\}, \qquad \forall t \in \mathbb{R}, \quad \forall m \in \mathbb{Z} \tag{4.3.10}$$

where $-\zeta \in \mathbb{C}$ is the symmetry point of order $\boldsymbol{n}$. From Table 2.6 and Eq. 2.4.3 it also follows that the tangent function $\dot{z}$ of $\gamma$ satisfies

$$\dot{z}(t) = \mathrm{e}^{\mathrm{i}m(2\pi/\boldsymbol{n})}\dot{z}(t - m(2\pi/\boldsymbol{n})), \qquad \forall t \in \mathbb{R}, \quad \forall m \in \mathbb{Z}. \tag{4.3.11}$$

In Eq. 4.3.7 we noted that the centroid of a region corresponds with the regional average. A complex formulation of Green's theorem reads (Spiegel [1964]):

$$\iint_R \frac{\partial Q(z, \bar{z})}{\partial z} \mathrm{d}A = \frac{\mathrm{i}}{2} \oint_\gamma Q(z, \bar{z}) \mathrm{d}\bar{z}. \tag{4.3.12}$$

If we apply Eq. 4.3.12 both to the numerator and to the denominator of the righthand side of Eq. 4.3.7 we obtain

$$\frac{1}{m_{00}}(m_{10} + \mathrm{i}m_{01}) = \frac{\frac{\mathrm{i}}{4}\oint_\gamma z^2 \mathrm{d}\bar{z}}{\frac{\mathrm{i}}{2}\oint_\gamma z \mathrm{d}\bar{z}} = \frac{1}{2}\frac{\int_0^{2\pi} z^2(t)\dot{\bar{z}}(t)\mathrm{d}t}{\int_0^{2\pi} z(t)\dot{\bar{z}}(t)\mathrm{d}t}. \tag{4.3.13}$$

Substitution of Eqs. 4.3.10 and 4.3.11 into the righthand side of Eq. 4.3.13 and rearranging summations and integrations leads to

$$\frac{1}{m_{00}}(m_{10} + \mathrm{i}m_{01})$$

$$= \frac{1}{2}\frac{\sum_{m=0}^{n-1}\int_0^{(2\pi/n)} \{\mathrm{e}^{\mathrm{i}m(2\pi/n)}z(t) + (\mathrm{e}^{\mathrm{i}m(2\pi/n)} - 1)\zeta\}^2 \mathrm{e}^{-\mathrm{i}m(2\pi/n)}\dot{\bar{z}}(t)\mathrm{d}t}{\sum_{m=0}^{n-1}\int_0^{(2\pi/n)} \{\mathrm{e}^{\mathrm{i}m(2\pi/n)}z(t) + (\mathrm{e}^{\mathrm{i}m(2\pi/n)} - 1)\zeta\} \mathrm{e}^{-\mathrm{i}m(2\pi/n)}\dot{\bar{z}}(t)\mathrm{d}t}$$

$$= \frac{1}{2}\frac{-2n\zeta\left\{\int_0^{(2\pi/n)} z(t)\dot{\bar{z}}(t)\mathrm{d}t + \zeta\int_0^{(2\pi/n)} \dot{\bar{z}}(t)\mathrm{d}t\right\}}{n\left\{\int_0^{(2\pi/n)} z(t)\dot{\bar{z}}(t)\mathrm{d}t + \zeta\int_0^{(2\pi/n)} \dot{\bar{z}}(t)\mathrm{d}t\right\}}$$

$$= -\zeta \tag{4.3.14}$$

where we have repeatedly used the property

$$\sum_{m=0}^{n-1} \mathrm{e}^{\mathrm{i}km(2\pi/n)} = 0, \qquad \text{if } k \neq 0 \bmod n. \tag{4.3.15}$$

Thus we have shown in Eq. 4.3.14 that $-\zeta$ in Eq. 4.3.10 indeed corresponds to the centroid of the region $R$.

Substitution of Eq. 4.3.10 into the equation of the average position along the contour yields (cf. Eq. 4.2.18)

$$\begin{aligned}\langle z \rangle &= \frac{1}{2\pi}\int_{2\pi} z(t)\mathrm{d}t \\ &= \frac{1}{2\pi}\sum_{m=0}^{\boldsymbol{n}-1}\int_0^{(2\pi/\boldsymbol{n})}\{\mathrm{e}^{im(2\pi/\boldsymbol{n})}z(t) + (\mathrm{e}^{im(2\pi/\boldsymbol{n})} - 1)\zeta\}\mathrm{d}t \\ &= -\frac{1}{2\pi}\cdot\boldsymbol{n}\cdot\frac{2\pi}{\boldsymbol{n}}\cdot\zeta \\ &= -\zeta \end{aligned} \tag{4.3.16}$$

where we have again applied Eq. 4.3.15. Combining Eqs. 4.3.14 and 4.3.16 completes the proof of the theorem.

□

It can be shown that the translation normalization parameter $\zeta_c^*(z) = -\langle z \rangle = -\hat{z}(0)$ minimizes $d^{(2)}(z_1, z_2)$ and $d^{(2)}(\hat{z}_1, \hat{z}_2)$ over all possible translation normalizations. The proof of this assertion follows immediately from the equation of $d^{(2)}(\hat{z}_1, \hat{z}_2)$ (cf. Eq. 4.2.37)

$$\begin{aligned} d^{(2)}(\hat{z}_1, \hat{z}_2) &= \min_{\alpha,\tau} \| \hat{z}_1^* - \mathcal{T}_\tau \mathcal{R}_\alpha \hat{z}_2^* \|_2 \\ &= \min_{\alpha,\tau}\Big[ |\beta_1^*\{\hat{z}_1(0) + \zeta_1^*\} - \mathrm{e}^{\mathrm{i}\alpha}\beta_2^*\{\hat{z}_2(0) + \zeta_2^*\}|^2 \\ &\quad + \sum_{k\in\mathbb{Z}-\{0\}} |\beta_1^*\hat{z}_1(k) - \mathrm{e}^{-\mathrm{i}(k\tau-\alpha)}\beta_2^*\hat{z}_2(k)|^2\Big]^{1/2} \end{aligned} \tag{4.3.17}$$

and from the equality of $d^{(2)}(\hat{z}_1, \hat{z}_2)$ and $d^{(2)}(z_1, z_2)$ by Parseval's formula (Eq. 4.2.42). Along different lines the optimality of $\zeta_c^*(z)$ as a translation normalization parameter in $d^{(2)}(z_1, z_2)$ has also been shown by Proffitt [1982].

*4.3.2 Normalization of contour size*

From Table 2.2 we know that of all contour representations that we proposed for dissimilarity measurement, $\psi$ is the only scale invariant representation. Therefore the scaling operator $\mathcal{S}_\beta$ applied to $\psi$ is equivalent to the identity operator (cf. Table 2.1). The contour representations $z$, $\dot{z}$, $\ddot{z}$ and $K$ are all variant under the scaling of a contour. Therefore we have to specify scale normalization parameters in the (mirror-) dissimilarity measures based on these contour representations. Let $\mathcal{D}_{\zeta^*}f$ denote any of these contour representations after translation normalization and let $\beta^*(\mathcal{D}_{\zeta^*}f)$ denote a scale normalization parameter for $\mathcal{D}_{\zeta^*}f$. By defining $\beta^*$ as a function of a translation-normalized contour representation we ensure that $\beta^*$ will be translation invariant. The necessity of this provision stems from the fact that the operators $\mathcal{D}_\zeta$ and $\mathcal{S}_\beta$ do not commute when $f$ corresponds to the position function $z$. In order to ensure that the normalization process has a unique solution, $\beta^*$ must be a single-valued function of $\mathcal{D}_{\zeta^*}f$. We also require that $\beta^*$ is a positive function of $\mathcal{D}_{\zeta^*}f$ since scaling is always performed by positive real-valued coefficients.

A proper scale normalization process has the property that the contour representations $\mathcal{D}_{\zeta^*}f$ and $\mathcal{S}_\beta\mathcal{D}_{\zeta^*}f$, the latter resulting from scaling $\mathcal{D}_{\zeta^*}f$ by a factor $\beta \in \mathbb{R}^+$, lead to the same contour representation after scale normalization, i.e.

$$\mathcal{S}_{\beta^*(\mathcal{D}_{\zeta^*}f)}(\mathcal{D}_{\zeta^*}f) = \mathcal{S}_{\beta^*(\mathcal{S}_\beta\mathcal{D}_{\zeta^*}f)}(\mathcal{S}_\beta\mathcal{D}_{\zeta^*}f) \qquad (4.3.18)$$

which leads, through the property $\mathcal{S}_{\beta_1}\mathcal{S}_{\beta_2} = \mathcal{S}_{\beta_1 \cdot \beta_2}$, to the requirement (cf. Table 2.1)

$$\beta^*(\mathcal{S}_\beta\mathcal{D}_{\zeta^*}f) = \beta^{-1} \cdot \beta^*(\mathcal{D}_{\zeta^*}f). \qquad (4.3.19)$$

Various propositions for scale normalization operators $\mathcal{S}_{\beta^*}$, that all satisfy the aforementioned requirements, can be found in the literature. These propositions occur mainly in relation to the translation-normalized position function $\mathcal{D}_{\zeta^*}z$ or to the corresponding Fourier representation $\mathcal{D}_{\zeta^*}\hat{z}$.

For example:

- $\beta^* = |\hat{z}(1)|^{-1}$ (Granlund [1972], Wallace and Wintz [1980], Mitchell, Reeves and Grogan [1982]).
- $\beta^* = \|\mathcal{D}_{\zeta^*}\hat{z}\|_{\infty}^{-1}$ (Burkhardt [1979]).
- $\beta^* = \|\mathcal{D}_{\zeta^*}\hat{z}\|_{1}^{-1}$ (Chen and Shi [1980]).
- $\beta^* = |\hat{z}(1) + \hat{z}(-1)|^{-1}$ (Sychra et al. [1976], Persoon and Fu [1977], Kuhl and Giardina [1982], Nguyen, Poulsen and Louis [1983]).
- $\beta^* = |\hat{z}(1) - \hat{z}(-1)|^{-1}$ (Tai, Li and Chiang [1982]).
- $\beta^* = \|\mathcal{D}_{\zeta^*}z\|_{2}^{-1}$ (Richard and Hemami [1974], Burkhardt [1979], Proffitt [1982]).
- $\beta^* = \|\mathcal{D}_{\zeta^*}z\|_{\infty}^{-1}$ (Freeman [1978a], Kuhl and Giardina [1982]).
- $\beta^* = L^{-1}$, where $L$ is the perimeter of the contour (Crimmins [1982]).
- $\beta^* = A^{-1/2}$, where $A$ is the area enclosed by the contour (Hu [1962], Alt [1962], Casey [1970], Reeves and Rostampour [1981], Tang [1982], Luerkens, Beddow and Vetter [1982a]). Note that $A = m_{00}$, the moment of order (0, 0) (cf. Eq. 4.3.4), which explains why this normalization coefficient is mainly proposed in shape analysis methods based on moments.

Though this constitutes quite a substantial list of scale normalization coefficients, many other coefficients, that all satisfy Eq. 4.3.19, can be formulated. The list also reveals that little agreement exists in the literature concerning scale normalization.

**Example 4.3.**
To obtain an impression of the effect of some of the aforementioned scale normalization coefficients, we have displayed in Figure 4.6 a

7-pointed star normalized by four different scale normalization coefficients. These scale normalizations were chosen such that, if they were applied to a circle, they would all yield a circle of unit radius. In Figure 4.6a we notice that rather thin shapes, that enclose relatively little area, virtually explode as a result of an area-based scale normalization. On the other hand, contours with a relatively large perimeter, will shrink to very small figures if we use a perimeter-based scale normalization (Figure 4.6b). Furthermore, varying signal-to-noise ratio conditions among contours may cause perimeter estimates to differ, even with otherwise congruent contours. These circumstances will lead to unrealistic values for $d^{(p)}(z_1, z_2)$ and $\tilde{d}^{(p)}(z_1, z_2)$ if we use a perimeter-based scale normalization for $z$, as proposed for example by Crimmins [1982]. The contour normalizations displayed in Figures 4.6c and 4.6d, based on $|\hat{z}(1)|$ and $\|\mathcal{D}_{\zeta^*}z\|_2$ respectively, both normalize the 7-pointed star to a size comparable to that of the circle. The reason for proposing $|\hat{z}(1)|^{-1}$ as a scale normalization coefficient is that for many simple closed con-

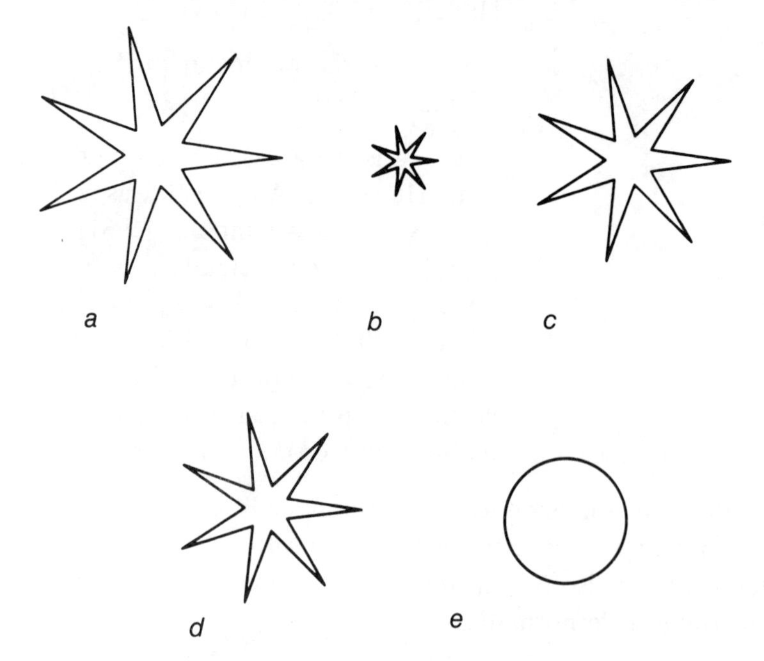


**Figure 4.6.** Illustration of the effect of various scale normalizations upon a 7-pointed star. In (*a*) the area has been normalized to $A = \pi$, in (*b*) the perimeter to $L = \pi$, in(*c*) the Fourier coefficient $\hat{z}(1)$ to $|\hat{z}(1)| = 1$ and in (*d*) the $\mathbf{L}^2$-norm to $\|\mathcal{D}_{\zeta^*}z\|_2 = 1$. For this 7-pointed star these normalizations lead to the values 2.083, 0.567, 1.643 and 1.536 for the radii of the circumscribed circles in the cases (*a*), (*b*), (*c*) and (*d*), respectively. For a circle, each of the four normalizations would yield a circle of radius 1, as displayed in (*e*).

tours $\hat{z}(1)$ is the Fourier coefficient of largest magnitude, not counting $\hat{z}(0)$. However, simple closed contours exist for which this is not true, as we will show in the next example.

□

**Example 4.4.**
In this example we consider a thin, strip-like object that loops around a center point (cf. Figure 4.7a). The magnitudes of the Fourier coeffi-

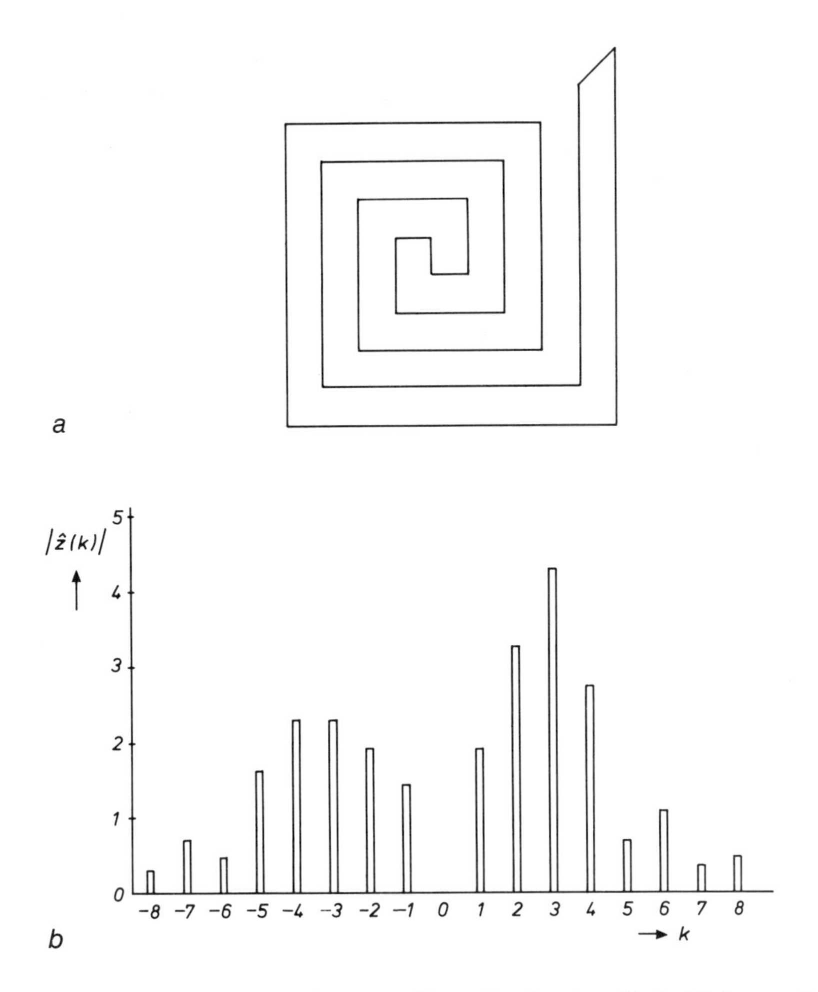


**Figure 4.7.** ($a$) shows a simple closed contour with position function $z(t)$. In ($b$) the magnitudes of the Fourier coefficients $\hat{z}(k)$, $k = -8, \ldots, 8$, generated by $z(t)$, are displayed. Clearly not $\hat{z}(1)$ but rather $\hat{z}(3)$ has the largest magnitude for this contour.

cients $\hat{z}(k)$ of this object are displayed in Figure 4.7b. It is clear that $\hat{z}(1)$ is not the Fourier coefficient of largest magnitude, but rather $\hat{z}(3)$. If we construct objects such as in Figure 4.7a with more loops, then it turns out that there is a direct correspondence between the number of loops and the index of the Fourier coefficient of largest magnitude.

□

Apart from satisfying the requirement in Eq. 4.3.19, a scale normalization parameter should be relatively insensitive to noise, thus limiting the noise sensitivity of the (mirror-)dissimilarity measures. It should also render scale-normalized contour representations that are comparable, in terms of the (mirror-)dissimilarity measure employed, for a wide range of shapes. The example just given shows that shapes exist for which $|\hat{z}(1)|^{-1}$, considered as a potential scale normalization parameter for $z$, does not satisfy the condition that it has a relatively low noise sensitivity. Experiments have shown that $\|\mathfrak{D}_{\zeta^*}z\|_2^{-1}$ is a better candidate for a scale normalization of $z$. More generally, we may state that $\mathcal{S}_{\beta^*} = \|\mathfrak{D}_{\zeta^*}z\|_p^{-1}$ is an appropriate scale normalization operator to be used in $d^{(p)}(z_1, z_2)$ and $\tilde{d}^{(p)}(z_1, z_2)$, since, apart from satisfying Eq. 4.3.19, it normalizes the size of objects by means of the same criterion that is also used to measure (mirror-)dissimilarity.

Continuing this argumentation, $\mathcal{S}_{\beta^*} = \|\mathfrak{D}_{\zeta^*}f\|_p^{-1}$ is an appropriate scale normalization operator in $d^{(p)}(f_1, f_2)$ and $\tilde{d}^{(p)}(f_1, f_2)$, where $f$ stands for any of the contour representations $z$, $\dot{z}$, $\ddot{z}$ and $K$. The effect of the scale normalization operator $\mathcal{S}_{\beta^*} = \|\mathfrak{D}_{\zeta^*}f\|_p^{-1}$ is that it maps the contour representation $\mathfrak{D}_{\zeta^*}f$ onto a unit sphere in the function space $\mathbf{L}^p(2\pi)$, endowed with the metric induced by the norm $\|\cdot\|_p$. It is easily verified that the scale normalization operator $\mathcal{S}_{\beta^*} = \|\mathfrak{D}_{\zeta^*}f\|_p^{-1}$ normalizes the representations $z$, $\dot{z}$, $\ddot{z}$ and $K$ of a circle to the corresponding representations of a unit circle.

In the following we evaluate the ranges of the (mirror-)dissimilarity measures proposed in Section 4.2. We also analyze the effect of scale normalization upon these ranges.

The effect of the scale normalization operator $\mathcal{S}_{\beta^*} = \|\mathfrak{D}_{\zeta^*}f\|_p^{-1}$ upon the range of the (mirror-)dissimilarity measures is easily established. Expanding $d^{(p)}(f_1, f_2)$, using Eqs. 4.2.1 and 4.2.2 and applying Minkowski's inequality, yields:

$$d^{(p)}(f_1, f_2) = \min_{\alpha,\tau} \|\mathcal{S}_{\beta_1^*}\mathcal{D}_{\zeta_1^*}f_1 - \mathcal{T}_\tau \mathcal{R}_\alpha \mathcal{S}_{\beta_2^*}\mathcal{D}_{\zeta_2^*}f_2\|_p$$

$$\leqslant \|\mathcal{S}_{\beta_1^*}\mathcal{D}_{\zeta_1^*}f_1\|_p + \|\mathcal{S}_{\beta_2^*}\mathcal{D}_{\zeta_2^*}f_2\|_p$$

$$= \mathcal{S}_{\beta_1^*}\|\mathcal{D}_{\zeta_1^*}f_1\|_p + \mathcal{S}_{\beta_2^*}\|\mathcal{D}_{\zeta_2^*}f_2\|_p. \tag{4.3.20}$$

With the proposed scale normalization operator $\mathcal{S}_{\beta^*} = \|\mathcal{D}_{\zeta^*}f\|_p^{-1}$, we find from Eq. 4.3.20 for the range of $d^{(p)}(f_1, f_2)$

$$0 \leqslant d^{(p)}(f_1, f_2) \leqslant 2. \tag{4.3.21}$$

Eq. 4.3.20 is also valid for $d^{(p)}(\psi_1, \psi_2)$ (cf. Eq. 4.2.17 and Table 2.1), since $\mathcal{S}_{\beta^*}$ is the identity operator when applied to $\psi$. However, for the same reason Eq. 4.3.21 is not valid for $d^{(p)}(\psi_1, \psi_2)$.

If $f$ stands for any of the contour representations $z$, $\dot{z}$ or $\ddot{z}$, we can find a more restricted range for $d^{(2)}(f_1, f_2)$. From Eqs. 4.2.1 and 4.2.35 we find

$$d^{(2)}(f_1, f_2) = \Big[\|\mathcal{S}_{\beta_1^*}\mathcal{D}_{\zeta_1^*}f_1\|_2^2 + \|\mathcal{S}_{\beta_2^*}\mathcal{D}_{\zeta_2^*}f_2\|_2^2 - 2\max_\tau |\varrho_{12}(\tau; \mathcal{S}_{\beta^*}\mathcal{D}_{\zeta^*}f)|\Big]^{1/2}$$

$$\leqslant \Big[\mathcal{S}_{\beta_1^*}^2\|\mathcal{D}_{\zeta_1^*}f_1\|_2^2 + \mathcal{S}_{\beta_2^*}^2\|\mathcal{D}_{\zeta_2^*}f_2\|_2^2\Big]^{1/2}. \tag{4.3.22}$$

Applying $\mathcal{S}_{\beta^*} = \|\mathcal{D}_{\zeta^*}f\|_2^{-1}$ we find for the range of $d^{(2)}(f_1, f_2)$

$$0 \leqslant d^{(2)}(f_1, f_2) \leqslant \sqrt{2}. \tag{4.3.23}$$

The range of $d^{(2)}(K_1, K_2)$ can be restricted even further. From Eqs. 2.2.31 and 2.2.34 it is easily verified that the average curvature along a simple closed curve equals

$$\langle K \rangle = \frac{1}{2\pi}\int_{2\pi} K(t)\mathrm{d}t = \frac{2\pi}{L} \tag{4.3.24}$$

where $L$ is the perimeter of the curve and $t$ the normalized arc length parameter. Consider $\varrho_{12}(\tau; K^*)$ as a function of $\tau$. Integration over $\tau$ yields:

$$\int_{2\pi} \varrho_{12}(\tau; K^*)\mathrm{d}\tau = \int_{2\pi} \left[ \frac{1}{2\pi} \int_{2\pi} K_1^*(t) K_2^*(t-\tau)\mathrm{d}t \right] \mathrm{d}\tau$$

$$= \frac{1}{2\pi} \int_{2\pi} K_1^*(t) \left[ \int_{2\pi} K_2^*(t-\tau)\mathrm{d}\tau \right] \mathrm{d}t$$

$$= \frac{(2\pi)^3 \mathcal{S}_{\beta_1^*} \mathcal{S}_{\beta_2^*}}{L_1 L_2}$$

$$\leqslant 2\pi \max_{\tau} \varrho_{12}(\tau; K^*) \tag{4.3.25}$$

where the last step follows from the fact that the maximum value of $\varrho_{12}(\tau; K^*)$ is always larger than its contour average. With $\mathcal{S}_{\beta^*} = \|K\|_2^{-1}$ we find from Eq. 4.3.25

$$\max_{\tau} \varrho_{12}(\tau; K^*) \geqslant \frac{(2\pi)^2 \mathcal{S}_{\beta_1^*} \mathcal{S}_{\beta_2^*}}{L_1 L_2} > 0 \tag{4.3.26a}$$

which gives through the substitution of $\mathcal{S}_{\beta^*} = \|K\|_2^{-1}$

$$\max_{\tau} \varrho_{12}(\tau; K^*) \geqslant \frac{(2\pi)^2 \|K_1\|_2^{-1} \|K_2\|_2^{-1}}{L_1 L_2} > 0. \tag{4.3.26b}$$

From Table 4.3 we find for $d^{(2)}(K_1, K_2)$, using Eq. 4.3.25

$$d^{(2)}(K_1, K_2) = \left[ \|\mathcal{S}_{\beta_1^*} K_1\|_2^2 + \|\mathcal{S}_{\beta_2^*} K_2\|_2^2 - 2 \max_{\tau} \varrho_{12}(\tau; \mathcal{S}_{\beta^*} K) \right]^{1/2}$$

$$\leqslant \left[ \mathcal{S}_{\beta_1^*}^2 \|K_1\|_2^2 + \mathcal{S}_{\beta_2^*}^2 \|K_2\|_2^2 - \frac{2(2\pi)^2 \mathcal{S}_{\beta_1^*} \mathcal{S}_{\beta_2^*}}{L_1 L_2} \right]^{1/2}. \tag{4.3.27}$$

Substituting again $\mathcal{S}_{\beta^*} = \|K\|_2^{-1}$ we find for the range of $d^{(2)}(K_1, K_2)$

$$0 \leqslant d^{(2)}(K_1, K_2) \leqslant \sqrt{2} \left[ 1 - \frac{(2\pi)^2 \|K_1\|_2^{-1} \|K_2\|_2^{-1}}{L_1 L_2} \right]^{1/2} < \sqrt{2}. \tag{4.3.28}$$

All results obtained in Eqs. 4.3.20-4.3.28 are equally valid for the corresponding mirror-dissimilarity measures.

If we use Fourier representations, instead of the contour representations themselves, to measure (mirror-)dissimilarity, then the requirement in Eq. 4.3.19 becomes

$$\beta^*(\mathcal{S}_\beta \mathcal{D}_{\zeta^*}\hat{f}) = \beta^{-1} \cdot \beta^*(\mathcal{D}_{\zeta^*}\hat{f}) \tag{4.3.29}$$

where $\hat{f}$ stands for any of the Fourier representations $\hat{z}$, $\hat{\dot{z}}$, $\hat{\ddot{z}}$ or $\hat{K}$. On the basis of the same argumentations as before we propose $\mathcal{S}_{\beta^*} = \|\mathcal{D}_{\zeta^*}\hat{f}\|_p^{-1}$ as a scale normalization operator in both $d^{(p)}(\hat{f}_1, \hat{f}_2)$ and $\tilde{d}^{(p)}(\hat{f}_1, \hat{f}_2)$. The effect of the scale normalization coefficient $\mathcal{S}_{\beta^*} = \|\mathcal{D}_{\zeta^*}\hat{f}\|_p^{-1}$ is that it maps the Fourier representation $\mathcal{D}_{\zeta^*}\hat{f}$ onto a unit sphere in the sequence space $\ell^p(\mathbb{Z})$, endowed with the metric induced by $\|\cdot\|_p$.

The proposed scale normalization operator $\mathcal{S}_{\beta^*} = \|\mathcal{D}_{\zeta^*}\hat{f}\|^{-1}$ normalizes the Fourier representations $\hat{z}$, $\hat{\dot{z}}$, $\hat{\ddot{z}}$ and $\hat{K}$ of a circle to the corresponding Fourier representations of a unit circle.

For the range of the (mirror-)dissimilarity measures $d^{(p)}(\hat{f}_1, \hat{f}_2)$ and $\tilde{d}^{(p)}(\hat{f}_1, \hat{f}_2)$ exactly the same results as obtained in Eqs. 4.3.20-4.3.28 can be derived.

Some remarks concerning the proposed scale normalization operators are still in order.

We recall from Section 2.2 that $|\dot{z}(t)| = L/2\pi$, $\forall t \in [0, 2\pi]$ if $t$ is the normalized arc length parameter. As a result we find $\|\dot{z}\|_p = L/2\pi$, $\forall p \geqslant 1$. Thus we see that the scale normalization operator $\mathcal{S}_{\beta^*} = \|\dot{z}\|_p^{-1}$ removes the effect of the noise-prone perimeter value from the (mirror-) dissimilarity measurement based on $\dot{z}$.

In Proffitt [1982] it is stated that $\mathcal{S}_{\beta^*} = \|\mathcal{D}_{\zeta^*}z\|_2^{-1}$ is an optimal choice as a scale normalization operator in $d^{(2)}(z_1, z_2)$ or $d^{(2)}(\hat{z}_1, \hat{z}_2)$, in the sense that the dissimilarity measure is minimized over all scale normalization operators that are chosen independently for each contour. However, if we use for example $\mathcal{S}_{\beta^*} = a\|\mathcal{D}_{\zeta^*}z\|_2^{-1}$, where $0 < a < 1$, then this will always yield a smaller value for the dissimilarity measure than if we use $\mathcal{S}_{\beta^*} = \|\mathcal{D}_{\zeta^*}z\|_2^{-1}$. Thus, we see that there does not exist an optimal scale normalization operator.

In one instance in the literature (Persoon and Fu [1977]), it is proposed to optimize the scale in one contour representation with respect to the other, instead of normalizing the scale in the individual contour representations. They proposed as a dissimilarity measure

$$d_{\mathrm{PF}}^{(2)}(\hat{z}_1, \hat{z}_2) = \min_{\beta,\tau,\alpha} \left[ \sum_{\substack{k \in \mathbb{Z} \\ k \neq 0}} |\hat{z}_1(k) - \beta e^{i(\alpha + k\tau)} \hat{z}_2(k)|^2 \right]^{1/2}. \qquad (4.3.30)$$

Note that $\zeta_c^*(\hat{z}) = -\hat{z}(0)$, which we found to be the optimal choice (cf. Section 4.3.1), has been chosen for translation normalization. Persoon and Fu [1977] used numerical techniques to find solutions for $\beta$, $\tau$ and $\alpha$ in Eq. 4.3.30. However, it is straightforward to find analytic solutions. Using the same methods of analysis as in Eqs. 4.2.32-4.2.35, we find for $\alpha$

$$\alpha = \arg \{ \varrho_{12}(\tau; \mathcal{D}_{\zeta_c^*} z) \} \qquad (4.3.31)$$

where the appropriate value of $\tau$ is found from

$$d_{\mathrm{PF}}^{(2)}(\hat{z}_1, \hat{z}_2) = \min_{\beta} \Big[ \| \mathcal{D}_{\zeta_{c1}^*} \hat{z}_1 \|_2^2 + \beta^2 \| \mathcal{D}_{\zeta_{c2}^*} \hat{z}_2 \|_2^2 - 2\beta \max_{\tau} | \varrho_{12}(\tau; \mathcal{D}_{\zeta_c^*} z) | \Big]^{1/2}. \qquad (4.3.32)$$

The solution for $\beta$ can be found by taking the partial derivative with respect to $\beta$ of $[d_{\mathrm{PF}}^{(2)}(\hat{z}_1, \hat{z}_2)]^2$, and equating the result to zero. This yields for $\beta$

$$\beta = \frac{\max_{\tau} | \varrho_{12}(\tau; \mathcal{D}_{\zeta_c^*} z) |}{\| \mathcal{D}_{\zeta_{c2}^*} \hat{z}_2 \|_2^2}. \qquad (4.3.33)$$

Substitution of this result into Eq. 4.3.32 leads to

$$\begin{aligned} d_{\mathrm{PF}}^{(2)}(\hat{z}_1, \hat{z}_2) &= \left[ \| \mathcal{D}_{\zeta_{c1}^*} \hat{z}_1 \|_2^2 - \frac{\max_{\tau} | \varrho_{12}(\tau; \mathcal{D}_{\zeta_c^*} z) |^2}{\| \mathcal{D}_{\zeta_{c2}^*} \hat{z}_2 \|_2^2} \right]^{1/2} \\ &= \| \mathcal{D}_{\zeta_{c1}^*} \hat{z}_1 \|_2 \left[ 1 - \max_{\tau} | \varrho_{12}(\tau; \mathcal{S}_{\beta^*} \mathcal{D}_{\zeta_c^*} z) |^2 \right]^{1/2} \end{aligned} \qquad (4.3.34)$$

where $\mathcal{S}_{\beta^*} = \left\| \mathcal{D}_{\zeta_c^*} \hat{z} \right\|_2^{-1}$, i.e. the scale normalization operator that we proposed earlier in this section.

If we use the translation and scale normalization operators $\mathcal{D}_{\zeta_c^*}$ and $\mathcal{S}_{\beta^*}$ in the dissimilarity measure $d^{(2)}(\hat{z}_1, \hat{z}_2)$, we can derive the expression (cf. Eq. 4.2.41, Table 4.3 and Eq. 4.2.1)

$$d^{(2)}(\hat{z}_1, \hat{z}_2) = \sqrt{2}\left[1 - \max_{\tau} \left| \varrho_{12}(\tau; \mathcal{S}_{\beta^*}\mathcal{D}_{\zeta_c^*} z) \right| \right]^{1/2}. \qquad (4.3.35)$$

With the aforementioned choice for $\mathcal{S}_{\beta^*}$ it is straightforward to show that

$$0 \leqslant \left| \varrho_{12}(\tau; \mathcal{S}_{\beta^*}\mathcal{D}_{\zeta_c^*} z) \right| \leqslant 1. \qquad (4.3.36)$$

Combining Eqs. 4.3.34-4.3.36 we can relate $d^{(2)}(\hat{z}_1, \hat{z}_2)$ and $d^{(2)}_{\mathrm{PF}}(\hat{z}_1, \hat{z}_2)$ by the inequality

$$d^{(2)}(\hat{z}_1, \hat{z}_2) \leqslant \sqrt{2} \left\| \mathcal{D}_{\zeta_c^*} \hat{z}_1 \right\|_2^{-1} d^{(2)}_{\mathrm{PF}}(\hat{z}_1, \hat{z}_2). \qquad (4.3.37)$$

Though $d^{(2)}_{\mathrm{PF}}(\hat{z}_1, \hat{z}_2)$ leads, through our analysis, to the same computational complexity as $d^{(2)}(\hat{z}_1, \hat{z}_2)$, it has some considerable drawbacks. The dissimilarity measure $d^{(2)}_{\mathrm{PF}}(\hat{z}_1, \hat{z}_2)$ is not symmetric, i.e. in general $d^{(2)}_{\mathrm{PF}}(\hat{z}_1, \hat{z}_2) \neq d^{(2)}_{\mathrm{PF}}(\hat{z}_2, \hat{z}_1)$. Therefore it does not constitute a metric over equivalence classes of geometrically similar contours. As a consequence of the asymmetry of $d^{(2)}_{\mathrm{PF}}(\hat{z}_1, \hat{z}_2)$ we have to take care that the sizes of the templates of shape classes are optimized with respect to the contour to be classified instead of the other way around, when this dissimilarity measure is used for shape classification. For shape clustering $d^{(2)}_{\mathrm{PF}}(\hat{z}_1, \hat{z}_2)$ seems inappropriate, since it lacks metric properties.

Concluding we can say that we have shown that $d^{(2)}_{\mathrm{PF}}(\hat{z}_1, \hat{z}_2)$ and $d^{(2)}(\hat{z}_1, \hat{z}_2)$ have comparable properties in view of dissimilarity measurement (cf. Eqs. 4.3.34, 4.3.35), but that $d^{(2)}(\hat{z}_1, \hat{z}_2)$ is preferable since it possesses metric properties (cf. Theorem 4.3).

### *4.3.3 Normalization of orientation and starting point*

In our formulation of (mirror-)dissimilarity measures in Section 4.2 we optimized for both the orientation and the starting point of one contour with respect to the other. As we can see in Tables 4.5-4.8, this accounts for most of the computational complexity of the (mirror-)dissimilarity measures, especially in measures with index $p \neq 2$. The definition of effective means for normalization of orientation and starting point in contours would greatly enhance the attractiveness of the proposed dissimilarity measures from a computational point of view. Clearly, a (mirror-)dissimilarity measure that is optimized for orientation and starting point yields a smaller value than a measure where the contour representations have been normalized with respect to these parameters. However, if the normalization procedures produce unique solutions, then a dissimilarity measure will still constitute a metric over the equivalence classes of representations of geometrically similar contours. In the latter case every equivalence class is represented by a unique normalized contour representation and a dissimilarity measure is a metric in the space of normalized contour representations.

If a normalization procedure does not always yield a unique solution, then, in order to preserve the metric properties of a dissimilarity measure, we can optimize the measure over the often limited set of candidate normalizations.

From Table 2.2 we know that all contour representations, on the basis of which we defined (mirror-)dissimilarity measures in Section 4.2, are variant under a shift of the parametric starting point and the representations $z$, $\dot{z}$ and $\ddot{z}$ are also variant under rotations of a contour.

Let $f$ denote any of the contour representations $z$, $\dot{z}$, $\ddot{z}$, $\psi^*$ and $K$. Since the operators $\mathcal{D}_\zeta$ and $\mathcal{R}_\alpha$ do not commute when $f$ stands for the position function $z$, we perform the normalization of orientation and starting point on $\mathcal{D}_{\zeta^*}f$. The operators $\mathcal{S}_\beta$, $\mathcal{R}_\alpha$ and $\mathcal{T}_\tau$ all commute. Therefore it is not necessary for orientation and starting point normalization that scale normalization has already taken place.

A proper orientation and starting point normalization process has the property that the contour representations $\mathcal{D}_{\zeta^*}f$ and $\mathcal{T}_\tau\mathcal{R}_\alpha\mathcal{D}_{\zeta^*}f$ lead to the same orientation- and starting point-normalized contour representation, i.e.

$$\mathcal{T}_{\tau^*(\mathcal{D}_{\zeta^*}f)}\mathcal{R}_{\alpha^*(\mathcal{D}_{\zeta^*}f)}(\mathcal{D}_{\zeta^*}f) = \mathcal{T}_{\tau^*(\mathcal{T}_\tau\mathcal{R}_\alpha\mathcal{D}_{\zeta^*}f)}\mathcal{R}_{\alpha^*(\mathcal{T}_\tau\mathcal{R}_\alpha\mathcal{D}_{\zeta^*}f)}(\mathcal{T}_\tau\mathcal{R}_\alpha\mathcal{D}_{\zeta^*}f) \tag{4.3.38}$$

which leads, through the commutativity of $\mathcal{R}_\alpha$ and $\mathcal{T}_\tau$ and the properties

$\mathcal{T}_{\tau_1}\mathcal{T}_{\tau_2} = \mathcal{T}_{\tau_1 + \tau_2}$ and $\mathcal{R}_{\alpha_1}\mathcal{R}_{\alpha_2} = \mathcal{R}_{\alpha_1 + \alpha_2}$, to the requirements

$$\tau^*(\mathcal{D}_{\zeta^*}f) = \tau^*(\mathcal{T}_\tau\mathcal{R}_\alpha\mathcal{D}_{\zeta^*}f) + \tau \tag{4.3.39}$$

independent of $\mathcal{R}_\alpha$ and

$$\alpha^*(\mathcal{D}_{\zeta^*}f) = \alpha^*(\mathcal{T}_\tau\mathcal{R}_\alpha\mathcal{D}_{\zeta^*}f) + \alpha \tag{4.3.40}$$

independent of $\mathcal{T}_\tau$.

The contour representations $\dot{z}$, $\ddot{z}$, $\psi^*$ and $K$ all involve differentiation (cf. Section 2.2) and are therefore more sensitive to noise than the position function $z$ of the object contour or the characteristic function $\chi_R(x, y)$ of the object (cf. Appendix B). As a consequence, it is natural to base orientation normalization either on features directly generated by $z$ or by $\chi_R(x, y)$ and to base starting point normalization on features generated by $z$. The orientation and/or starting point normalization parameters thus obtained can then be used to normalize the contour representations $z$, $\dot{z}$, $\ddot{z}$, $\psi^*$ or $K$, as desired.

In the literature we find two main approaches towards orientation and/or starting point normalization:

- Orientation normalization based on the gravitational moments $m_{pq}$ (cf. Definition 4.9) of the region that is occupied by the object.

- Orientation and starting point normalization based on the Fourier coefficients $\hat{z}(k)$ generated by the position function $z$.

In the following we will evaluate these two approaches.

The oldest technique for orientation normalization in pictorial pattern recognition is based upon the moments $m_{pq}$ (Hu [1962], Alt [1962]), which are features generated by the characteristic function $\chi_R(x, y)$ of the object (cf. Definition 4.9).

In the orientation normalization technique based on moments, translation normalization of the object is achieved by the moment-based translation normalization parameter $\zeta_\text{r}^*$ (cf. Eq. 4.3.6). The orientation normalization process uses the central moments $\mu_{pq}(R)$ of the region $R$.

**Definition 4.10.** *Central moment $\mu_{pq}$.*
The central moment $\mu_{pq}$ of a region $R$ in the plane is defined as

$$\mu_{pq} = \mu_{pq}(R) = \iint_R \left(x - \frac{m_{10}}{m_{00}}\right)^p \left(y - \frac{m_{01}}{m_{00}}\right)^q \mathrm{d}x\,\mathrm{d}y$$

$$p, q = 0, 1, \ldots . \tag{4.3.41}$$

The central moment $\mu_{pq}$ is said to be of order $(p + q)$.

□

Note that $\mu_{pq}(R) \equiv m_{pq}(\mathcal{D}_{\zeta_r^*}R)$, $\forall p,q = 0, 1, \ldots$ .

The first step in the orientation normalization method based on moments is to determine an orientation $\alpha$ such that, after a rotation of $\mathcal{D}_{\zeta_r^*}R$ over $-\alpha$, we obtain $\mu_{11}(\mathcal{R}_{-\alpha}\mathcal{D}_{\zeta_r^*}R) = 0$, i.e.

$$\begin{aligned}
\mu_{11}(\mathcal{R}_{-\alpha}\mathcal{D}_{\zeta_r^*}R) &= \iint_R \{(\mathcal{D}_{\zeta_r^*}x)\cos\alpha + (\mathcal{D}_{\zeta_r^*}y)\sin\alpha\} \\
&\qquad \cdot \{-(\mathcal{D}_{\zeta_r^*}x)\sin\alpha + (\mathcal{D}_{\zeta_r^*}y)\cos\alpha\}\mathrm{d}x\,\mathrm{d}y \\
&= -\sin\alpha\cos\alpha\,\mu_{20}(R) + (\cos^2\alpha - \sin^2\alpha)\mu_{11}(R) \\
&\qquad + \sin\alpha\cos\alpha\,\mu_{02}(R) \\
&= 0.
\end{aligned} \tag{4.3.42}$$

Solving $\alpha$ from Eq. 4.3.42 gives

$$\tan 2\alpha = \frac{2\mu_{11}(R)}{\mu_{20}(R) - \mu_{02}(R)} \tag{4.3.43}$$

from which we find

$$\alpha = \frac{1}{2}\arctan\left(\frac{2\mu_{11}(R)}{\mu_{20}(R) - \mu_{02}(R)}\right) + n \cdot \frac{\pi}{2}, \qquad n \in \mathbb{Z}. \tag{4.3.44}$$

This result shows that the constraint in Eq. 4.3.42 yields four solutions for $\alpha$ in a range of length $2\pi$. In order to obtain a unique solution for $\alpha$, additional constraints are needed. Hu [1962] and Reeves and Rostampour [1981] have proposed to choose that value of $\alpha$ from the solutions in Eq. 4.3.44, such that

$$\mu_{20}(\mathcal{R}_{-\alpha}\mathcal{D}_{\zeta_r^*}R) > \mu_{02}(\mathcal{R}_{-\alpha}\mathcal{D}_{\zeta_r^*}R) \tag{4.3.45a}$$

and

$$\mu_{30}(\mathcal{R}_{-\alpha}\mathcal{D}_{\zeta_r^*}R) > 0. \tag{4.3.45b}$$

However, under rather general conditions it may happen that

$$\mu_{20}(\mathcal{R}_{-\alpha}\mathcal{D}_{\zeta_r^*}R) = \mu_{02}(\mathcal{R}_{-\alpha}\mathcal{D}_{\zeta_r^*}R) \tag{4.3.46a}$$

and/or

$$\mu_{30}(\mathcal{R}_{-\alpha}\mathcal{D}_{\zeta_r^*}R) = 0 \tag{4.3.46b}$$

(cf. Casey [1970], Nagy and Tuong [1970]) or that these equations are almost valid. In these cases even more constraints are needed to arrive at a unique value for $\alpha$. Another possible approach is to optimize (mirror-)dissimilarity measures over the four solutions that we found for $\alpha$ in Eq. 4.3.44.

In the special case of $\boldsymbol{n}$-fold rotational symmetry, with $\boldsymbol{n} > \mathbf{2}$, it is not possible to use Eq. 4.3.42 as a constraint on $\alpha$, since in that case $\mu_{11}(\mathcal{R}_{-\alpha}\mathcal{D}_{\zeta_r^*}R) = 0$, for any of the values of $\alpha$ in Eq. 4.3.44. Then moments of order $(p + q) = \boldsymbol{n}$ are required to define a useful constraint on the orientation normalization parameter $\alpha$ (cf. Hu [1962]). In that case $\alpha$ must be solved from constraints on $\mu_{pq}(\mathcal{R}_{-\alpha}\mathcal{D}_{\zeta_r^*}R)$, which is related to $\mu_{pq}(\mathcal{D}_{\zeta_r^*}R)$ as

$$\mu_{pq}(\mathcal{R}_{-\alpha}\mathcal{D}_{\zeta_r^*}R)$$

$$= \sum_{n=0}^{p} \sum_{m=0}^{q} (-1)^{q-m} \binom{p}{n} \binom{q}{m} (\sin\alpha)^{q+n-m} (\cos\alpha)^{p-n+m}$$

$$\cdot \mu_{p+q-n-m,\, n+m}(\mathcal{D}_{\zeta_r^*}R). \tag{4.3.47}$$

The central moments $\mu_{pq}(\mathcal{D}_{\zeta_r^*}R)$ can be computed directly from a segmented image.

The concept of a parametric starting point on the contour of a region in the plane does not naturally arise if we are dealing with its moments. This explains why starting point normalization is not mentioned in the pattern recognition literature on moments. However, if we apply orientation normalization to the translation-normalized position function $\mathcal{D}_{\zeta_r^*}z$, it is very well possible to define constraints on the basis of this contour representation to arrive at a starting point normalization. Examples of such constraints are the following.

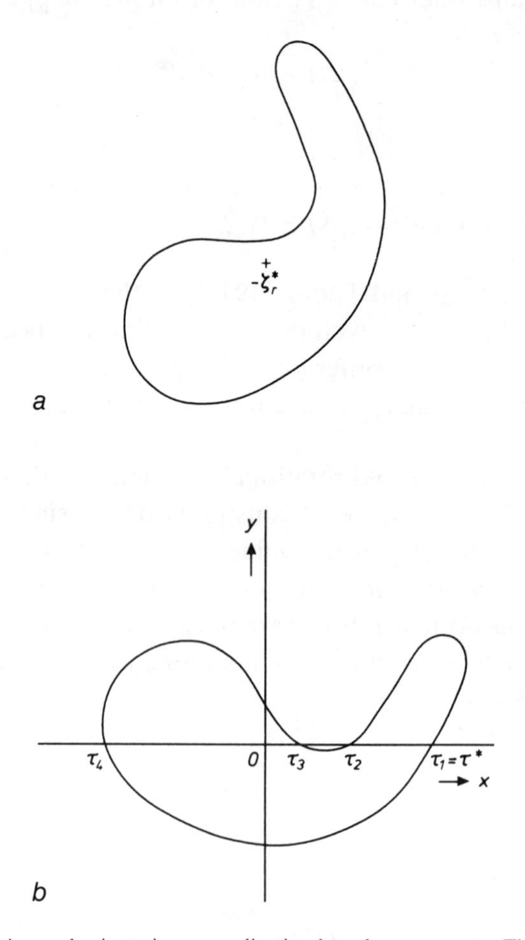


**Figure 4.8.** Translation and orientation normalization based on moments. The contour in ($a$) has position function $z(t)$ and centroid $-\zeta_r^* = (m_{10} + im_{01})/m_{00}$. The contour, resulting from the normalization, is displayed in ($b$) and has position function $\mathcal{R}_{\alpha^*}\mathcal{D}_{\zeta_r^*}z(t)$. The set of parameter values $t$, for which the $x$-axis in ($b$) intersects $\mathcal{R}_{\alpha^*}\mathcal{D}_{\zeta_r^*}z(t)$, is $T_0 = \{\tau_1, \tau_2, \tau_3, \tau_4\}$. The set $T_0$ may be used for starting point normalization, as described in the text.

Orientation normalization of $\mathcal{D}_{\zeta_r^*}z$ yields $\mathcal{R}_{\alpha^*}\mathcal{D}_{\zeta_r^*}z$, where $-\alpha^*$ is a unique solution for $\alpha$. Next we determine the set $T_0$ of values $\tau$ for which $\mathcal{R}_{\alpha^*}\mathcal{D}_{\zeta_r^*}z(\tau)$ lies on the $x$-axis, i.e. $\text{Im}\{\mathcal{R}_{\alpha^*}\mathcal{D}_{\zeta_r^*}z(\tau)\} = 0$, and choose $\tau^* \in T_0$ such that

$$\mathcal{R}_{\alpha^*}\mathcal{D}_{\zeta_r^*}z(\tau^*) = \max_{\tau \in T_0}\{\mathcal{R}_{\alpha^*}\mathcal{D}_{\zeta_r^*}z(\tau)\} \tag{4.3.48}$$

cf. Figure 4.8.

Another possibility for starting point normalization in $\mathcal{R}_{\alpha^*}\mathcal{D}_{\zeta_r^*}z$ is to use the phase of one of the Fourier coefficients generated by $\mathcal{R}_{\alpha^*}\mathcal{D}_{\zeta_r^*}z$. For example, let the index $k_1$ be such that

$$\left|\mathcal{R}_{\alpha^*}\mathcal{D}_{\zeta_r^*}\hat{z}(k_1)\right| = \sup_{k \in \mathbb{Z} - \{0\}} \left|\mathcal{R}_{\alpha^*}\mathcal{D}_{\zeta_r^*}\hat{z}(k)\right|. \tag{4.3.49}$$

A reason to make this choice is the fact that the phase of the most significant Fourier coefficient may be expected to be less sensitive to noise.

We denote

$$\eta(k) = \arg\{\mathcal{R}_{\alpha^*}\mathcal{D}_{\zeta_r^*}\hat{z}(k)\} \tag{4.3.50a}$$

and

$$\eta_\tau(k) = \arg\{\mathcal{T}_\tau\mathcal{R}_{\alpha^*}\mathcal{D}_{\zeta_r^*}\hat{z}(k)\}. \tag{4.3.50b}$$

Then we have the relation (cf. Table 3.2)

$$\eta_\tau(k) = \eta(k) - k\tau. \tag{4.3.51}$$

If the constraint is that the starting point normalization parameter $\tau^*$ is chosen such that $\mathcal{T}_{\tau^*}\mathcal{R}_{\alpha^*}\mathcal{D}_{\zeta_r^*}\hat{z}(k_1)$ is a positive real value, we find for $\tau^*$:

$$\tau^* = \frac{1}{k_1}\eta(k_1) + n\frac{2\pi}{k_1}, \qquad n \in \mathbb{Z}. \tag{4.3.52}$$

Only for $|k_1| = 1$ does Eq. 4.3.52 define a unique solution for $\tau^*$. Otherwise additional constraints are needed to select a $\tau^*$ from the $|k_1|$ candidates defined by Eq. 4.3.52.

The latter method for starting point normalization is quite similar to the methods for orientation and starting point normalization based entirely on the Fourier coefficients generated by $\mathcal{D}_{\zeta^*}z$, that we will discuss next.

In methods for orientation and starting point normalization entirely based on the Fourier coefficients generated by the position function $z$ it is natural to use $\zeta_c^*(z) = -\hat{z}(0)$ (cf. Eq. 4.3.3) for translation normalization. We observe from the literature that these methods have the following characteristics in common:

- Select two indices $k_1, k_2 \in \mathbb{Z} - \{0\}$, $k_1 \neq k_2$.
- Determine an angle of rotation $\alpha^*$ and a starting point shift $\tau^*$ such that $\mathcal{T}_{\tau^*}\mathcal{R}_{\alpha^*}\mathcal{D}_{\zeta_c^*}\hat{z}(k_i)$, $i = 1, 2$, both are positive real values.
- If the latter constraints do not lead to a unique solution for $\alpha^*$ and $\tau^*$, then determine additional constraints to arrive at a unique pair of solutions.

We will now derive the solutions for $\alpha^*$ and $\tau^*$ from the phases of the selected pair of Fourier coefficients $\hat{z}(k_1)$ and $\hat{z}(k_2)$ and determine how the number of solutions depends upon the values of $k_1$ and $k_2$. We denote

$$\eta(k) = \arg\{\mathcal{D}_{\zeta_c^*}\hat{z}(k)\} \tag{4.3.53a}$$

and

$$\eta_{\tau,\alpha}(k) = \arg\{\mathcal{T}_\tau\mathcal{R}_\alpha\mathcal{D}_{\zeta_c^*}\hat{z}(k)\}. \tag{4.3.53b}$$

Then we have the relation (cf. Table 3.2)

$$\eta_{\tau,\alpha}(k) = \eta(k) - k\tau + \alpha. \tag{4.3.54}$$

Note that $\eta(k)$ and $\eta_{\tau,\alpha}(k)$ can only be determined up to a multiple of $2\pi$. With the aid of Eq. 4.3.54 the constraint that $\mathcal{T}_{\tau^*}\mathcal{R}_{\alpha^*}\mathcal{D}_{\zeta_c^*}\hat{z}(k_i)$, $i = 1, 2$, are positive real values yields the equations

$$\eta(k_i) - k_i\tau^* + \alpha^* = n_i 2\pi, \qquad n_i \in \mathbb{Z}, \quad i = 1, 2 \tag{4.3.55}$$

from which $\tau^*$ and $\alpha^*$ can be solved:

$$\tau^* = \frac{\eta(k_1) - \eta(k_2)}{k_1 - k_2} + n\frac{2\pi}{k_1 - k_2} \tag{4.3.56a}$$

$$\alpha^* = \frac{k_2\eta(k_1) - k_1\eta(k_2)}{k_1 - k_2} + n\frac{k_1 2\pi}{k_1 - k_2}, \qquad n \in \mathbb{Z}. \tag{4.3.56b}$$

The solutions for $\tau^*$ and $\alpha^*$ in Eqs. 4.3.56a and 4.3.56b need not necessarily be in the range $(0, 2\pi)$, though we note that solutions $\tau^*$ mod $2\pi$ and $\alpha^*$ mod $2\pi$ are equally valid. Keeping this in mind, it is clear that in Eqs. 4.3.56a and 4.3.56b we have obtained $|k_1 - k_2|$ pairs of solutions for $\tau^*$ and $\alpha^*$ in ranges of length $2\pi$. Only in the case of $|k_1 - k_2| = 1$ do we obtain a unique solution for the orientation and starting point normalization. In all other cases additional constraints are needed to determine a unique pair $\tau^*$ and $\alpha^*$. Another possibility is to minimize (mirror-)dissimilarity measures over the $|k_1 - k_2|$ candidate normalizations.

The orientation and starting point normalization techniques based on Fourier coefficients that have been published so far differ in the selection of the indices $k_1$ and $k_2$ and in the definition of additional constraints, if any, to arrive at a unique solution for $\tau^*$ and $\alpha^*$. These techniques are reviewed in the following.

Persoon and Fu were the first to present such normalization techniques (Persoon and Fu [1974], Persoon and Fu [1977]). They choose the indices $k_1 = 1$ and $k_2 = -1$ and determine $\tau^*$ and $\alpha^*$ such that $\mathcal{T}_{\tau^*}\mathcal{R}_{\alpha^*}\mathcal{D}_{\zeta_c^*}\hat{z}(k_i)$, $i = 1, 2$, become purely imaginary values, instead of positive real values. Since they do not specify the signs of the imaginary Fourier coefficients after normalization, their normalization constraints lead to the equation:

$$\eta(k_i) - k_i\tau^* + \alpha^* = \frac{\pi}{2} + n_i\pi, \qquad n_i \in \mathbb{Z}, \quad i = 1, 2. \tag{4.3.57}$$

With $k_1 = 1$ and $k_2 = -1$, $\tau^*$ and $\alpha^*$ are solved from Eq. 4.3.57 as

$$\tau^* = \frac{1}{2}\{\eta(1) - \eta(-1)\} + n\frac{\pi}{2} \tag{4.3.58a}$$

$$\alpha^* = -\frac{1}{2}\{\eta(1) + \eta(-1)\} + (n \pm 1)\frac{\pi}{2}, \qquad n \in \mathbb{Z}. \tag{4.3.58b}$$

From these equations we see that the normalization procedure proposed by Persoon and Fu yields eight distinct pairs of $\tau^*$ and $\alpha^*$, instead of two, as claimed by Wallace [1981]. Persoon and Fu do not give any additional constraints to arrive at a unique solution for $\tau^*$ and $\alpha^*$.

A problem with the choice of the index $k_2 = -1$ is that $|\mathcal{D}_{\zeta_c^*}\hat{z}(-1)|$ can be zero or close to zero under fairly general conditions (see Figure 4.9), which causes $\eta(-1)$ either to be undetermined or more likely to be corrupted by noise. In Section 3.6 we already found that, if the contour to be normalized has $\boldsymbol{n}$-fold rotational symmetry, with $\boldsymbol{n} > \mathbf{2}$, this constitutes a sufficient condition such that $|\mathcal{D}_{\zeta_c^*}\hat{z}(-1)| = 0$.

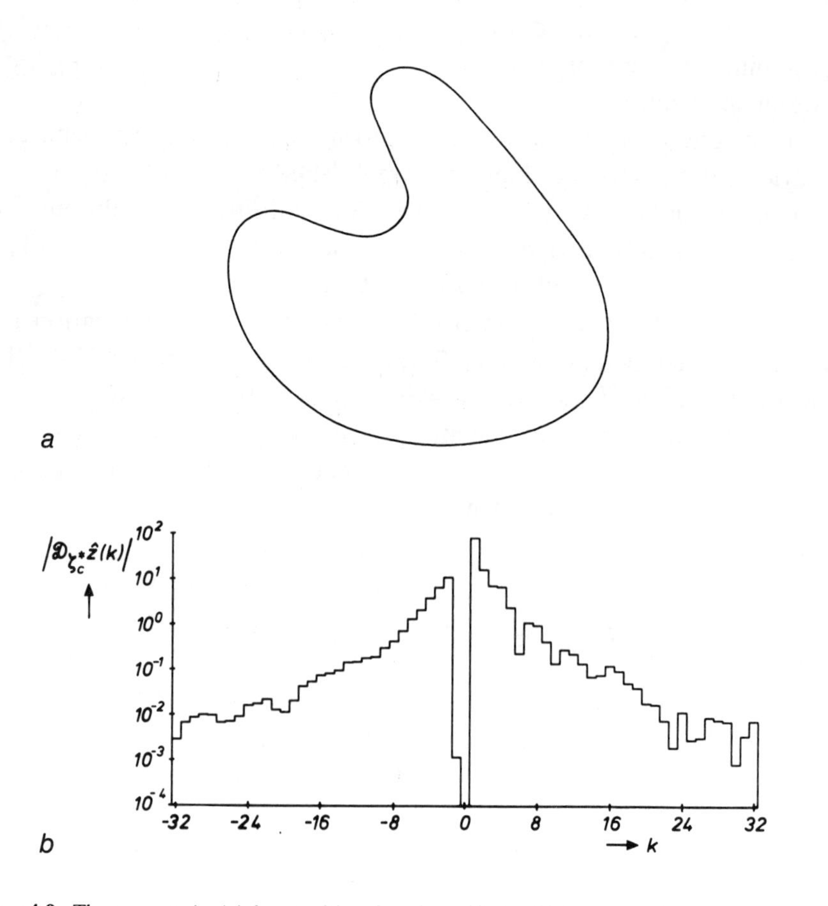


**Figure 4.9.** The contour in ($a$) has position function $z(t)$. In ($b$) the magnitudes of the Fourier coefficients $\mathcal{D}_{\zeta_c^*}\hat{z}(k)$, $k = -32, \ldots, 32$, generated by $\mathcal{D}_{\zeta_c^*}z(t)$, are displayed. Note that $|\mathcal{D}_{\zeta_c^*}\hat{z}(-1)|$ is close to zero.

Kuhl and Giardina [1982] presented an orientation and starting point normalization technique based on 'elliptic Fourier features'. Some straightforward analysis reveals that all their results can be expressed in terms of the Fourier coefficients $\hat{z}(k)$. In particular their orientation and starting point normalization turn out to be virtually identical to the one proposed by Persoon and Fu. Kuhl and Giardina also chose the indices $k_1 = 1$ and $k_2 = -1$, but they determine $\tau^*$ and $\alpha^*$ such that $\mathcal{T}_{\tau^*}\mathcal{R}_{\alpha^*}\mathcal{D}_{\zeta_c^*}\hat{z}(k_i)$, $i = 1, 2$ become positive real values, as we did earlier (cf. Eqs. 4.3.55, 4.3.56a, b), instead of purely imaginary values, as Persoon and Fu did. From Eqs. 4.3.56a and 4.3.56b we observe that, with $|k_1 - k_2| = 2$, Kuhl and Giardina obtain two pairs of solutions for $\tau^*$ and $\alpha^*$ which are found by substitution of $k_1 = 1$ and $k_2 = -1$ into these equations. In Figure 4.10 we have displayed an example of the results of the orientation and starting point normalization according to Kuhl and Giardina. They do not specify additional constraints to arrive at a unique solution for $\tau^*$ and $\alpha^*$.

In the orientation and starting point normalization procedure proposed by Exel [1978] and Burkhardt [1979] $k_1$ is chosen such that

$$\left|\mathcal{D}_{\zeta_c^*}\hat{z}(k_1)\right| = \sup_{k \in \mathbb{Z}} \left|\mathcal{D}_{\zeta_c^*}\hat{z}(k)\right| \tag{4.3.59}$$

i.e. $k_1$ is the index of the most significant Fourier coefficient in the sequence $\mathcal{D}_{\zeta_c^*}\hat{z}$. In their reports the authors argue that in most cases $k_1 = 1$. In Example 4.3 we have shown circumstances for which this is not true. For contours with a counterclockwise sense we conjecture that $k_1 > 0$ in Eq. 4.3.59.

Exel and Burkhardt determine $k_2 > k_1$ such that $k_2 - k_1$ is the order of rotational symmetry of the contour. The authors state that, in order to reduce the noise sensitivity of the normalization procedure, the most significant Fourier coefficients must be chosen. However, the criterion they use for the selection of $k_2$ is not always in accordance with this statement since $(\boldsymbol{k}_2 - \boldsymbol{k}_1)$-fold rotational symmetry does not guarantee that $\mathcal{D}_{\zeta_c^*}\hat{z}(k_2)$ is significant, as shown for example in Figure 4.11. This figure also shows that there is no clear reason to exclude the indices $k < k_1$ for the selection of $k_2$.

Once the indices $k_1$ and $k_2$ have been selected, Exel and Burkhardt determine $\tau^*$ and $\alpha^*$ such that $\mathcal{T}_{\tau^*}\mathcal{R}_{\alpha^*}\mathcal{D}_{\zeta_c^*}\hat{z}(k_i)$, for $i = 1, 2$, become positive real values. Thus they obtain $k_2 - k_1$ pairs of solutions for $\tau^*$

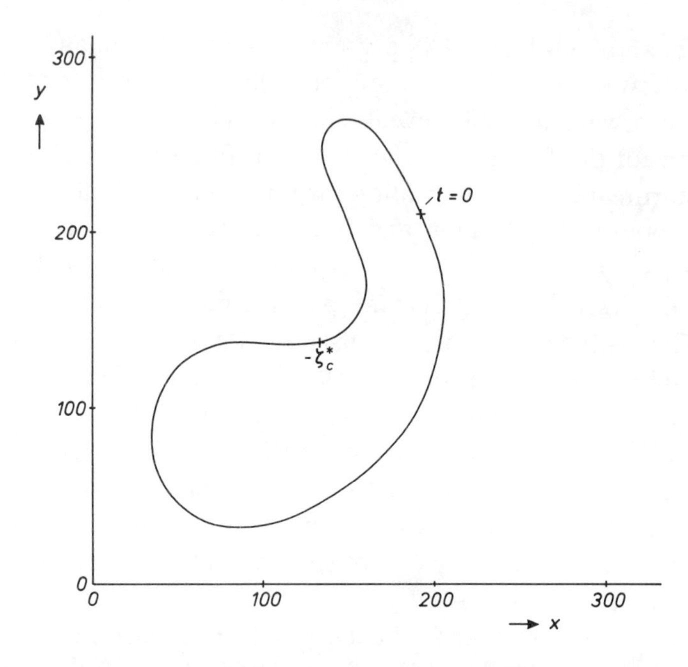


**Figure 4.10.** Normalization of orientation and starting point of the contour in ($a$) using the constraint that $\mathcal{T}_{\tau^*}\mathcal{R}_{\alpha^*}\mathcal{D}_{\zeta_c^*}\hat{z}(k_i)$, $i = 1, 2$, are positive real values. For $k_1 = 1$ and $k_2 = -1$ (Kuhl and Giardina [1982]) this constraint yields two solutions, displayed in ($b$) and ($c$). In this example $\hat{z}(1)$ and $\hat{z}(-1)$ are the two most significant Fourier coefficients: $|\hat{z}(1)| = 78.133$ and $|\hat{z}(-1)| = 30.663$. In ($b$) and ($c$) the position and orientation of the coordinate axes with respect to the contour and the location of the starting point on the contour after normalization have been indicated.

and $\alpha^*$ (cf. Eqs. 4.3.56a, b). No additional constraints are given to obtain a unique pair of $\tau^*$ and $\alpha^*$ in case $k_2 - k_1 > 1$. However, if the contour has $(\boldsymbol{k}_2 - \boldsymbol{k}_1)$-fold rotational symmetry, then each of the $k_2 - k_1$ candidate normalizations will lead to the same normalized contour representation. Since rotational symmetry was indeed chosen as a criterion for choosing $k_2$, ambiguity resolving criteria are not needed in this procedure. In Section 4.5, where we will treat the subject of symmetry measurement in detail, we will return to Exel and Burkhardt's proposal for symmetry measurement.

Finally, Wallace and Mitchell [1979] and Wallace and Wintz [1980] chose $k_1 = 1$ in Eqs. 4.3.56a and 4.3.56b. The index $k_2$ is chosen such that

$$\left|\mathcal{D}_{\zeta_c^*}\hat{z}(k_2)\right| = \sup_{k \in \mathbb{Z} - \{1\}} \left|\mathcal{D}_{\zeta_c^*}\hat{z}(k)\right| \tag{4.3.60}$$

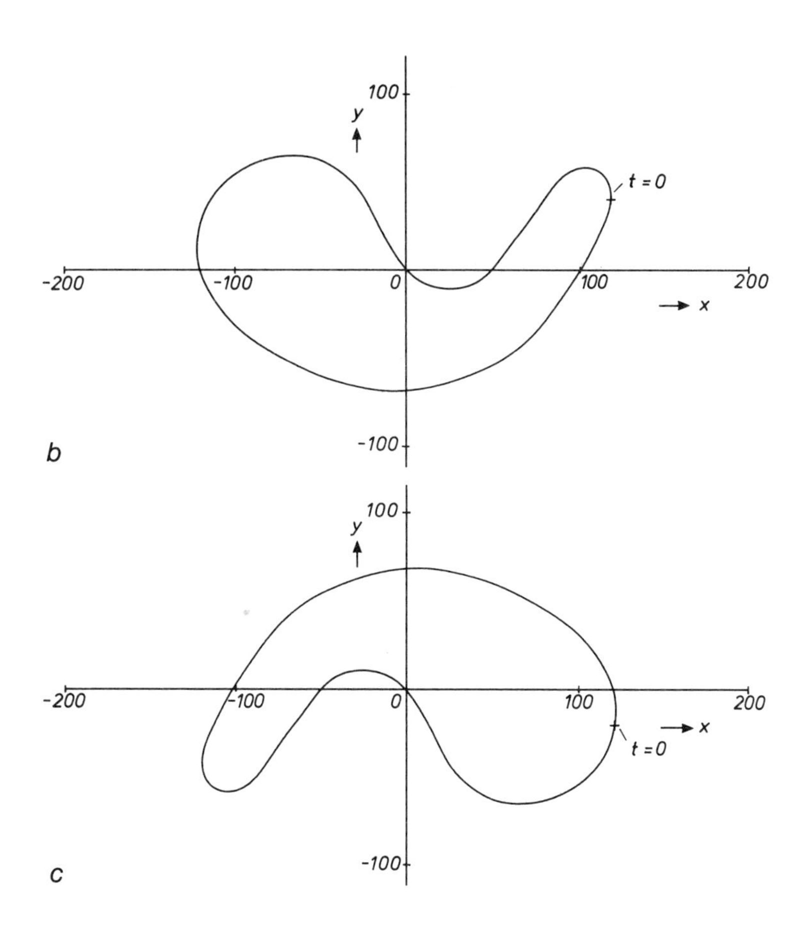


i.e. $\mathscr{D}_{\zeta_c^*}\hat{z}(k_2)$ is the most significant Fourier coefficient, not considering $k = 1$. Note that $\mathscr{D}_{\zeta_c^*}\hat{z}(0) \equiv 0$ (cf. Eq. 4.3.2). Wallace et al. also determine $\tau^*$ and $\alpha^*$ such that $\mathscr{T}_{\tau^*}\mathscr{R}_{\alpha^*}\mathscr{D}_{\zeta_c^*}\hat{z}(k_i)$, $i = 1, 2$, become positive real values, leading to $|k_1 - k_2|$ pairs of candidate solutions for $\tau^*$ and $\alpha^*$ (cf. Eqs. 4.3.56a, b). To arrive at a unique solution for $\tau^*$ and $\alpha^*$ if $|k_1 - k_2| > 1$, Wallace, Mitchell and Wintz present two methods.

The first method can be summarized as follows:

- Determine the set $\boldsymbol{K}$ of indices such that

$$\boldsymbol{K} \cap \{0, k_1, k_2\} = \varnothing \tag{4.3.61a}$$

and

$$k \in \boldsymbol{K} \quad \text{iff} \quad \text{GCD}\left(|k - k_1|, |k_2 - k_1|\right) = 1 \tag{4.3.61b}$$

where GCD( ) stands for *greatest common divisor*.

- Select an index $k_3 \in \boldsymbol{K}$ such that

$$\left|\mathcal{D}_{\zeta_c^*}\hat{z}(k_3)\right| = \sup_{k \in \boldsymbol{K}} \left|\mathcal{D}_{\zeta_c^*}\hat{z}(k)\right|. \tag{4.3.62}$$

- Choose that pair $\tau^*$ and $\alpha^*$ from the $|k_1 - k_2|$ candidate normalizations for which

$$\operatorname{Re}\left\{\mathcal{T}_{\tau^*}\mathcal{R}_{\alpha^*}\mathcal{D}_{\zeta_c^*}\hat{z}(k_3)\right\} \tag{4.3.63}$$

is maximum.

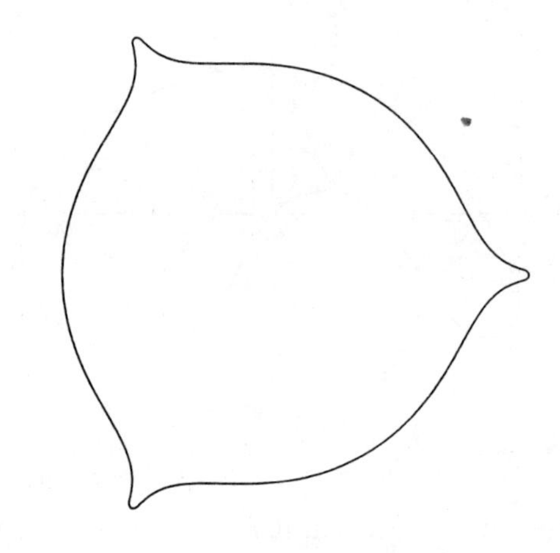


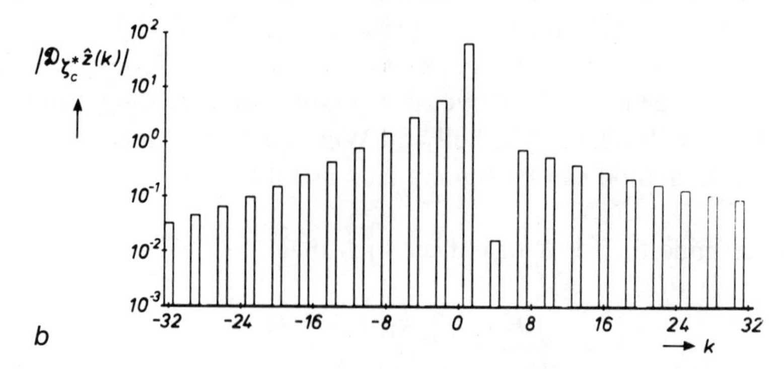


**Figure 4.11.** In ($a$) we have a contour with **3**-fold rotational symmetry. The position function of this contour is $z(t)$. In ($b$) the magnitudes of the Fourier coefficients $\mathcal{D}_{\zeta_c^*}\hat{z}(k)$, $k = -32, \ldots, 32$, generated by $\mathcal{D}_{\zeta_c^*}z(t)$, are displayed. $\mathcal{D}_{\zeta_c^*}\hat{z}(1)$ is the most significant Fourier coefficient, whereas $\mathcal{D}_{\zeta_c^*}\hat{z}(4)$ is relatively small. Also note that $\mathcal{D}_{\zeta_c^*}\hat{z}(-2)$ is the second most significant Fourier coefficient.

The second ambiguity resolving criterion proposed by Wallace, Mitchell and Wintz simply consists of selecting that pair $\tau^*$ and $\alpha^*$ from the $|k_1 - k_2|$ candidate normalizations for which

**Table 4.9.** Survey of orientation and starting point normalization techniques based on Fourier coefficients $\hat{z}(k)$, $k \in \mathbb{Z} - \{0\}$.

| References | Indices of $\mathcal{D}_{\zeta_c^*}\hat{z}(k)$ | Constraints on $\mathcal{T}_{\tau^*}\mathcal{R}_{\alpha^*}\mathcal{D}_{\zeta_c^*}\hat{z}(k_i)$, $i = 1, 2$ | Remarks |
|---|---|---|---|
| Persoon and Fu [1974]<br>Persoon and Fu [1977] | $k_1 = 1$;<br>$k_2 = -1$ | imaginary values | 8 pairs of solutions for $\tau^*$ and $\alpha^*$ in Eqs. 4.3.58a, b;<br>no additional ambiguity resolving constraints |
| Kuhl and Giardina [1982] | $k_1 = 1$;<br>$k_2 = -1$ | positive real values | 2 pairs of solutions for $\tau^*$ and $\alpha^*$ in Eqs. 4.3.56a, b;<br>no additional ambiquity resolving constraints |
| Exel [1978]<br>Burkhardt [1979] | $k_1$ such that: $|\mathcal{D}_{\zeta_c^*}\hat{z}(k_1)| = \sup_{k \in \mathbb{Z}} |\mathcal{D}_{\zeta_c^*}\hat{z}(k)|$;<br>$k_2$ such that: $k_2 > k_1$ and $\boldsymbol{k}_2 - \boldsymbol{k}_1$ is the order of rotational symmetry | positive real values | $k_2 - k_1$ pairs of solutions for $\tau^*$ and $\alpha^*$ in Eqs. 4.3.56a, b;<br>no additional ambiguity resolving constraints (not needed, see text) |
| Wallace and Mitchell [1979]<br>Wallace and Wintz [1980] | $k_1 = 1$;<br>$k_2$ such that: $|\mathcal{D}_{\zeta_c^*}\hat{z}(k_2)| = \sup_{k \in \mathbb{Z} - \{k_1\}} |\mathcal{D}_{\zeta_c^*}\hat{z}(k)|$ | positive real values | $|k_2 - 1|$ pairs of solutions for $\tau^*$ and $\alpha^*$ in Eqs. 4.3.56a, b;<br>unique solution for $\tau^*$ and $\alpha^*$ obtained through additional constraints |

$$\sum_{k \in \mathbb{Z}} \mathrm{Re}\left\{\mathcal{T}_{\tau^*}\mathcal{R}_{\alpha^*}\mathcal{D}_{\zeta_c^*}\hat{z}(k)\right\} \cdot \left|\mathrm{Re}\left\{\mathcal{T}_{\tau^*}\mathcal{R}_{\alpha^*}\mathcal{D}_{\zeta_c^*}\hat{z}(k)\right\}\right| \tag{4.3.64}$$

is maximum.

A survey of the characteristics of the orientation and starting point normalization techniques based on the Fourier coefficients $\hat{z}$, as discussed in the foregoing, is presented in Table 4.9.

We now propose a slightly modified technique for orientation and starting point normalization:

- Select $k_1$ such that

$$\left|\mathcal{D}_{\zeta_c^*}\hat{z}(k_1)\right| = \sup_{k \in \mathbb{Z}} \left|\mathcal{D}_{\zeta_c^*}\hat{z}(k)\right| \tag{4.3.65}$$

and $k_2$ such that

$$\left|\mathcal{D}_{\zeta_c^*}(k_2)\right| = \sup_{k \in \mathbb{Z} - \{k_1\}} \left|\mathcal{D}_{\zeta_c^*}(k)\right| \tag{4.3.66}$$

i.e. $k_1$ and $k_2$ are the indices of the two most significant Fourier coefficients.

- Determine values of $\tau^*$ and $\alpha^*$ such that $\mathcal{T}_{\tau^*}\mathcal{R}_{\alpha^*}\mathcal{D}_{\zeta_c^*}\hat{z}(k_i)$, $i = 1, 2$, become positive real values. There are $|k_2 - k_1|$ candidate pairs of $\tau^*$ and $\alpha^*$, given by Eqs. 4.3.56a, b. If $|k_2 - k_1| = 1$, then we have obtained a unique solution for $\tau^*$ and $\alpha^*$, else we perform the next step.

- Apply additional ambiguity resolving constraints to find a single pair $\tau^*$ and $\alpha^*$ from the $|k_2 - k_1|$ candidate pairs.

Concerning this normalization scheme we make the following remarks:

- The most significant pair of Fourier coefficients is chosen in order to reduce the noise sensitivity of the method. It may happen that there are other almost as significant Fourier coefficients. In the latter case it is advisable to consider also normalizations that use these Fourier coefficients and to optimize (mirror-)dissimilarity measurement over these normalizations, since the selection of a contour normalization, that is incomparable to that of another contour, will lead to a poor (mirror-)dissimilarity measurement.

- If the third step in the proposed normalization procedure is necessary in order to arrive at a unique solution for $\tau^*$ and $\alpha^*$, either of the two methods, defined by Wallace, Mitchell and Wintz (Eqs. 4.3.61a-4.3.64), can be used. An alternative ambiguity resolving constraint that we propose here is to select that pair $\tau^*$ and $\alpha^*$ from the $|k_2 - k_1|$ candidate pairs such that

$$\sum_{k \in \mathbb{Z}} \mathrm{Re}\left\{\mathcal{T}_{\tau^*}\mathcal{R}_{\alpha^*}\mathcal{D}_{\zeta_c^*}\hat{z}(k)\right\} \tag{4.3.67}$$

is maximum. Experiments have shown that the second method to obtain a unique solution for $\tau^*$ and $\alpha^*$ proposed by Wallace, Mitchell and Wintz (Eq. 4.3.64) is more robust than the first method they proposed (Eqs. 4.3.61a-4.3.63). These experiments have also shown that the alternative method we propose here in Eq. 4.3.67 has properties that are comparable to the method in Eq. 4.3.64. However, our method requires less computational effort than the second method of Wallace, Mitchell and Wintz (compare Eq. 4.3.64 and Eq. 4.3.67).

- An alternative for the third step in our procedure simply consists of minimizing (mirror-)dissimilarity measures over $|k_2 - k_1|$ candidate normalizations obtained in the second step.

### *4.3.4 Discussion of normalization versus optimization in dissimilarity and mirror-dissimilarity measures*

In the previous sections we have discussed methods for the normalization of contour representations with respect to the contour position, size, orientation and location of the parametric starting point. The normalizations with respect to position and size are expected by the (mirror-)dissimilarity measures, as defined in Section 4.2. The normalizations with respect to orientation and parametric starting point were investigated in order to be able to modify the previously defined dissimilarity measures to measures with comparable metric properties, but with a much lower computational complexity. Assume that a unique normalized version $\mathcal{T}_{\tau^*}\mathcal{R}_{\alpha^*}\mathcal{S}_{\beta^*}\mathcal{D}_{\zeta^*}f$ of a contour representation $f$ has been obtained, where $f$ stands for any of the contour representations $z$, $\dot{z}$, $\ddot{z}$, $\psi^*$ and $K$. Then a normalized dissimilarity measure of index $p$ is defined as follows.

**Definition 4.11.** *Normalized dissimilarity measure of index p.*
Let $f$ act as a generic symbol for any of the contour representations $z$, $\dot{z}$, $\ddot{z}$, $\psi^*$ and $K$. Then a normalized measure of dissimilarity of index $p$ between a pair of contours $\gamma_1$ and $\gamma_2$, with contour representations $f_1$ and $f_2$ respectively, $f_1, f_2 \in \mathbf{L}^p(2\pi)$, is defined as

$$d_*^{(p)}(f_1, f_2) = \| \mathcal{T}_{\tau_1^*}\mathcal{R}_{\alpha_1^*}\mathcal{S}_{\beta_1^*}\mathcal{D}_{\zeta_1^*}f_1 - \mathcal{T}_{\tau_2^*}\mathcal{R}_{\alpha_2^*}\mathcal{S}_{\beta_2^*}\mathcal{D}_{\zeta_2^*}f_2 \|_p, \qquad 1 \leqslant p \leqslant \infty \tag{4.3.68}$$

where $\mathcal{T}_{\tau_i^*}\mathcal{R}_{\alpha_i^*}\mathcal{S}_{\beta_i^*}\mathcal{D}_{\zeta_i^*}f_i$, $i = 1, 2$, are the uniquely normalized versions of $f_i$.

□

The discrete normalized dissimilarity measure of index $p$, $d_*^{(p)}[f_1, f_2]$, based on $N$ equidistant samples of both $f_1$ and $f_2$, is defined similarly (compare with Eq. 4.2.49). If we do not take the computational complexities of the normalization procedures into account, the computational complexity of $d_*^{(p)}[f_1, f_2]$ is $O(N)$ for any of the contour representations and for any value of the index $p$. This is considerably less than the computational complexities of $d^{(p)}[f_1, f_2]$, mentioned in Table 4.5, and also less than those of $d^{(2)}[f_1, f_2]$, mentioned in Table 4.7.

All methods proposed for translation and scale normalization, in Sections 4.3.1 and 4.3.2 respectively, lead to unique solutions. The methods for orientation and starting point normalization, however, may lead to multiple candidate solutions. In order to preserve the metric properties of a dissimilarity measure, we must find a unique solution from these candidates.

If we decide to minimize a dissimilarity measure over the candidate normalizations of the contour representations, instead of obtaining a unique normalization through an ambiguity resolving criterion, then this has the following effect upon the computational complexity of the dissimilarity measure.

Consider for example the orientation and starting point normalization based on Fourier coefficients. Let there be $|k_{11} - k_{12}|$ candidate normalizations for contour $\gamma_1$ and $|k_{21} - k_{22}|$ for contour $\gamma_2$, where $k_{i1}$ and $k_{i2}$, $i = 1, 2$, are the indices of Fourier coefficients $\hat{z}_i$, selected for the normalization of contour $\gamma_i$. Since

$$\| \mathcal{T}_{\tau_1^*}\mathcal{R}_{\alpha_1^*}\mathcal{S}_{\beta_1^*}\mathcal{D}_{\zeta_1^*}f_1 - \mathcal{T}_{\tau_2^*}\mathcal{R}_{\alpha_2^*}\mathcal{S}_{\beta_2^*}\mathcal{D}_{\zeta_2^*}f_2 \|_p$$

$$= \| \mathcal{S}_{\beta_1^*}\mathcal{D}_{\zeta_1^*}f_1 - \mathcal{T}_{\tau_2^* - \tau_1^*}\mathcal{R}_{\alpha_2^* - \alpha_1^*}\mathcal{S}_{\beta_2^*}\mathcal{D}_{\zeta_2^*}f_2 \|_p \qquad (4.3.69)$$

it is easily verified, using Eqs. 4.3.56a and 4.3.56b, that there exist LCM $(|k_{11} - k_{12}|, |k_{21} - k_{22}|)$ distinct normalized pairs of contour representations for the contours $\gamma_1$ and $\gamma_2$, where LCM( ) stands for *least common multiple*. As a consequence, the computational complexity of $d_*^{(p)}[f_1, f_2]$, when optimized over the candidate normalizations, is $O(N \cdot \text{LCM}(|k_{11} - k_{12}|, |k_{21} - k_{22}|))$. This complexity can still be considerably lower than the complexities reported in Tables 4.5 and 4.7 for $d^{(p)}[f_1, f_2]$ and $d^{(2)}[f_1, f_2]$ respectively. Optimizing dissimilarity measurement over a limited set of candidate contour normalizations may be an acceptable compromise that offers a reduced computational complexity and that limits the danger of arriving at inappropriate normalizations.

In the same way as we defined $d_*^{(p)}(f_1, f_2)$ in Eq. 4.3.68, a normalized mirror-dissimilarity measure $\tilde{d}_*^{(p)}(f_1, f_2)$ can be defined (cf. Definition 4.4). The discrete normalized version $\tilde{d}_*^{(p)}[f_1, f_2]$ has the same properties with regard to computational complexity as $d_*^{(2)}[f_1, f_2]$. Along the same lines normalized (mirror-)dissimilarity measures $d_*^{(p)}(\hat{f}_1, \hat{f}_2)$ and $\tilde{d}_*^{(p)}(\hat{f}_1, \hat{f}_2)$, based on the Fourier representation $\hat{f}$ of contours, can be defined. If $N$ Fourier coefficients are used in the discrete normalized (mirror-)dissimilarity measures $d_*^{(p)}[\hat{f}_1, \hat{f}_2]$ and $\tilde{d}_*^{(p)}[\hat{f}_1, \hat{f}_2]$, then their computational complexity is $O(N)$, if we disregard the computational complexity of the normalization procedure itself and that of the computation of the Fourier coefficients. Likewise, if we minimize $d_*^{(p)}[\hat{f}_1, \hat{f}_2]$ and $\tilde{d}_*^{(p)}[\hat{f}_1, \hat{f}_2]$ over multiple candidate normalizations, generated by Eqs. 4.3.56a and 4.3.56b, then the computational complexity of these measures is $O(N \cdot \text{LCM}(|k_{11} - k_{12}|, |k_{21} - k_{22}|))$.

As we already observed, the normalization procedures for translation, size, orientation and starting point, discussed in the previous sections, can be distinguished into two main classes:

- those based on the gravitational moments $m_{pq}$ of the region enclosed by the contour,
- those based on the Fourier coefficients generated by the position function $z$ of the contour.

From a theoretical point of view normalization procedures based on Fourier coefficients are somewhat better adapted to the mathematical form of the (mirror-)dissimilarity measures defined in Section 4.2 (cf. for example Section 4.3.1 concerning the optimality of $\zeta_c^* = -\hat{z}(0)$ in $d^{(2)}(z_1, z_2)$). However, there is no indication that normalization procedures based on moments $m_{pq}$ perform less well than those based on Fourier coefficients (apart from the absence of the concept of a starting point on a contour in moment-based methods). Since the moments in $m_{pq}$ are defined in Definition 4.9 as region integrals, the insensitivity for noise of the moment-based normalization methods may be expected to be at least as good as that of methods based on Fourier coefficients generated by the position function $z$.

A drawback, at first sight, of the normalization methods based on moments is that the computation of moments of gravity requires integration over a region in the plane, or, in the context of the discrete geometry of digital pictures, summation over a region. However, through the application of Green's theorem in the plane, by means of which the surface integral becomes a contour integral, the computational complexity of moments can be highly reduced (cf. e.g. Tang [1982], Cyganski and Orr [1985]). In Appendix B we show that, if the contour can be approximated by a polygon with $N$ vertices, $m_{pq}$ can be computed exactly with a computational complexity of $O(pqN)$. Recently Bamieh and de Figueiredo [1986] published a similar result, by means of which the computational complexity can even be reduced to $O(\min(p, q) \cdot N)$.

## 4.4 Theoretical and experimental evaluation of the behavior of dissimilarity measures

In Section 4.4.1 we derive some further theoretical properties of the dissimilarity measures defined in Section 4.2. In particular we establish a number of mathematical relations between various dissimilarity measures. We also relate a special case of the dissimilarity measures, based on the curvature function $K$, to the bending energy necessary to deform one contour, considered as a thin elastic beam, into an other contour.

In Section 4.4.2 we discuss some experimental results, obtained with a number of dissimilarity measures. Such analyses are needed in order to enable the choice of an appropriate dissimilarity measure, or of a combination of dissimilarity measures, in a particular application. First we evaluate through clustering techniques the relative behavior of the dissimilarity measures used in the experiments. Thus we obtain insight into which dissimilarity measures perform similarly and which perform differently. Next we use clustering techniques to analyze the performance of individual dissimilarity measures. We evaluate which aspects of geometric dissimilarity are emphasized by a particular dissimilarity measure.

Since each mirror-dissimilarity measure constitutes a special case of a corresponding dissimilarity measure (cf. Eq. 4.2.29), we limit the discussions to dissimilarity measures.

### *4.4.1 Further analysis of theoretical properties of dissimilarity measures*

First we consider the effect of the index $p$ upon the dissimilarity measures based on contour representations. The lower the value of $p$ the more globally the differences between a pair of contour representations are measured. For $p = 1$ the average deviation between contour representations is measured and for $p = 2$ the square root of the mean square deviation (cf. Table 4.1). The larger the value of $p$ the more sensitive a dissimilarity measure becomes for local deviations between contour representations. In the limit, for $p = \infty$, the measure expresses the maximum deviation between a pair of contour representations.

It is a well-known fact from mathematical analysis (cf. e.g. Edwards [1979], p. 28) that, if $f \in \mathbf{L}^p$, then

$$\|f\|_q \leqslant \|f\|_p, \qquad 0 < q < p \leqslant \infty. \tag{4.4.1}$$

Since dissimilarity measures have been defined in Section 4.2 as the norm of the difference of a pair of contour representations, we might expect this inequality also to be valid for dissimilarity measures. However, in Section 4.3 we proposed $\mathcal{S}_{\beta^*} = \|\mathcal{D}_\zeta f\|_p^{-1}$ as a scale normalization operator for a contour representation $f$ in $d^{(p)}(f_1, f_2)$, when $f$ stands for any of the scale variant contour representations $z$, $\dot{z}$, $\ddot{z}$ and $K$.

As a result, the inequality in Eq. 4.4.1 is only valid for a dissimilarity measure based on $f$ if $\|\mathscr{D}_\zeta f\|_p$ is independent of $p$ or if $f$ is scale invariant. This is true for the tangent function $\dot{z}$, since $\|\dot{z}\|_p = L/2\pi$, $\forall p \geqslant 1$ (cf. Section 4.3), and for the scale invariant periodic cumulative angular function $\psi$. Thus we obtain the inequalities

$$d^{(q)}(\dot{z}_1, \dot{z}_2) \leqslant d^{(p)}(\dot{z}_1, \dot{z}_2) \tag{4.4.2a}$$

$$d^{(q)}(\psi_1, \psi_2) \leqslant d^{(p)}(\psi_1, \psi_2) \tag{4.4.2b}$$

for $1 \leqslant p < q \leqslant \infty$.

For dissimilarity measures based on $z$, $\ddot{z}$ and $K$ similar inequalities are not valid.

The interpretation of dissimilarity measurement based on Fourier representations of contours is different from that of the measurement based on the contour representations themselves. Fourier coefficients have been defined as contour averages (cf. Definition 3.1). Therefore each Fourier coefficient expresses a global feature of a contour: local shape information is not present in Fourier representations. As a consequence, in the dissimilarity measures based on Fourier representations, the deviation between an individual pair of Fourier coefficients with corresponding indices still expresses a global difference between the pair of contours under consideration. This constitutes a fundamental difference in the way dissimilarity measures based on contour representations and those based on Fourier representations, operate.

However, the effect of varying the value of the index $p$ in dissimilarity measures based on Fourier representations is fairly similar to that in dissimilarity measures based on contour representations. Also in the former measures we observe that the lower the value of $p$ the more globally the differences between a pair of Fourier representations are measured. For $p = 1$ the total deviation between a pair of Fourier representations is measured and for $p = 2$ the square root of the total squared deviation. The larger the value of $p$ the more sensitive a dissimilarity measure becomes for deviations between individual Fourier coefficients in a pair of Fourier representations. In the limit, for $p = \infty$, the measure expresses the largest deviation between any individual pair of corresponding Fourier coefficients.

In analogy with Eq. 4.4.1, we have for norms of a sequence $\hat{f}$ the inequality (cf. Edwards [1979], p. 29)

$$\|\hat{f}\|_\infty \leq \|\hat{f}\|_p \leq \|\hat{f}\|_q, \qquad 0 < q < p < \infty. \tag{4.4.3}$$

In Section 4.3 we proposed $\mathcal{S}_{\beta^*} = \|\mathcal{D}_{\zeta^*}\hat{f}\|_p^{-1}$ as a scale normalization operator for a Fourier representation $\hat{f}$ in $d^{(p)}(\hat{f}_1, \hat{f}_2)$, when $\hat{f}$ stands for any of the scale variant Fourier representations $\hat{z}$, $\hat{\dot{z}}$, $\hat{\ddot{z}}$ and $\hat{K}$.

Therefore the inequality in Eq. 4.4.3 applies again only to those dissimilarity measures, based on Fourier representations, for which $\|\mathcal{D}_{\zeta^*}\hat{f}\|_p$ is independent of the value of $p$ or for which $\hat{f}$ is scale invariant. It turns out that none of the scale variant Fourier representations satisfies this condition. Thus Eq. 4.4.3 applies only to $d^{(p)}(\hat{\psi}_1, \hat{\psi}_2)$, which is based on the scale invariant Fourier representation $\hat{\psi}$

$$d^{(p)}(\hat{\psi}_1, \hat{\psi}_2) \leq d^{(q)}(\hat{\psi}_1, \hat{\psi}_2), \qquad \text{for } 1 \leq p < q \leq \infty. \tag{4.4.4}$$

It is also possible to relate norms of periodic functions to norms of Fourier series. In Eq. 4.2.42 we already noted Parseval's identity

$$\|f\|_2 = \|\hat{f}\|_2, \qquad \forall f \in \mathbf{L}^2(2\pi) \tag{4.4.5}$$

which causes a dissimilarity measure $d^{(2)}(f_1, f_2)$, to be isometric with $d^{(2)}(\hat{f}_1, \hat{f}_2)$ (cf. Eq. 4.2.41). For more general values of the index $p$, inequalities exist, known as the Hausdorff-Young inequalities (cf. e.g. Edwards [1982], pp. 153-157).

If $f \in \mathbf{L}^p(2\pi)$, then

$$\|\hat{f}\|_{p'} \leq \|f\|_p, \qquad 1 \leq p \leq 2 \tag{4.4.6a}$$

and, if $\hat{f} \in \ell^p(\mathbb{Z})$, then

$$\|f\|_{p'} \leq \|\hat{f}\|_p, \qquad 1 \leq p \leq 2 \tag{4.4.6b}$$

where $1/p + 1/p' = 1$.

Since we proposed in Section 4.3 $\mathcal{S}_{\beta^*} = \|\mathcal{D}_{\zeta^*}f\|_p^{-1}$ as a scale normalization operator for the scale variant contour representation $f$ in $d^{(p)}(f_1, f_2)$ and $\mathcal{S}_{\beta^*} = \|\mathcal{D}_{\zeta^*}\hat{f}\|_p^{-1}$ for $\hat{f}$ in $d^{(p)}(\hat{f}_1, \hat{f}_2)$, the validity of the inequalities in Eqs. 4.4.6a and 4.4.6b does not extend to these dissimilarity measures. For other choices of scale normalization operators the inequalities in Eqs. 4.4.6a and 4.4.6b may apply to dissimilarity measures. For dissimilarity measures based on the scale invariant contour representation $\psi$ and Fourier representation $\hat{\psi}$ the following inequalities are valid

If $\psi_1, \psi_2 \in \mathbf{L}^p(2\pi)$, then

$$d^{(p')}(\hat{\psi}_1, \hat{\psi}_2) \leqslant d^{(p)}(\psi_1, \psi_2), \qquad 1 \leqslant p \leqslant 2 \tag{4.4.7}$$

and, if $\hat{\psi}_1, \hat{\psi}_2 \in \ell^p(\mathbb{Z})$, then

$$d^{(p')}(\psi_1, \psi_2) \leqslant d^{(p)}(\hat{\psi}_1, \hat{\psi}_2), \qquad 1 \leqslant p \leqslant 2. \tag{4.4.8}$$

The analysis of the effects of varying the value of the index $p$ in dissimilarity measures just given and the inequalities mentioned give some insight into what is measured by a dissimilarity measure and can be helpful to guide the choice for a particular dissimilarity measure in a given application.

Another means of analyzing the relative behavior of dissimilarity measures may be provided by inclusion relations that exist for $\mathbf{L}^p$-spaces of functions and their derivatives (cf. Beckenbach and Bellman [1971], Ch. 5). In some cases even explicit inequalities have been derived between the norm of a function and the norm of a derivative of that function: e.g. Wirtinger's inequality (cf. Hardy, Littlewood and Pólya [1952], Beckenbach and Bellman [1971], pp. 177-178), relating $||f||_2$ and $||\dot{f}||_2$ of a zero-mean function $f$, and the Northcott-Bellman inequalities (cf. Beckenbach and Bellman [1971], p. 182), relating $||f||_p$ and $||f^{(k)}||_p$ for $p = \infty$ and for arbitrary values of $p > 0$, where $f^{(k)}$ is the $k$-th derivative of a zero-mean function $f$.

Since these inequalities relate only norms of functions with norms of their derivatives we may only consider the representations $z$, $\dot{z}$ and $\ddot{z}$. Unfortunately these inequalities do not directly apply to the dissimilarity measures based on these representations as a result of the scale normalization coefficients proposed in Section 4.3. For different choices of scale normalization operators, that are not dependent upon the particular contour representation to be scale-normalized itself, these inequalities may be valid though.

More powerful guidance for the choice of an appropriate dissimilarity measure may be obtained through an interpretation of the role of each of the contour representations in dissimilarity measurement. The general properties of the dissimilarity measures based on contour representations are that they are functions of pointwise absolute differences between contour representations. We have just seen above that the value of the index $p$ in these measures determines how locally or globally these absolute differences are expressed in the measure. The absolute differences are measured after the appropriate translation and scale normalizations and contour orientation and starting point optimization or normalization, if the contour representation is variant for these parameters.

Dissimilarity measures $d^{(p)}(z_1, z_2)$ are functions of pointwise distances between a contour pair. As a consequence this dissimilarity measure tends to operate rather coarsely. Relatively small protrusions or intrusions in contours, which may be very important for a proper distinction between objects, will in general have little effect on the value of $d^{(p)}(z_1, z_2)$. Dissimilarity measures $d^{(p)}(\dot{z}_1, \dot{z}_2)$ are functions of pointwise absolute differences between complex tangents along a pair of contours.

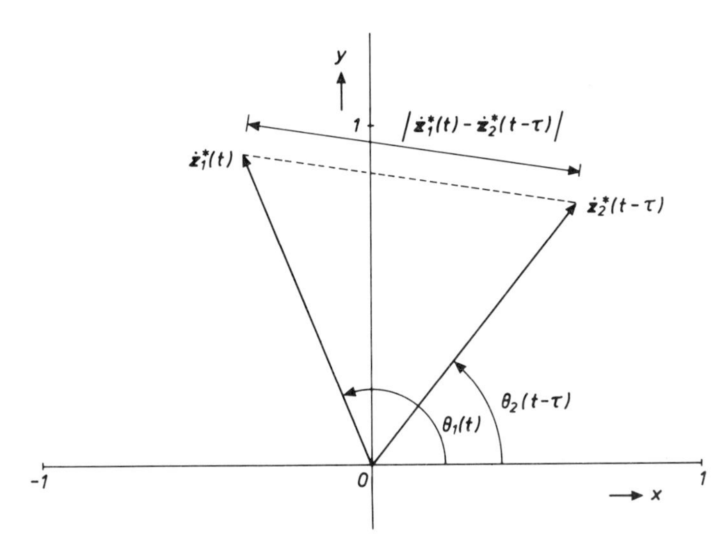


**Figure 4.12.** Vector representation of the tangent functions of two contours as an illustration of dissimilarity measurement by means of $d^{(p)}(\dot{z}_1, \dot{z}_2)$. Apart from the value of the index $p$, the local contribution to $d^{(p)}(\dot{z}_1, \dot{z}_2)$ is determined by the length of the difference vector:

$$\left|\dot{z}^*(t) - \dot{z}^*(t-\tau)\right| = 2\left|\sin\left(\tfrac{1}{2}\{\theta_1(t) - \theta_2(t-\tau)\}\right)\right|.$$

Representing the tangents as two-dimensional vectors is very illustrative in this context (cf. Figure 4.12). The scale normalization proposed in Section 4.3 leads to $|\dot{z}^*(t)| = 1$, everywhere where $\dot{z}(t)$ exists, for every contour. Therefore $d^{(p)}(\dot{z}_1, \dot{z}_2)$ is really a function of pointwise differences in tangent direction between a pair of contours (cf. Figure 4.12).

Dissimilarity measurement by $d^{(p)}(\dot{z}_1, \dot{z}_2)$ is much less coarse than by $d^{(p)}(z_1, z_2)$. Small protrusions or intrusions in a contour, for example, can have a significant effect on the value of $d^{(p)}(\dot{z}_1, \dot{z}_2)$, especially for larger values of $p$. On the one hand this reveals an increased noise sensitivity of $d^{(p)}(\dot{z}_1, \dot{z}_2)$ with respect to $d^{(p)}(z_1, z_2)$, but on the other hand it shows the sensitivity of $d^{(p)}(\dot{z}_1, \dot{z}_2)$ for fine shape detail.

Dissimilarity measures $d^{(p)}(\dot{z}_1, \dot{z}_2)$ are closely related to $d^{(p)}(\psi_1, \psi_2)$. Through Eq. 2.2.32 and Eqs. 4.2.11, 4.2.12 and 4.2.14 we can rewrite $d^{(p)}(\psi_1, \psi_2)$, as defined in Eq. 4.2.17, as

$$d^{(p)}(\psi_1, \psi_2) = \min_{\tau} \left[ \frac{1}{2\pi} \int_{2\pi} |\varphi_1(t) - \varphi_2(t-\tau) + \lambda(\psi_1) - \lambda(\psi_2) - \tau|^p \mathrm{d}t \right]^{1/p}. \tag{4.4.9}$$

For a given pair of contours $\gamma_1$ and $\gamma_2$, represented by $\psi_1$ and $\psi_2$ respectively, $\lambda(\psi_1)$ and $\lambda(\psi_2)$ in Eq. 4.4.9 are constants. Furthermore, the value of $\varphi_1(t) - \varphi_2(t - \tau)$ indicates the local difference in tangent direction at $t$ on $\gamma_1$ and $t - \tau$ on $\gamma_2$. If we apply Eq. 2.2.30 to Eq. 4.4.9 and compare the result with Figure 4.12, then the resemblance of $d^{(p)}(\dot{z}_1, \dot{z}_2)$ and $d^{(p)}(\psi_1, \psi_2)$ becomes clear immediately.

We can derive an inequality between $d^{(p)}(\dot{z}_1, \dot{z}_2)$ and $d^{(p)}(\psi_1, \psi_2)$. For the proof of this inequality we need the following lemma.

**Lemma 4.1.**

$$|e^{it} - e^{is}| \leqslant |t - s|. \tag{4.4.10}$$

*Proof*

$$|e^{it} - e^{is}| = \left| i \int_s^t e^{iu} \mathrm{d}u \right| \leqslant \left| \int_s^t \mathrm{d}u \right| = |t - s|.$$

□

**Theorem 4.5.**
If $\psi_1, \psi_2 \in \mathbf{L}^p(2\pi)$, then

$$d^{(p)}(\dot{z}_1, \dot{z}_2) \leqslant d^{(p)}(\psi_1, \psi_2). \tag{4.4.11}$$

*Proof*

Through Eqs. 2.2.9, 2.2.30, 2.2.32 and Eq. 4.2.12 we can relate $\dot{z}^*(t)$ and $\psi^*(t)$ as

$$\dot{z}^*(t) = \left|\dot{z}^*(t)\right| e^{i\{\psi^*(t) - \lambda(\psi) + t + \theta(0)\}}, \tag{4.4.12}$$

where we proposed in Section 4.3 for $\dot{z}^*(t)$

$$\dot{z}^*(t) = \mathcal{S}_{\beta^*(\dot{z})}\dot{z}(t) = \frac{\dot{z}(t)}{\|\dot{z}\|_p}. \tag{4.4.13}$$

As a result of Eq. 2.2.5 and $\|\dot{z}\|_p = L/2\pi, \forall p \geqslant 1$ (cf. Section 4.3), we obtain

$$\left|\dot{z}^*(t)\right| = 1, \qquad \forall t \in [0, 2\pi]. \tag{4.4.14}$$

The dissimilarity measure $d^{(p)}(\dot{z}_1, \dot{z}_2)$ has been defined as (cf. Table 4.1)

$$\begin{aligned} d^{(p)}(\dot{z}_1, \dot{z}_2) \\ &= \min_{\alpha,\tau} \|\dot{z}_1^* - \mathcal{T}_\tau \mathcal{R}_\alpha \dot{z}_2^*\|_p \\ &= \min_{\alpha,\tau} \left[\frac{1}{2\pi}\int_{2\pi} \left|\dot{z}_1^*(t) - e^{i\alpha}\dot{z}_2^*(t-\tau)\right|^p dt\right]^{1/p}. \end{aligned} \tag{4.4.15}$$

The substitution of Eqs. 4.4.12 and 4.4.14 into Eq. 4.4.15 and the application of Lemma 4.1, give

$$d^{(p)}(\dot{z}_1, \dot{z}_2) = \min_{\alpha,\tau} \left[ \frac{1}{2\pi} \int_{2\pi} \left| e^{i\{\psi_1^*(t) - \lambda(\psi_1) + t + \theta_1(0)\}} \right.\right.$$

$$\left.\left. - e^{i\alpha} e^{i\{\psi_2^*(t-\tau) - \lambda(\psi_2) + t - \tau + \theta_2(0)\}} \right|^p dt \right]^{1/p}$$

$$\leqslant \min_{\alpha,\tau} \left[ \frac{1}{2\pi} \int_{2\pi} \left| \psi_1^*(t) - \lambda(\psi_1) + \theta_1(0) \right.\right.$$

$$\left.\left. - \alpha - \psi_2^*(t-\tau) + \lambda(\psi_2) + \tau - \theta_2(0) \right|^p dt \right]^{1/p} . \qquad (4.4.16)$$

Through the application of Minkowski's inequality (cf. Appendix A) to the latter expression in Eq. 4.4.16, we obtain, for $p > 1$

$$\min_{\alpha,\tau} \left[ \frac{1}{2\pi} \int_{2\pi} \left| \psi_1^*(t) - \lambda(\psi_1) + \theta_1(0) \right.\right.$$

$$\left.\left. - \alpha - \psi_2^*(t-\tau) + \lambda(\psi_2) + \tau - \theta_2(0) \right|^p dt \right]^{1/p}$$

$$\leqslant \min_{\tau} \left\{ \left[ \frac{1}{2\pi} \int_{2\pi} \left| \psi_1^*(t) - \psi_2^*(t-\tau) \right|^p dt \right]^{1/p} \right.$$

$$\left. + \min_{\alpha} \left| -\lambda(\psi_1) + \lambda(\psi_2) + \theta_1(0) - \theta_2(0) - \alpha + \tau \right| \right\}$$

$$= \min_{\tau} \left[ \frac{1}{2\pi} \int_{2\pi} \left| \psi_1^*(t) - \psi_2^*(t-\tau) \right|^p dt \right]^{1/p} . \qquad (4.4.17)$$

The transition from the second to the third expression in Eq. 4.4.17 can be made because for any value of $\tau$ we can find a value for $\alpha$ such that $|-\lambda(\psi_1) + \lambda(\psi_2) + \theta_1(0) - \theta_2(0) - \alpha + \tau| = 0$. The third expression in Eq. 4.4.17 equals $d^{(p)}(\psi_1, \psi_2)$ (cf. Table 4.1). By combining Eqs. 4.4.16 and 4.4.17 the proof of this theorem is complete.

□

The proof of this theorem clearly depends upon the fact that $|\dot{z}^*(t)|$ is a contour-independent constant, which is a result of the choice of the scale normalization operator $\mathcal{S}_{\beta^*} = \|\dot{z}\|_p^{-1}$, that was proposed in Section 4.3. For other scale normalization operators this theorem may no longer be valid.

The dissimilarity measures $d^{(p)}(\ddot{z}_1, \ddot{z}_2)$ and $d^{(p)}(K_1, K_2)$ also show great resemblance. This becomes clear if we rewrite $d^{(p)}(\ddot{z}_1, \ddot{z}_2)$. First we examine the scale normalization operator $\mathcal{S}_{\beta^*(\ddot{z})}$, proposed in Section 4.3, more closely by using the relation in Eq. 2.2.25

$$
\begin{aligned}
\mathcal{S}_{\beta^*(\ddot{z})} &= \|\ddot{z}\|_p^{-1} \\
&= \left[\frac{1}{2\pi}\int_{2\pi}\left|\mathrm{i}\left(\frac{L}{2\pi}\right)^2 K(t)\mathrm{e}^{\mathrm{i}\theta(t)}\right|^p \mathrm{d}t\right]^{-1/p} \\
&= \left(\frac{2\pi}{L}\right)^2 \|K\|_p^{-1}.
\end{aligned}
\tag{4.4.18}
$$

Next we rewrite $d^{(p)}(\ddot{z}_1, \ddot{z}_2)$, by using Eqs. 2.2.25 and 4.4.18

$$
\begin{aligned}
d^{(p)}(\ddot{z}_1, \ddot{z}_2) &= \min_{\alpha,\tau}\left[\frac{1}{2\pi}\int_{2\pi}\left|\ddot{z}_1^*(t) - \mathrm{e}^{\mathrm{i}\alpha}\ddot{z}_2^*(t-\tau)\right|^p \mathrm{d}t\right]^{1/p} \\
&= \min_{\alpha,\tau}\left[\frac{1}{2\pi}\int_{2\pi}\Big|\|K_1\|_p^{-1}K_1(t)\mathrm{e}^{\mathrm{i}\theta_1(t)}\right. \\
&\qquad \left. - \mathrm{e}^{\mathrm{i}\alpha}\|K_2\|_p^{-1}K_2(t-\tau)\mathrm{e}^{\mathrm{i}\theta_2(t-\tau)}\Big|^p \mathrm{d}t\right]^{1/p}.
\end{aligned}
\tag{4.4.19}
$$

From Table 4.1 it is immediately clear that $d^{(p)}(K_1, K_2)$ is a function of pointwise differences in curvature between a pair of contours. From Eq. 4.4.19 we observe that $d^{(p)}(\ddot{z}_1, \ddot{z}_2)$ is a function of pointwise differences both in curvature and in tangent direction between a pair of contours. For the latter reason $d^{(p)}(\ddot{z}_1, \ddot{z}_2)$ also exhibits some resemblance with $d^{(p)}(\dot{z}_1, \dot{z}_2)$.

For the dissimilarity measure $d^{(2)}(K_1, K_2)$ we can also find a physical interpretation from elasticity theory. In Section 3.2 we already mentioned the concept of bending energy in relation with shape analysis. Over a number of years elastic energy or bending energy has been proposed by various authors as a feature to characterize the shape of a curve or contour (e.g. Freeman and Glass [1969], Young, Walker and Bowie [1974], Freeman [1979], Horn [1983]).

It is a well-known fact from elasticity theory (Den Hartog [1949], Landau and Lifschitz [1970]) that the elastic energy or bending energy per unit length $U$, stored in a thin elastic beam, is proportional to the squared curvature of the beam

$$U(s) = \frac{1}{2}E(s)I(s)K^2(s) \tag{4.4.20}$$

where:

$s$ – arc length parameter,
$E(s)$ – Young's modulus at $s$,
$I(s)$ – moment of inertia at $s$.

If we assume the cross-section of the beam to be circular, with constant diameter, then the moment of inertia is constant, i.e. $I(s) = I$. For homogeneous isotropic media Young's modulus is also a constant, i.e. $E(s) = E$. Further we assume the elastic beam to be very thin and its elastic properties to be such that Hooke's Law, which is the basis for Eq. 4.4.20, is valid over a wide range of curvatures. Then the average elastic energy per unit length $\langle U \rangle$ is proportional to $\|K\|_2^2$

$$\langle U \rangle = \frac{1}{2} \cdot \frac{EI}{L} \int_0^L K^2(s)\mathrm{d}s = \frac{1}{2} \cdot \frac{EI}{2\pi} \int_0^{2\pi} K^2(t)\mathrm{d}t = \frac{1}{2}EI\,\|K\|_2^2 \tag{4.4.21}$$

where $K(t)$ serves as a shorthand notation for $K(s(t))$.

We now consider two thin elastic beams of equal arc length. Landau and Lifschitz [1970], pp. 78-82, derive that the total bending energy $U_{tot}$, necessary to deform one beam, with curvature function $K_1$, into the other, with curvature function $K_2$, is

$$U_{tot}(K_1, K_2) = \frac{1}{2} EI \int_0^L \{K_1(s) - K_2(s)\}^2 ds. \tag{4.4.22}$$

Thus the average bending energy per unit length $\langle E \rangle$ for a pair of equal length curves is proportional to $\| K_1 - K_2 \|_2^2$

$$\langle U(K_1, K_2) \rangle$$

$$= \frac{1}{2} \cdot \frac{EI}{L} \int_0^L \{K_1(s) - K_2(s)\}^2 ds = \frac{1}{2} EI \| K_1 - K_2 \|_2^2. \tag{4.4.23}$$

If we use arc length as a scale normalization parameter for the curvature function, i.e.

$$\mathcal{S}_{\beta^*(K)} = |\hat{K}(0)|^{-1} = \frac{L}{2\pi} \tag{4.4.24}$$

thus normalizing curvature functions to those of contours with perimeter $2\pi$, then we find for $d^{(2)}(K_1, K_2)$

$$d^{(2)}(K_1, K_2) = \min_{\tau} \| K_1^* - \mathcal{T}_\tau K_2^* \|_2. \tag{4.4.25}$$

Comparing Eq. 4.4.23 and Eq. 4.4.25 we find that $d^{(2)}(K_1, K_2)$ is proportional to the square root of the average bending energy per unit of arc length, needed to deform the contour $\gamma_1^*$ with curvature function $K_1^*$ to the contour $\gamma_2^*$ with curvature function $K_2^*$, with optimization of the starting point of $\gamma_2^*$ with respect to that of $\gamma_1^*$

$$d^{(2)}(K_1, K_2) = \left(\frac{2}{EI}\right)^{1/2} \cdot \min_{\tau} \langle U(K_1^*, \mathcal{T}_\tau K_2^*) \rangle^{1/2}. \tag{4.4.26}$$

Apart from the theoretical analyses and interpretations of the behavior of dissimilarity measures, a practical example may provide more insight. Such an example will be described in the next section.

### *4.4.2 Experimental analysis of the behavior of dissimilarity measures*

Though the analysis of the theoretical properties of dissimilarity measures has shed some light on their relative behavior, further insight is desirable. To this end we have performed an experiment, which is described in the following.

In this experiment we use 18 different contours, $\{\gamma_1, \ldots, \gamma_{18}\}$. These contours have been depicted in Figure 4.13. The fact that these contours, apart from $\gamma_{10}$, the circle, all have **5**-fold rotational symmetry does not really constitute a limitation: in this example we have interpreted the performance of dissimilarity measures on the basis of $^1/_5$-th of each contour.

As for the smoothness of the contours in this example, there are two distinct groups: the subset $\{\gamma_1, \gamma_2, \gamma_3, \gamma_4, \gamma_5, \gamma_9, \gamma_{14}, \gamma_{18}\}$ of piecewise regular contours and the subset $\{\gamma_6, \gamma_7, \gamma_8, \gamma_{10}, \gamma_{11}, \gamma_{12}, \gamma_{13}, \gamma_{15}, \gamma_{16}, \gamma_{17}\}$ of regular contours.

Though we have defined $\ddot{z}$ and $K$ as distributions when they represent contours that belong to $\boldsymbol{\Gamma}_{ps}$ or to $\boldsymbol{\Gamma}_{pr}$ (cf. Section 2.2), these distributions cannot be used in dissimilarity measures with index $p > 1$. Therefore some smoothing is performed in the neighborhood of nonsmooth points ('corners') on contours. This smoothing is a natural phenomenon in the process of estimating contour representations from a finite number of contour samples.

The contours in this experiment are all known in analytic form: polygons, hypocycloids, circle, epicycloids and limaçons of Pascal (cf. e.g. Wieleitner [1908], Lawrence [1972]). The contour parameter in these analytic expressions does not necessarily correspond to (normalized) arc length. As a result, dissimilarity measurement on the basis of the analytic expressions that describe these contours is not feasible in view of the conventions introduced in Section 2.2. Therefore we perform dissimilarity measurement on the basis of the discrete versions of the dissimilarity measures, which were introduced in Section 4.2.4. To do this we need $N$ equidistant (in the sense of arc length) contour representation samples for every contour. In this experiment we used the following procedure to obtain these samples.

Step 1. Let $u$ be the parameter of the analytic form of the position function of a contour. In general, $u$ is not a (normalized) arc length parameter.

Using the analytic form, compute $M$ samples of the position function, taken uniformly in the parameter $u$. We take $M$ substantially larger than $N$. In this experiment $M = 4096$ and $N = 512$.

Step 2. We consider the $M$ position function samples, resulting from Step 1, as the vertices of a polygon. This polygon is resampled, taking $N$ equidistant samples along its perimeter. These samples are used as (approximately) equidistant position function samples $z[n]$.

Step 3. In order to estimate $N$ equidistant samples of $\dot{z}$, $\ddot{z}$, $\varphi$ or $K$ we determine at each position $z[n]$ a polynomial fit, in least squares sense. For each fit we use a window containing an odd number of position function samples, with $z[n]$ as the central sample. The resulting polynomial is differentiated once or twice, as is required for the contour representation to be estimated and is evaluated at the central sample.

Details of this procedure and efficient implementations as FIR filters are presented in Appendix C. Some resulting estimates of contour representations have been depicted in Figures 4.14a-f.

**Remarks.**
For the contours $\gamma_1$, $\gamma_2$, $\gamma_3$, $\gamma_4$, $\gamma_5$ and $\gamma_{10}$ the parameter $u$ in the analytic form of the position function can be related in a straightforward manner to arc length. Therefore for these contours Step 1 of our procedure has been discarded and the $N$ equidistant position function samples $\{z[n]\}$ have been computed directly from the analytic form.

In this experiment we are interested in the behavior of the dissimilarity measures as a function of the contour representation used and of the value of the index $p$. We wish to minimize the effects of using discrete dissimilarity measures based on a finite number of contour representation samples. Therefore we have chosen $N = 512$, which may be larger than what is necessary in most practical situations for a set of contours such as $\{\gamma_1, \ldots, \gamma_{18}\}$.

The piecewise polynomial fitting procedures, used for the estimation of contour representations in this experiment, may not be optimal.

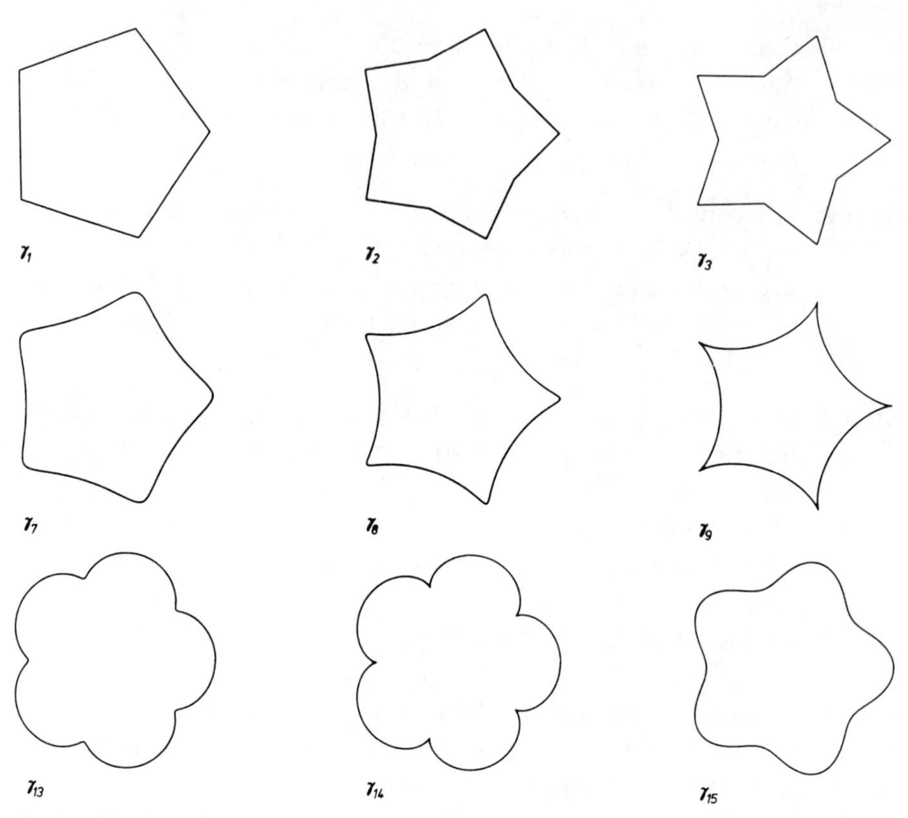


**Figure 4.13.** The set of contours $\{\gamma_1, \ldots, \gamma_{18}\}$ used in the dissimilarity measurement experiments in this section.

Many different options are available as estimation procedures. For example, (differentiating) low-pass filters (cf. McClellan, Parks and Rabiner [1973], Rabiner, McClellan and Parks [1975], Hamming [1977]), (differentiating) Gaussian filters (cf. Marr and Hildreth [1980], Hodson, Thayer and Franklin [1981], Marr [1982], Witkin [1983], Asada and Brady [1986], Babaud et al. [1986], Mokhtarian and Mackworth [1986], Yuille and Poggio [1986]), approximating splines (cf. De Boor [1978], Faux and Pratt [1979], Pavlidis [1982]).

The criterion for optimality of contour representation estimation in the context of dissimilarity measurement is quite obvious: that estimation procedure should be selected which may be expected to yield the smallest distortion, in the sense of the dissimilarity measure used in the estimate with respect to the original representation. An in-depth investigation of this important issue is outside the scope of this book.

□

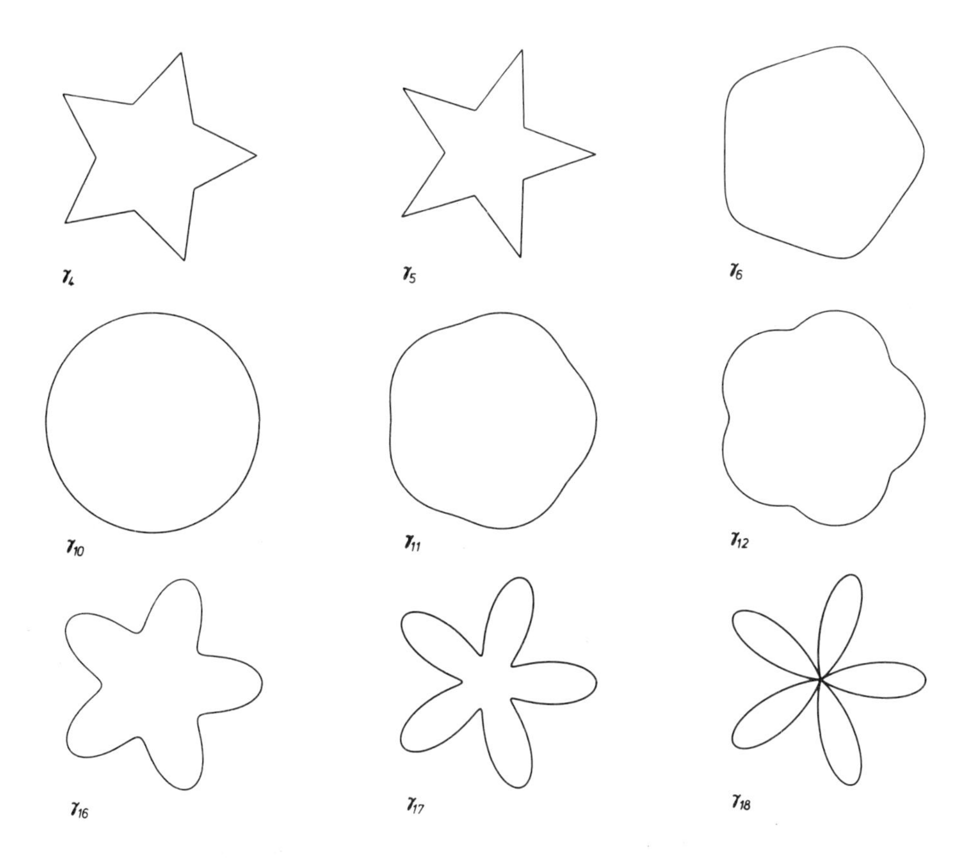


For each of the 5 contour representations of the 18 test figures we have computed pairwise discrete dissimilarities $d^{(p)}[\ ]$ for $p = 1, 2, 3, 5$ and $\infty$. Thus we obtained 25 symmetric $18 \times 18$ matrices of dissimilarity values. From these matrices we wish to learn two things:

1. What is the relative performance of the 25 dissimilarity measures? Are there dissimilarity measures among these 25 that can be grouped together because they give similar results?

2. What is the performance of individual dissimilarity measures. Which aspects of similarity or dissimilarity are emphasized or neglected by a given measure?

We begin by considering the first question in more detail. In order to compare the relative performance of the 25 dissimilarity measures we compute for each pair of dissimilarity matrices $\boldsymbol{D}_i$ and $\boldsymbol{D}_j$ a dissimilarity matrix correlation coefficient $R(\boldsymbol{D}_i, \boldsymbol{D}_j)$, which is defined as (cf.

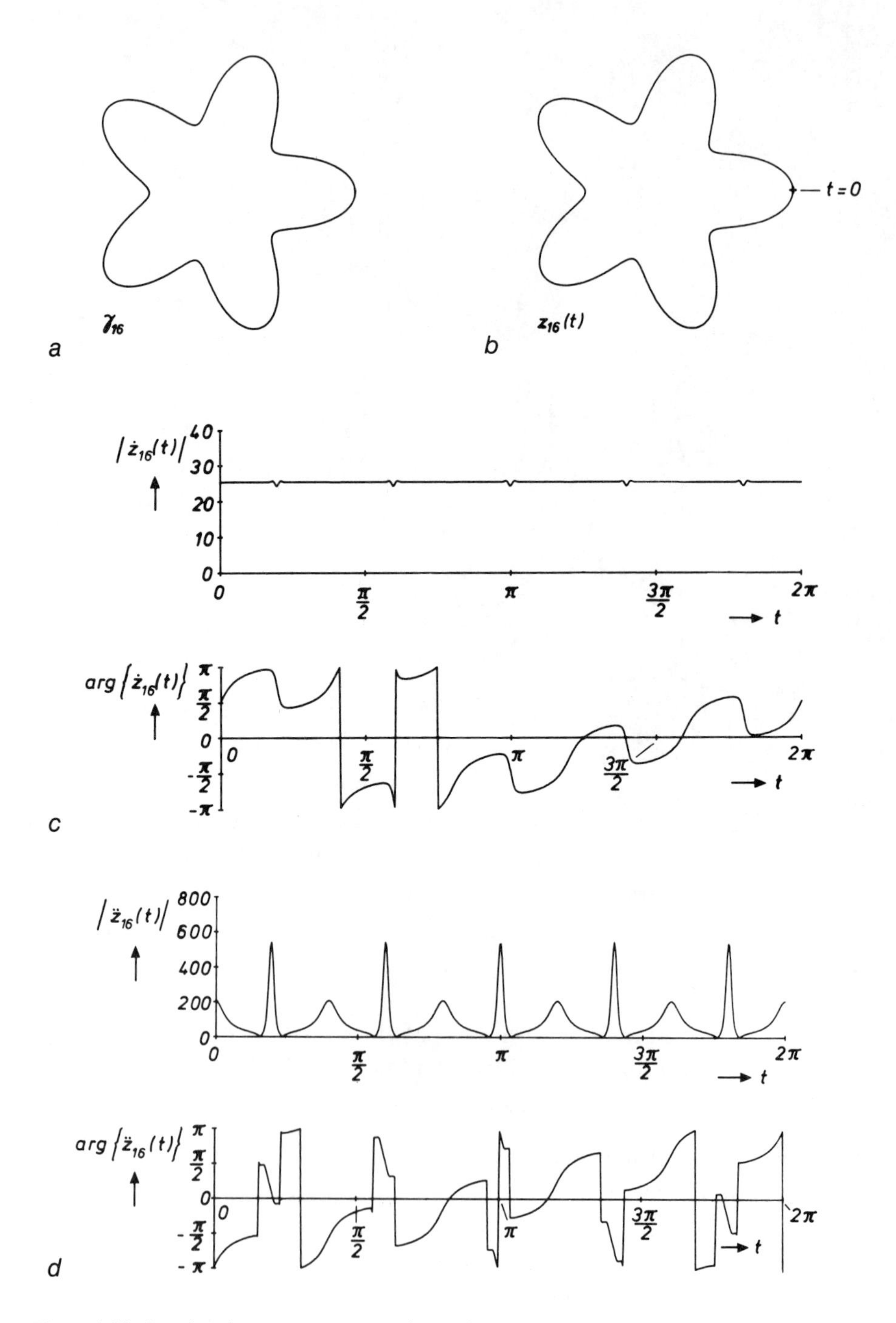


**Figure 4.14.** Results of contour representation estimation by means of piecewise polynomial fitting for the contour in (*a*), which corresponds to contour $\gamma_{16}$ in Figure 4.13. The order of the polynomials in this experiment is 4 and the fitting window contains 11 position function samples. In (*b*), (*c*), (*d*), (*e*) and (*f*) the resulting representations $z_{16}$, $\dot{z}_{16}$, $\ddot{z}_{16}$, $\psi_{16}$ and $K_{16}$ of $\gamma_{16}$ are displayed.

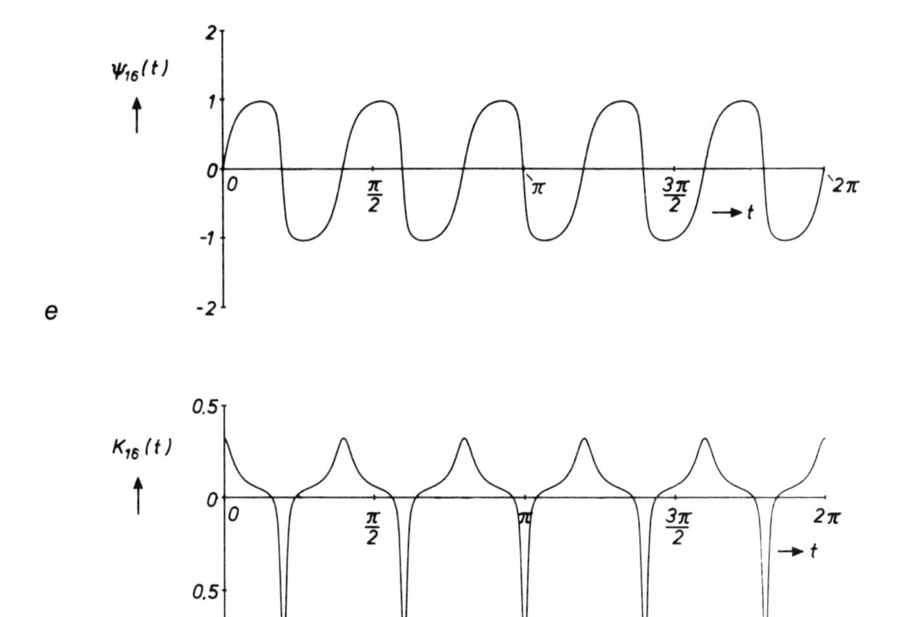


Sneath and Sokal [1973], p. 138 and pp. 279-280, or Dubes and Jain [1980], p. 159)

$R(\boldsymbol{D}_i, \boldsymbol{D}_j)$

$$= \frac{\frac{1}{P^2} \sum_{m=1}^{P} \sum_{n=1}^{P} D_i(m,n) \cdot D_j(m,n) - \langle \boldsymbol{D}_i \rangle \langle \boldsymbol{D}_j \rangle}{\left[\frac{1}{P^2} \sum_{m=1}^{P} \sum_{n=1}^{P} D_i^2(m,n) - \langle \boldsymbol{D}_i \rangle^2\right]^{1/2} \left[\frac{1}{P^2} \sum_{m=1}^{P} \sum_{n=1}^{P} D_j^2(m,n) - \langle \boldsymbol{D}_j \rangle^2\right]^{1/2}} \tag{4.4.27}$$

where $\langle \boldsymbol{D}_i \rangle$ and $\langle \boldsymbol{D}_j \rangle$ are matrix averages

$$\langle \boldsymbol{D}_k \rangle = \frac{1}{P^2} \sum_{m=1}^{P} \sum_{n=1}^{P} D_k(m,n). \tag{4.4.28}$$

In our example the number of test figures $P = 18$. Since the dissimilarity matrices $\boldsymbol{D}_k$ are symmetric and all elements on their main diagonal are equal to zero, the summations in Eq. 4.4.28 can be reduced to

$$\langle \boldsymbol{D}_k \rangle = \frac{2}{P^2} \sum_{m=2}^{P} \sum_{n=1}^{m-1} D_k(m,n). \tag{4.4.29}$$

The indices $i$ and $j$ of the dissimilarity matrices $\boldsymbol{D}_i$ and $\boldsymbol{D}_j$ in Eq. 4.4.27 are in the range $\{1, \ldots, 25\}$. The correspondence between these indices and the dissimilarity measures is indicated in Table 4.10.

The computation of $R(\boldsymbol{D}_i, \boldsymbol{D}_j)$ for each pair of dissimilarity matrices yields a $25 \times 25$ matrix $\boldsymbol{R}$ of matrix correlation coefficients. Note from Eq. 4.4.27 that $\boldsymbol{R}$ is also a symmetric matrix and that all elements on its main diagonal are equal to one. Though in theory the entries $R(\boldsymbol{D}_i, \boldsymbol{D}_j)$ in $\boldsymbol{R}$ are in the range $[-1, 1]$, we found in practice only positive values, as might be expected when comparing dissimilarity matrices generated through a single set of contours (the smallest entry we found was $R(\boldsymbol{D}_1, \boldsymbol{D}_{12}) = 0.055$). Therefore the conversion

$$D(\boldsymbol{D}_i, \boldsymbol{D}_j) = 1 - R(\boldsymbol{D}_i, \boldsymbol{D}_j) \tag{4.4.30}$$

defines a dissimilarity coefficient (cf. Section 4.1 and Späth [1980]). Thus we obtain a $25 \times 25$ matrix $\boldsymbol{D}$ of dissimilarity coefficients, resulting

**Table 4.10.** Correspondence between dissimilarity measures and the indexed dissimilarity matrices; e.g. dissimilarity matrix $\boldsymbol{D}_{13}$ has been generated by $d^{(3)}(\ddot{z}_m, \ddot{z}_n)$.

| Contour representation | Index | | | | |
|---|---|---|---|---|---|
| | $p = 1$ | $p = 2$ | $p = 3$ | $p = 5$ | $p = \infty$ |
| $z$ | $\boldsymbol{D}_1$ | $\boldsymbol{D}_2$ | $\boldsymbol{D}_3$ | $\boldsymbol{D}_4$ | $\boldsymbol{D}_5$ |
| $\dot{z}$ | $\boldsymbol{D}_6$ | $\boldsymbol{D}_7$ | $\boldsymbol{D}_8$ | $\boldsymbol{D}_9$ | $\boldsymbol{D}_{10}$ |
| $\ddot{z}$ | $\boldsymbol{D}_{11}$ | $\boldsymbol{D}_{12}$ | $\boldsymbol{D}_{13}$ | $\boldsymbol{D}_{14}$ | $\boldsymbol{D}_{15}$ |
| $\psi$ | $\boldsymbol{D}_{16}$ | $\boldsymbol{D}_{17}$ | $\boldsymbol{D}_{18}$ | $\boldsymbol{D}_{19}$ | $\boldsymbol{D}_{20}$ |
| $K$ | $\boldsymbol{D}_{21}$ | $\boldsymbol{D}_{22}$ | $\boldsymbol{D}_{23}$ | $\boldsymbol{D}_{24}$ | $\boldsymbol{D}_{25}$ |

from a pairwise comparison of the dissimilarity matrices $\boldsymbol{D}_i$, $i = 1, \ldots, 25$, using Eqs. 4.4.27-4.4.30. The matrix $\boldsymbol{D}$ is symmetric and has zeroes on its main diagonal. In order to determine similarities in the performance of the 25 dissimilarity measures in this experiment we performed cluster analysis on the matrix $\boldsymbol{D}$.

We used three clustering methods that all belong to the family of sequential, agglomerative, hierarchic, nonoverlapping clustering methods (SAHN methods) (cf. Sneath and Sokal [1973], Anderberg [1973], Dubes and Jain [1980], Späth [1980]):

1. Single Linkage Clustering,
2. Complete Linkage Clustering,
3. Average Linkage Clustering (UPGMA: unweighted pair groups using metric averages).

Since we lack the space to go into very much detail on these methods we refer the reader to the literature just mentioned.

The main reason for using three different clustering methods was to enable the detection of dependencies of clustering results upon the particular method. In our experiments, however, the characteristics of the results were quite similar, irrespective of the clustering method that was employed.

The result of a clustering is a *dendrogram* or a *phenogram* (cf. Sneath and Sokal [1973]), which is a graphical representation of the dissimilarity values at which items to be clustered or clusters of such items are merged into a new, larger cluster. A dendrogram reflects in what way the clustering procedure has embedded transitivity between the items to be clustered. In Figures 4.15a-c the dendrograms are shown that result from applying the three clustering methods to the matrix $\boldsymbol{D}$. The correspondence between the dissimilarity matrices $\boldsymbol{D}_i$, $i = 1, \ldots, 25$, and the dissimilarity measures that generated these matrices can be found in Table 4.10.

Before analyzing these dendrograms in detail we first investigate how accurately these dendrograms reflect the mutual relations between the performances of dissimilarity measures, as specified in the input matrix $\boldsymbol{D}$. To this end we generate for each dendrogram a cophenetic matrix $\boldsymbol{C}_k$ of dimensions $25 \times 25$. In the cophenetic matrix $\boldsymbol{C}_k$ each entry indicates at which dissimilarity value the corresponding pair of dissimilarity measures were merged into a single cluster by the $k$-th clustering method. Obviously the information contained in a dendrogram and that in a cophenetic matrix is identical. In order to determine the correspondence between $\boldsymbol{D}$ and $\boldsymbol{C}_k$, the cophenetic correlation coefficient (CPCC) (cf. Sneath and Sokal [1973], Dubes and Jain [1980]) is computed. The CPCC has the same form as the dissimilarity matrix correlation coefficient, as defined in Eq. 4.4.27. The results obtained in our experiment were:

$$\text{CPCC}(\boldsymbol{D}, \boldsymbol{C}_1) = 0.887 \quad \text{(Single Linkage)},$$

$$\text{CPCC}(\boldsymbol{D}, \boldsymbol{C}_2) = 0.905 \quad \text{(Complete Linkage)},$$

$$\text{CPCC}(\boldsymbol{D}, \boldsymbol{C}_3) = 0.911 \quad \text{(Average Linkage)}.$$

These values are sufficiently high in all three cases to be confident that the cophenetic matrices $\boldsymbol{C}_k$ and, equivalently, the corresponding dendrograms properly reflect the mutual relations in the matrix $\boldsymbol{D}$.

Analyzing the three dendrograms more closely, we observe that the clustering structure in each case is virtually the same.

First we note that the dissimilarity measures based on $z$ ($\boldsymbol{D}_1$-$\boldsymbol{D}_5$, cf. Table 4.10) all cluster together at low dissimilarity values and that in each of the three cases $d^{(\infty)}[z_m, z_n]$ ($\boldsymbol{D}_5$) is the last measure to join the cluster. This means that the dissimilarity measures based on $z$, for varying values of the index $p$, all measure similar aspects of geometric dissimilarity; only the measure $d^{(\infty)}[z_m, z_n]$ exhibits a slightly different character.

The dendrograms in Figures 4.15a-c also show that, with the exception of $d^{(\infty)}[\dot{z}_m, \dot{z}_n]$ and $d^{(\infty)}[\psi_m, \psi_n]$ ($\boldsymbol{D}_{10}$ and $\boldsymbol{D}_{20}$), the dissimilarity measures based on $\dot{z}$ ($\boldsymbol{D}_6$-$\boldsymbol{D}_9$) and on $\psi$ ($\boldsymbol{D}_{16}$-$\boldsymbol{D}_{19}$) are merged into a single cluster at low dissimilarity levels. Looking at this cluster in greater detail we we note that $d^{(1)}[\dot{z}_m, \dot{z}_n]$ and $d^{(1)}[\psi_m, \psi_n]$ ($\boldsymbol{D}_6$ and $\boldsymbol{D}_{16}$) and $d^{(2)}[\dot{z}_m, \dot{z}_n]$ and $d^{(2)}[\psi_m, \psi_n]$ ($\boldsymbol{D}_7$ and $\boldsymbol{D}_{17}$) behave almost identically. In view of the lower computational complexity of the dissimilarity measures based on $\psi$ for arbitrary values of the index $p$ (cf. Table 4.5), we draw the conclusion that the contour representation $\psi$ is to be preferred over $\dot{z}$ with the set of contours used in this experiment. Though the performance of $d^{(\infty)}[\dot{z}_m, \dot{z}_n]$ and $d^{(\infty)}[\psi_m, \psi_n]$ ($\boldsymbol{D}_{10}$ and $\boldsymbol{D}_{20}$) is similar, the behavior of these dissimilarity measures differs considerably from that of the measures based on $z$ ($\boldsymbol{D}_1$-$\boldsymbol{D}_5$) and from that of the other measures

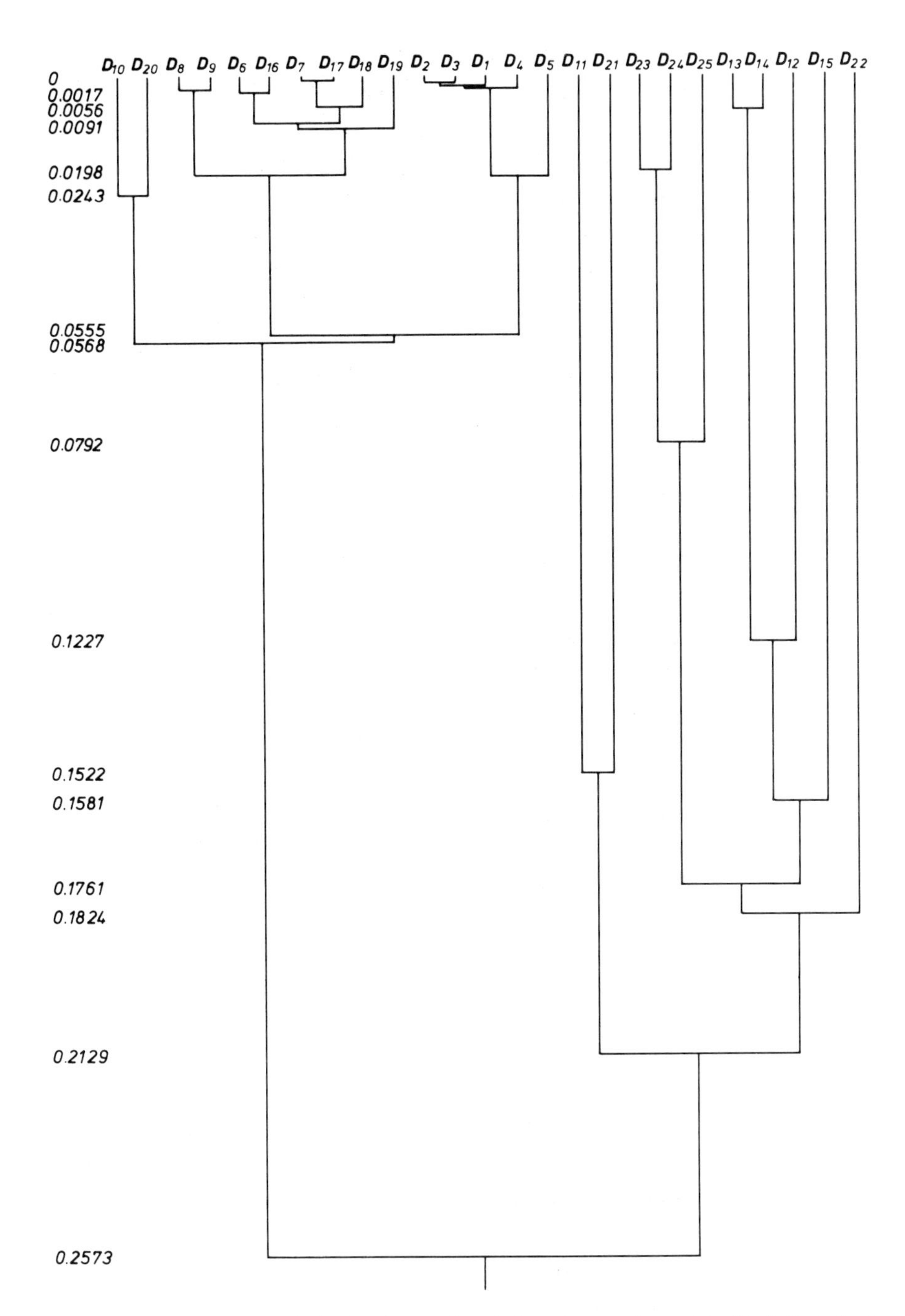


**Figure 4.15a.** Dendrogram displaying the results of clustering on the matrix $\boldsymbol{D}$ of dissimilarity coefficients, resulting from a pairwise comparison of the dissimilarity matrices $\boldsymbol{D}_k$, $k = 1, \ldots, 25$, by means of Eqs. 4.4.27-4.4.30. The dendrogram in this figure is the result of the Single Linkage Clustering Scheme. The numbers on the left of the dendrogram mark the dissmilarity values at which clusters merge.

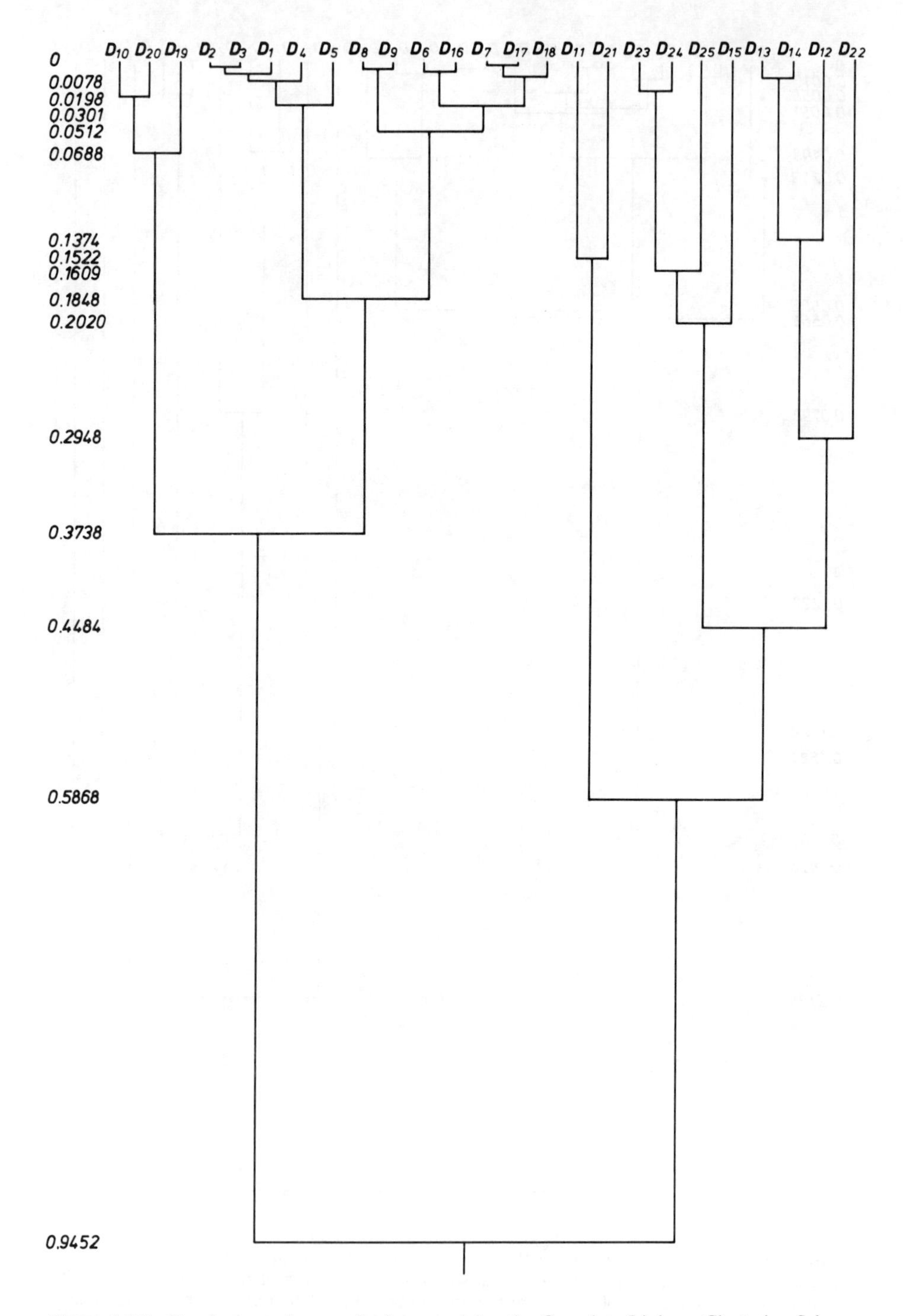


**Figure 4.15b.** Dendrogram that results from applying the Complete Linkage Clustering Scheme to ***D***.

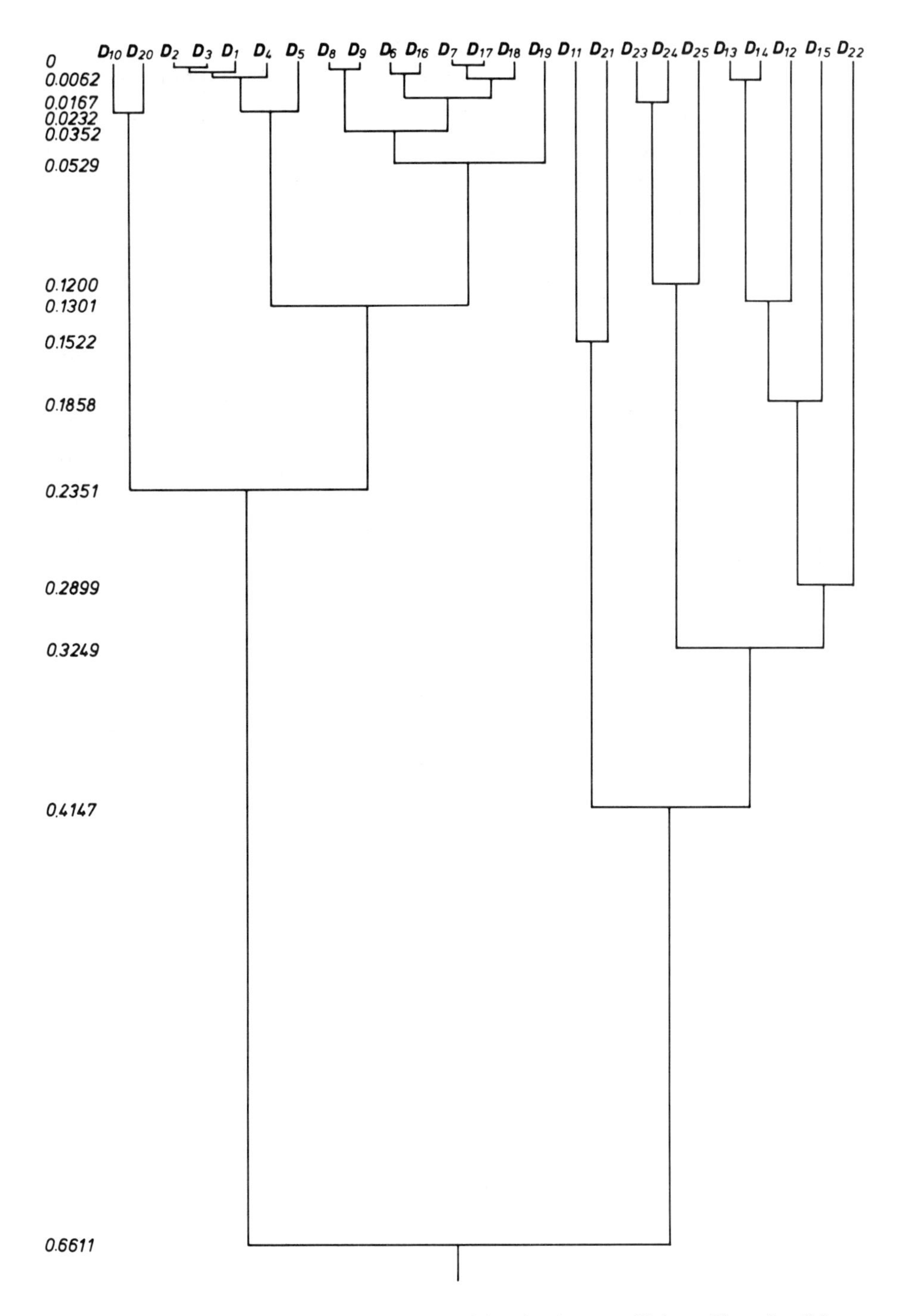


**Figure 4.15c**. Dendrogram that results from applying the Average Linkage Clustering Scheme (UPGMA) to $\boldsymbol{D}$.

based on $\dot{z}$ ($\boldsymbol{D}_6$-$\boldsymbol{D}_9$) and $\psi$ ($\boldsymbol{D}_{16}$-$\boldsymbol{D}_{19}$) (with the exception of $d^{(5)}[\psi_m, \psi_n]$ ($\boldsymbol{D}_{19}$) under the Complete Linkage Clustering Scheme, where this dissimilarity measure is closer to $d^{(\infty)}[\dot{z}_m, \dot{z}_n]$ and $d^{(\infty)}[\psi_m, \psi_n]$, cf. Figure 4.15b). This is not surprising since both $d^{(\infty)}[\dot{z}_m, \dot{z}_n]$ and $d^{(\infty)}[\psi_m, \psi_n]$ measure the maximum directional difference between a pair of contours, minimized over orientation and/or starting point, which is a local and noise-sensitive property. Though these measures are legitimate dissimilarity measures, their usefulness in practical applications is doubtful.

The dissimilarity measures based on $\ddot{z}$ ($\boldsymbol{D}_{11}$-$\boldsymbol{D}_{15}$) and $K$ ($\boldsymbol{D}_{21}$-$\boldsymbol{D}_{25}$) also cluster together in a single cluster, though at much higher dissimilarity values. The fact that the measures based on $\ddot{z}$ and $K$ are merged into a single cluster shows that they behave similarly, although this similarity is not as pronounced as with the measures based on $z$ or with those based on $\dot{z}$ and $\psi$. The large dispersion of this cluster is an indication of the high noise sensitivity of dissimilarity measures based on $\ddot{z}$ and $K$. Therefore we draw the conclusion that dissimilarity measures based on these contour representations can only be useful if a number of conditions are satisfied:

1. The curves under consideration must at least belong to the class $\boldsymbol{\Gamma}_{\mathrm{wr}}$ of weakly regular simple closed curves (cf. Definition 3.3). In the present experiment this condition is clearly not met by the contours $\gamma_1, \ldots, \gamma_5, \gamma_9, \gamma_{14}$ and $\gamma_{18}$, which all belong to the class $\boldsymbol{\Gamma}_{\mathrm{pr}}$ of piecewise regular curves.

2. The value of the index $p$ must be kept sufficiently low.

3. An appropriate and noise-resistent method for estimating $\ddot{z}$ and $K$ from the (segmented) input picture must be used. As remarked earlier, the polynomial filters (cf. Appendix C) used for that purpose in this experiment may not be optimal in this respect.

If one has to make a choice between dissimilarity measures based on $\ddot{z}$ and those on $K$, then dissimilarity measures based on $K$ are to be preferred because they have a lower computational complexity (cf. Table 4.5).

After the foregoing analysis of the relative performance of the dissimilarity measures, we arrive at the second question raised in this exper-

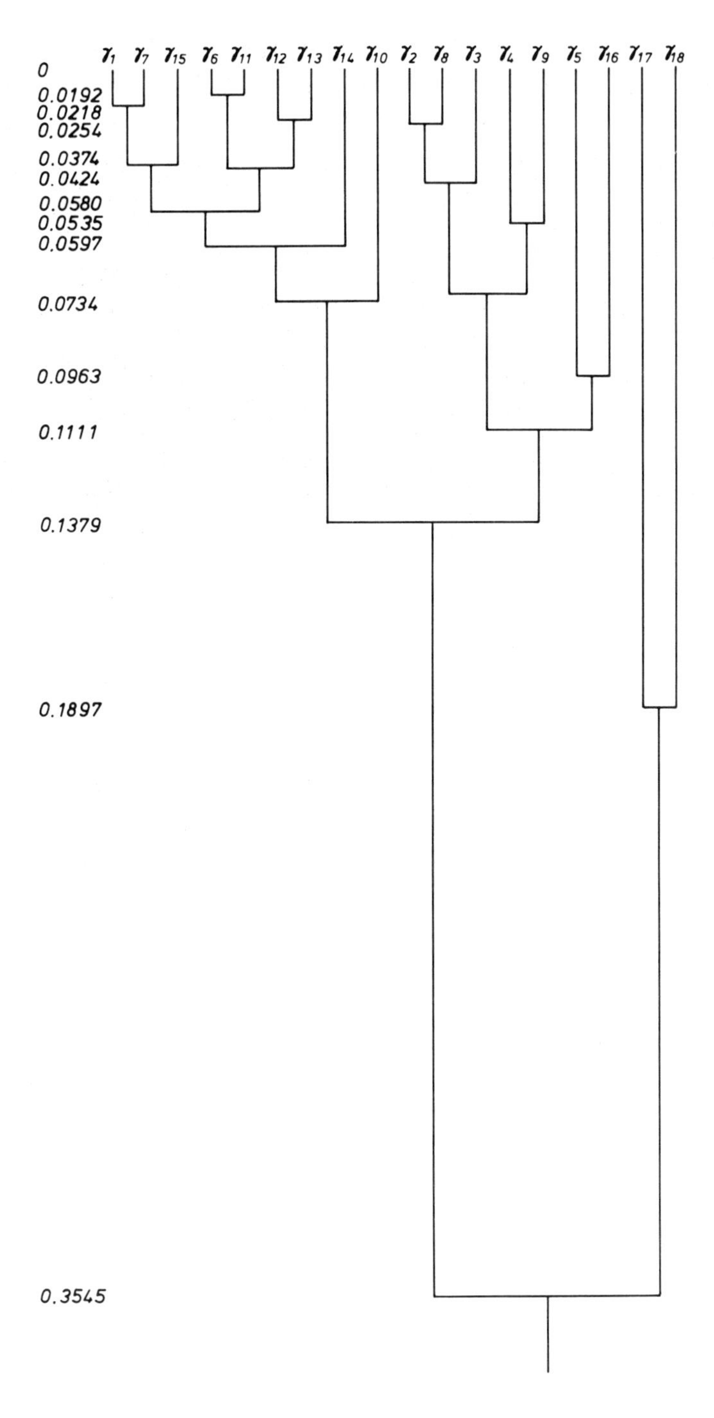


**Figure 4.16.** Dendrogram resulting from UPGMA clustering on the matrix of dissimilarity coefficients $\boldsymbol{D}_3$, generated by $d^{(3)}(z_m, z_n)$, $m, n = 1, \ldots, 18$, for the set of contours $\{\gamma_1, \ldots, \gamma_{18}\}$.

iment: what is the performance of individual dissimilarity measures and what aspects of geometrical dissimilarity do they measure. Obviously we lack the space for an exhaustive evaluation of the performance of all 25 dissimilarity measures used in this experiment. Therefore we present a global evaluation, using one of the major observations from the foregoing analysis of the relative performance of the dissimilarity measures, i.e. that the main distinction in behaviour is not brought about by a variation of the value of the index $p$, but mainly by the contour representation used and, more in particular, by the order of differentiation to which the contour representation belongs.

We evaluate the performance of individual dissimilarity measures through clustering experiments on the $18 \times 18$ matrices of dissimilarity coefficients $\boldsymbol{D}_i$, $i = 1, \ldots, 25$, generated by these dissimilarity measures (cf. Table 4.10). For these experiments we have used again a clustering technique from the family of SAHN methods: Average Linkage Clustering (UPGMA). We have chosen a clustering technique from the family of SAHN methods because these methods enable us to distinguish structures at multiple levels of clustering, a possibility that is lacking in partitional (or non-hierarchic) clustering methods. From the family of SAHN methods we have chosen UPGMA because it takes, in many respects, a middle position between Single Linkage Clustering and Complete Linkage Clustering. Thus we avoid the extreme properties of the latter two methods (cf. Anderberg [1973], Sneath and Sokal [1973]).

To analyze dissimilarity measures based on the position function $z$ we clustered on $\boldsymbol{D}_1$, $\boldsymbol{D}_3$ and $\boldsymbol{D}_5$, generated by $d^{(p)}[z_m, z_n]$ for $p = 1$, 3 and $\infty$, respectively (cf. Table 4.10). Figure 4.16 shows the dendrogram resulting from UPGMA clustering on $\boldsymbol{D}_3$. We observe in this figure two major clusters of contours: $(\gamma_1, \gamma_6, \gamma_7, \gamma_{10}, \gamma_{11}, \gamma_{12}, \gamma_{13}, \gamma_{14}, \gamma_{15})$ and $(\gamma_2, \gamma_3, \gamma_4, \gamma_5, \gamma_8, \gamma_9, \gamma_{16})$, plus an isolated pair of contours: $(\gamma_{17}, \gamma_{18})$ (cf. Figure 4.13 for a display of the contours). The clustering resulting from $\boldsymbol{D}_5$ is virtually identical to that of $\boldsymbol{D}_3$. For $\boldsymbol{D}_1$ there are some differences in the sense that $\gamma_2$, $\gamma_3$ and $\gamma_8$ are merged into the first cluster instead of the second. Evaluating the clusterings based on $\boldsymbol{D}_1$, $\boldsymbol{D}_3$ and $\boldsymbol{D}_5$ jointly, we find the clusters $(\gamma_1, \gamma_7, \gamma_{15})$, $(\gamma_2, \gamma_3, \gamma_8)$, $(\gamma_4, \gamma_5, \gamma_9, \gamma_{16})$, $(\gamma_6, \gamma_{10}, \gamma_{11}, \gamma_{12}, \gamma_{13}, \gamma_{14})$ and the isolated pair $(\gamma_{17}, \gamma_{18})$. Analyzing this result qualitatively we see in Figure 4.13 that contours of substantially different smoothness properties are merged together into a single cluster (e.g. $\gamma_1$, $\gamma_7$ and $\gamma_{15}$). For the set of contours $\{\gamma_1, \ldots, \gamma_{18}\}$ used in this

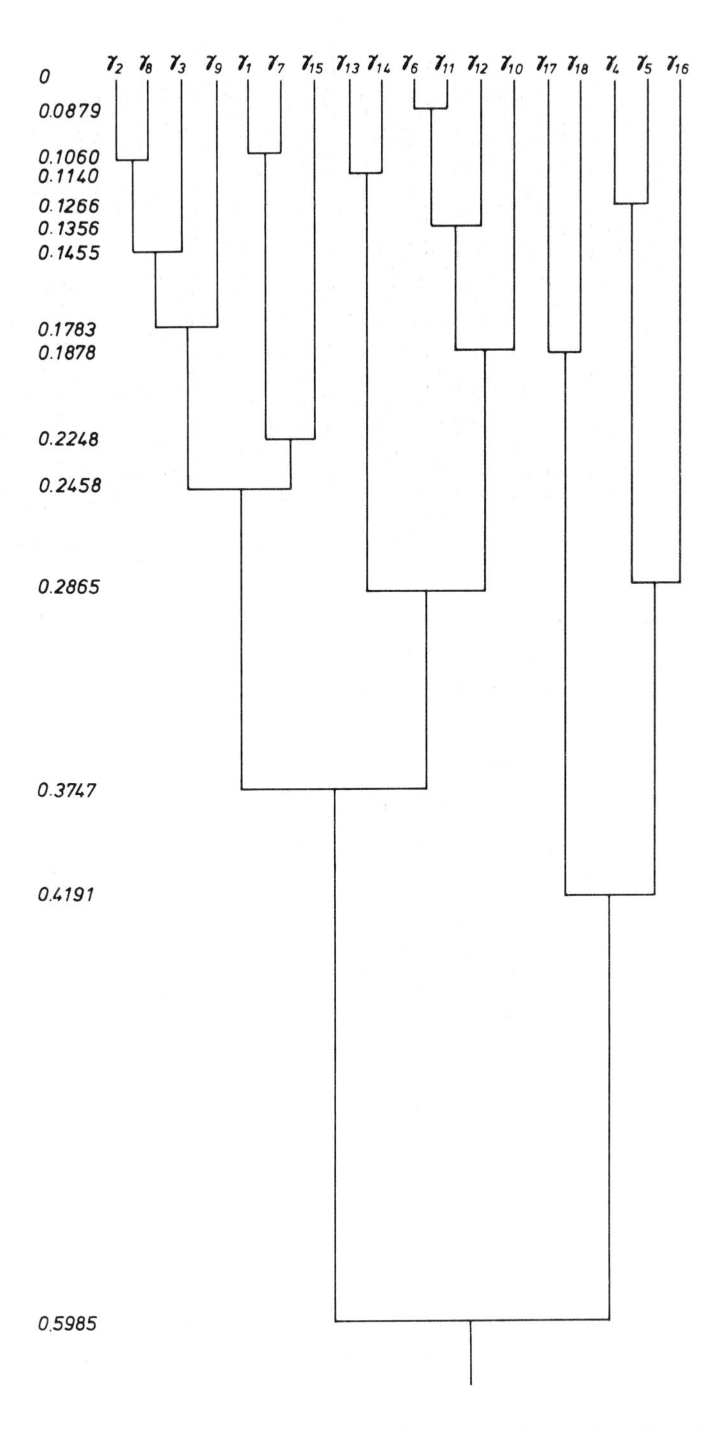


**Figure 4.17.** Dendrogram resulting from UPGMA clustering on the matrix of dissimilarity coefficients $\boldsymbol{D}_6$, generated by $d^{(1)}(\dot{z}_m, \dot{z}_n)$, $m, n = 1, \ldots, 18$, for the set of contours $\{\gamma_1, \ldots, \gamma_{18}\}$.

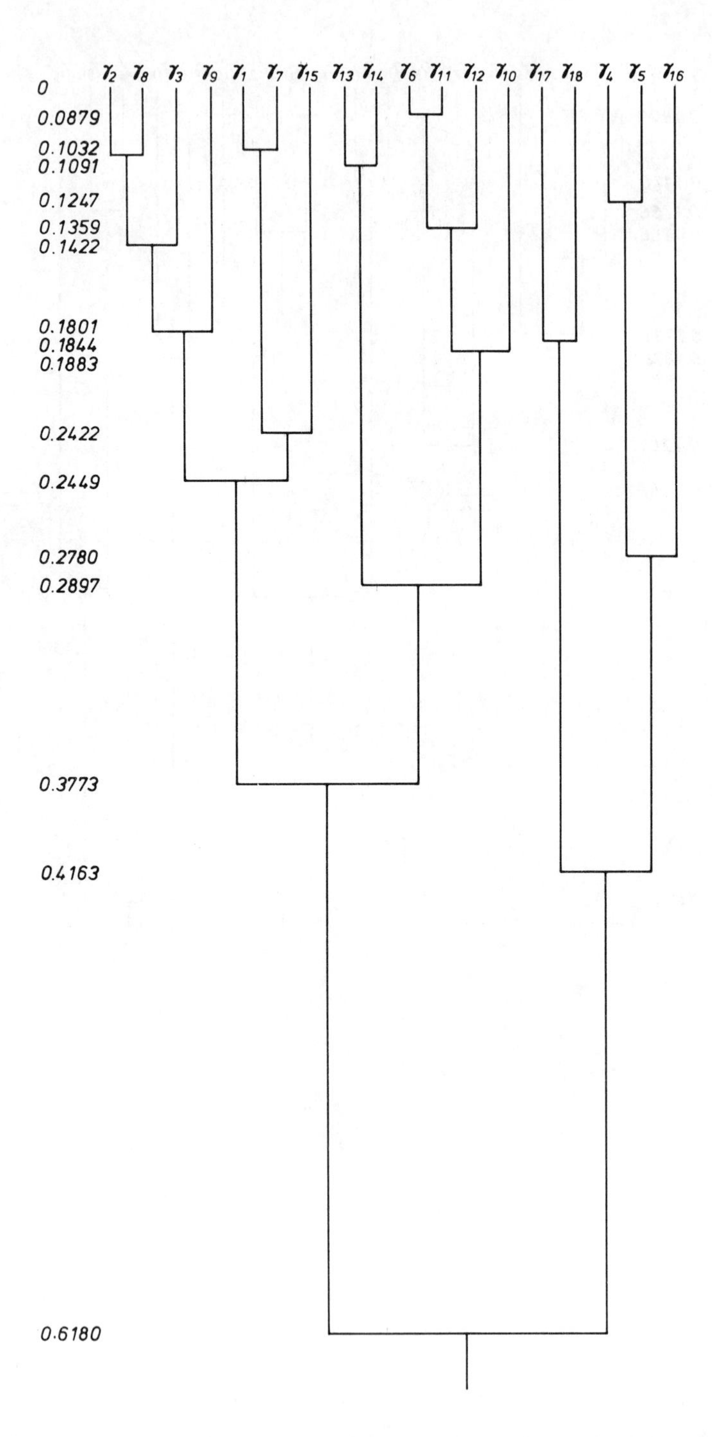


**Figure 4.18.** Dendrogram resulting from UPGMA clustering on the matrix of dissimilarity coefficients $\boldsymbol{D}_{16}$, generated by $d^{(1)}(\psi_m, \psi_n)$, $m, n = 1, \ldots, 18$, for the set of contours $\{\gamma_1, \ldots, \gamma_{18}\}$.

experiment, the clustering on the basis of $d^{(p)}[z_m, z_n]$, for varying values of $p$, seems to generate a grouping of contours according to a similarity of their concavity properties.

With dissimilarity measures based on the tangent function $\dot{z}$ we clustered on $\boldsymbol{D}_6$, $\boldsymbol{D}_8$ and $\boldsymbol{D}_{10}$, generated by $d^{(p)}[\dot{z}_m, \dot{z}_n]$ for $p = 1, 3$ and $\infty$, respectively. Likewise, with dissimilarity measures based on the periodic cumulative angular function $\psi$, we clustered on $\boldsymbol{D}_{16}$, $\boldsymbol{D}_{18}$ and $\boldsymbol{D}_{20}$, generated by $d^{(p)}[\psi_m, \psi_n]$ for the same values of $p$. Figures 4.17 and 4.18 show the dendrograms resulting from clustering on $\boldsymbol{D}_6$ and $\boldsymbol{D}_{16}$. These clustering results are clearly virtually identical, which confirms our previous observation concerning the similarity in performance of dissimilarity measures based on $\dot{z}$ and $\psi$ with the set of contours used in this experiment. In Figures 4.17 and 4.18 we can identify two major clusters, each consisting of two smaller clusters. The two clusters in the first major cluster consist of $(\gamma_1, \gamma_2, \gamma_3, \gamma_7, \gamma_8, \gamma_9, \gamma_{15})$ and $(\gamma_6, \gamma_{10}, \gamma_{11}, \gamma_{12}, \gamma_{13}, \gamma_{14})$, while the two clusters in the second major cluster consist of $(\gamma_4, \gamma_5, \gamma_{16})$ and $(\gamma_{17}, \gamma_{18})$.

Looking globally at the clustering results for various values of $p$ we find that the dissimilarity measures based on $\dot{z}$ give rise to the clusters $(\gamma_1, \gamma_2, \gamma_3, \gamma_7, \gamma_8, \gamma_9)$, $(\gamma_4, \gamma_5)$, $(\gamma_6, \gamma_{10}, \gamma_{11}, \gamma_{12})$, $(\gamma_{13}, \gamma_{14})$ and $(\gamma_{17}, \gamma_{18})$. Comparing this result with the contours in Figure 4.13, we observe that contours that are merged into a single cluster generally have similar smoothness characteristics. These clustering results also correspond fairly well with our subjective notion of shape similarity.

For the dissimilarity measures based on $\psi$ similar results are found.

In order to evaluate the similarity in performance of dissimilarity measures based on $\dot{z}$ and on $\psi$ further, we performed another experiment with a different set of contours. Apart from a circle and a square this set contained six contours of thin, strip-like objects, such as in Figure 4.7. The latter six contours had an increasing number of loops.

For contours with loops the contour representations $\dot{z}$ and $\psi$ behave differently. The reason for this difference is the fact that the phase of $\dot{z}$ is restricted to a range of length $2\pi$, whereas in $\psi$ the directional changes along a contour are accumulated.

As a result the dissimilarity measures based on $\dot{z}$ and on $\psi$ also behave differently for contours with loops and for other strongly non-holomorphic contours. In this additional experiment we frequently found substantially higher values for $d^{(p)}[\psi_m, \psi_n]$ than for $d^{(p)}[\dot{z}_m, \dot{z}_n]$.

However, these differences in dissimilarity values did not lead to substantial differences in clustering results. We performed both a Single Linkage Clustering and a Complete Linkage Clustering on the dissimilarity matrices, generated for this set of contours by $d^{(2)}[\dot{z}_m, \dot{z}_n]$ and $d^{(2)}[\psi_m, \psi_n]$. The structure of the Single Linkage Clustering results was identical for both dissimilarity measures. The structure of the Complete Linkage Clustering results differed only for one contour. More differences occurred between the results of Single Linkage Clustering and those of Complete Linkage Clustering on the same dissimilarity matrix. These differences are the result of the difference in behavior of these two clustering schemes (cf. Sneath and Sokal [1973], Anderberg [1973], Dubes and Jain [1980], Späth [1980]).

From this experiment, and from previously obtained evidence, we draw the conclusion that, with contour clustering and classification, $d^{(p)}(\dot{z}_m, \dot{z}_n)$ and $d^{(p)}(\psi_m, \psi_n)$ have a similar performance. In view of its lower computational complexity in practice for general values of the index $p$ (cf. Table 4.5), $d^{(p)}(\psi_m, \psi_n)$ is to be preferred as a dissimilarity measure to $d^{(p)}(\dot{z}_m, \dot{z}_n)$.

Finally, with dissimilarity measures based on $\ddot{z}$ we clustered on $\boldsymbol{D}_{11}$ and $\boldsymbol{D}_{15}$, generated by $d^{(p)}[\ddot{z}_m, \ddot{z}_n]$ for $p = 1$ and $\infty$, respectively. With dissimilarity measures based on $K$ we clustered on $\boldsymbol{D}_{21}$ and $\boldsymbol{D}_{25}$, generated by $d^{(p)}[K_m, K_n]$, for the same values of $p$. Figures 4.19 and 4.20 show the dendrograms resulting from clustering on $\boldsymbol{D}_{11}$ and $\boldsymbol{D}_{21}$.

First we note a significantly increased dispersion in these clusterings: clustering generally takes place at higher dissimilarity levels than in the previous clusterings with dissimilarity measures based on $z$, $\dot{z}$. This is remarkable since we found in Section 4.3.2 that $d^{(p)}(z_m, z_n)$, $d^{(p)}(\dot{z}_m, \dot{z}_n)$, $d^{(p)}(\ddot{z}_m, \ddot{z}_n)$ and $d^{(p)}(K_m, K_n)$ are in the same range.

Secondly we see in Figures 4.19 and 4.20 that there are many local differences between these two clusterings, though at a more global level the results are similar.

A global analysis of the results from clustering on $\boldsymbol{D}_{11}$ and $\boldsymbol{D}_{15}$ leads to four clusters of contours: $(\gamma_1, \gamma_2, \gamma_3, \gamma_4, \gamma_5, \gamma_8, \gamma_9)$, $(\gamma_6, \gamma_{15})$, $(\gamma_{12}, \gamma_{13}, \gamma_{14})$ and $(\gamma_{16}, \gamma_{17})$. Analyzing the results from clustering on $\boldsymbol{D}_{21}$ and $\boldsymbol{D}_{25}$ jointly, we find only two clusters: $(\gamma_1, \gamma_2, \gamma_3, \gamma_4, \gamma_5, \gamma_7, \gamma_8, \gamma_9)$ and $(\gamma_{11}, \gamma_{12}, \gamma_{13}, \gamma_{15}, \gamma_{16}, \gamma_{17})$. In the clustering on $\boldsymbol{D}_{25}$ the circle, $\gamma_{10}$, behaves as a complete outlier, which is understandable for the dissimilarity measure $d^{(\infty)}(K_m, K_n)$.

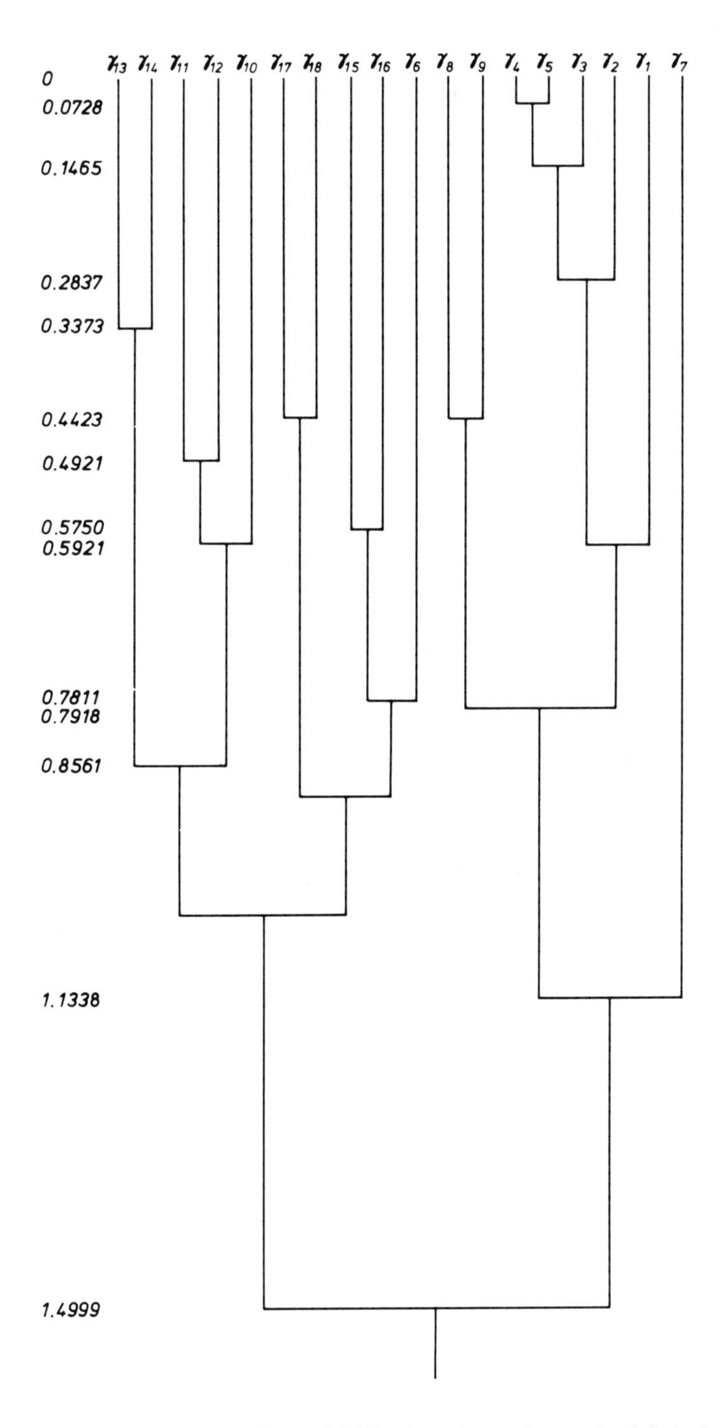


**Figure 4.19.** Dendrogram resulting from UPGMA clustering on the matrix of dissimilarity coefficients $\boldsymbol{D}_{11}$, generated by $d^{(1)}(\ddot{z}_m, \ddot{z}_n)$, $m, n = 1, \ldots, 18$, for the set of contours $\{\gamma_1, \ldots, \gamma_{18}\}$.

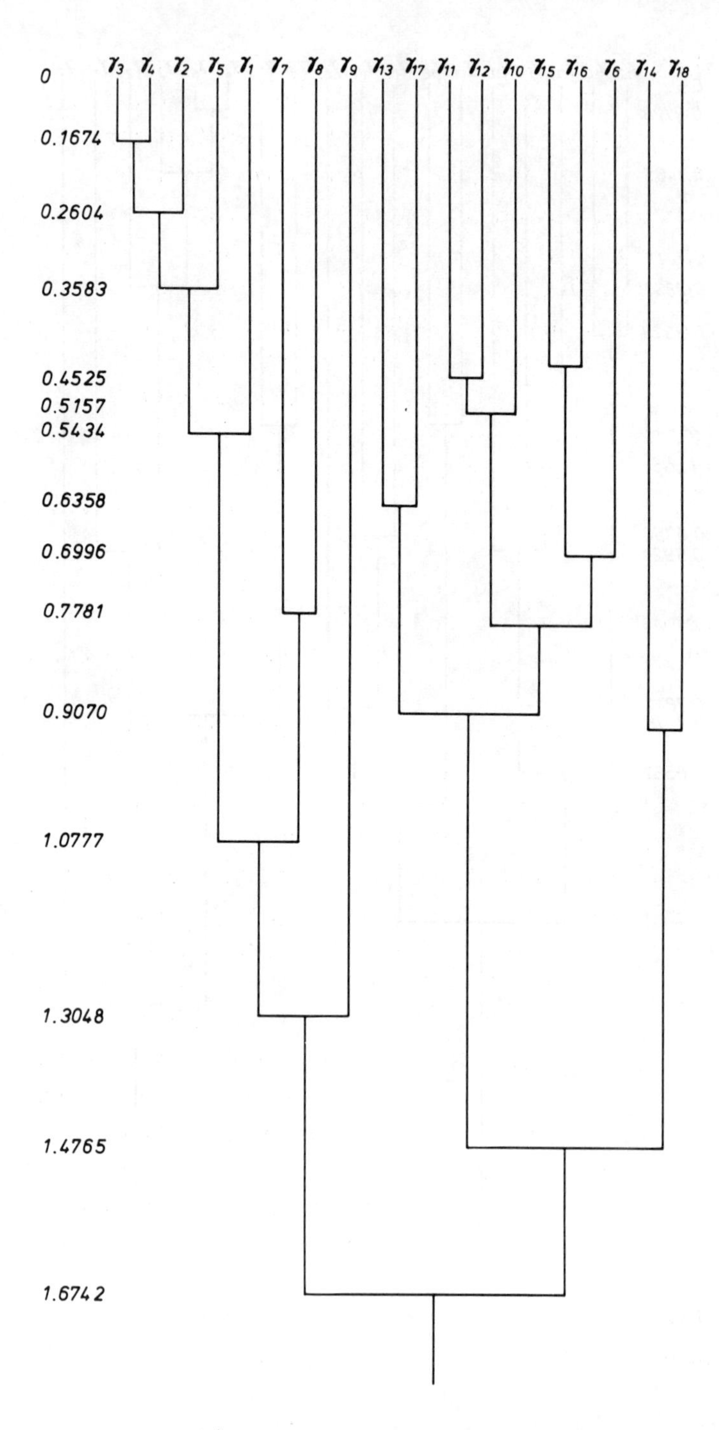


**Figure 4.20.** Dendrogram resulting from UPGMA clustering on the matrix of dissimilarity coefficients $\boldsymbol{D}_{21}$, generated by $d^{(1)}(K_m, K_n)$, $m, n = 1, \ldots, 18$, for the set of contours $\{\gamma_1, \ldots, \gamma_{18}\}$.

From the clustering results it appears that there are analogies in the behavior of dissimilarity measures based on $\ddot{z}$ and those based on $K$, as might be expected from the relation that exists between $\ddot{z}$ and $K$ (cf. Eq. 2.2.25). However, the analogies in behavior are far less pronounced than those between dissimilarity measures based on $\dot{z}$ and on $\psi$. Though there is no theoretical reason for this weaker analogy, there are some practical reasons. Due to the high noise sensitivity of double differentiation it is far more difficult to obtain a reliable estimate of the contour representations $\ddot{z}$ and $K$ than of the contour representations $\dot{z}$ and $\psi$. Also the class of curves for which $\ddot{z}$ and $K$ can be usefully estimated in practice is more restricted than that for which $\dot{z}$ and $\psi$ can be estimated.

Contour representation estimation processes, as described for example in Appendix C, necessarily exercise a smoothing operation on the contour data. Thus we obtain also estimates for $\ddot{z}$ and $K$ of the piecewise regular contours $\gamma_1$, $\gamma_2$, $\gamma_3$, $\gamma_4$, $\gamma_5$, $\gamma_9$, $\gamma_{14}$ and $\gamma_{18}$ for which $\ddot{z}$ and $K$ are defined as distributions (cf. Section 2.2). Therefore the estimates obtained have a limited reliability, which is one of the causes of the large dispersion in the clustering results. Since 8 out of the 18 contours in this experiment are piecewise regular, with unreliable estimates for $\ddot{z}$ and $K$, it is hard to judge the clustering results properly.

Analyzing the clustering results qualitatively, we find that the contours in the large cluster ($\gamma_1$, $\gamma_2$, $\gamma_3$, $\gamma_4$, $\gamma_5$, $\gamma_8$, $\gamma_9$), resulting from clustering on $\boldsymbol{D}_{11}$ and $\boldsymbol{D}_{15}$, all have five pronounced convex corners. Each of the other three clusters, ($\gamma_6$, $\gamma_{15}$), ($\gamma_{12}$, $\gamma_{13}$, $\gamma_{14}$) and ($\gamma_{16}$, $\gamma_{17}$), indeed contains contours that may be called similar from a subjective point of view. The contours in each of the latter two clusters belong to the same family of plane curves: epicycloids and limaçons of Pascal, respectively.

The contours in the first cluster that we found from clustering on $\boldsymbol{D}_{21}$ and $\boldsymbol{D}_{25}$, ($\gamma_1$, $\gamma_2$, $\gamma_3$, $\gamma_4$, $\gamma_5$, $\gamma_7$, $\gamma_8$, $\gamma_9$), all have pronounced convex corners. With the exception of $\gamma_7$ and $\gamma_8$, these contours are not twice differentiable. Most contours in the second cluster, ($\gamma_{11}$, $\gamma_{12}$, $\gamma_{13}$, $\gamma_{15}$, $\gamma_{16}$, $\gamma_{17}$), have more or less pronounced concavities (an exception is $\gamma_{11}$), while all contours in this cluster are at least twice differentiable.

## 4.5 Measures of mirror-symmetry and of *n*-fold rotational symmetry

In Section 2.4 we formulated for each of the contour representations, introduced in Sections 2.1 and 2.2, the conditions that these representations must satisfy in order to represent a contour that possesses mirror-symmetry (or symmetry ***m***) or ***n***-fold rotational symmetry (or symmetry ***n***). Likewise, in Section 3.6 we formulated these conditions in terms of the Fourier coefficients generated by each of the aforementioned contour representations.

As we remarked in Section 2.5, symmetry in objects will only be approximate in real life. Thus, if we want to establish quantitatively the extent to which a certain type of symmetry is present in objects, we need measures of mirror-symmetry and of ***n***-fold rotational symmetry, or, equivalently, measures of mirror-dissymmetry and of ***n***-fold rotational dissymmetry. The purpose of this section is to define such measures and to investigate some of their theoretical properties. Since we consider an object to be symmetric if it comes into coincidence with itself upon a symmetry transformation, the concept of dissimilarity naturally arises in dissymmetry measurement.

We will make use of the (mirror-)dissimilarity measures, defined in Section 4.2, in order to define measures of mirror-dissymmetry and of ***n***-fold rotational dissymmetry.

In Section 4.5.1 we define measures of mirror-dissymmetry as special cases of measures of mirror-dissimilarity, which were defined earlier in Section 4.2.

In Section 4.5.2 we define measures of ***n***-fold rotational dissymmetry which are closely related to the previously defined dissimilarity measures. The measures of ***n***-fold rotational dissymmetry, based on Fourier representations, will be compared with existing proposals for such measures.

To obtain insight into the behavior and performance of the measures of dissymmetry we also present and evaluate the results of an experiment.

### *4.5.1 Measures of mirror-dissymmetry or dissymmetry* ***m***

There are various reasons why the quantitative assessment of the degree of symmetry ***m*** has been given considerable attention in the literature on digital shape analysis, even more than that of symmetry ***n***. Symmetry ***m*** plays an important role in human perception of shape orientation (Rock [1973]), and can also be of significance in a theory of shape understanding (Davis [1977b]). Furthermore, the detection of symmetry ***m*** can be useful for the normalization of orientation and starting point of contours. It can also lead to a compaction of shape description since we need to store only half of the contour representation of a mirror-symmetric object. The detection of symmetry ***m*** in partially occluded shapes can give clues as to how shape completion should be performed. In the latter case we are dealing with nonclosed curves or contour segments. Though we study in this thesis only simple closed contours, the (mirror-)dissimilarity measures based on contour representations can be adapted to open curves or contour segments in a straightforward manner.

In the literature on shape analysis various propositions for the measurement of symmetry ***m*** can be found.

Zahn and Roskies [1972] derive a property between the phases of the Fourier coefficients generated by $\psi$ when the contour is mirror-symmetric. This property forms the basis for a measure of dissymmetry ***m***, where dissymmetry is complemetary to the concept of symmetry.

Davis [1977b] uses a hierarchical model of contour segments and angles between these segments to represent shape. Local mirror-symmetries are detected at a low level in the hierarchical representation. Through clustering of local symmetries and through the definition of relations between the clusters, a global axis of symmetry is found.

Parui and Dutta Majumder [1983] also use a hierarchical shape representation, similar to that of Davis (Davis [1977a], Davis [1977b]). At each level in the hierarchy a shape is represented as a polygon. Their method starts at a high, i.e. coarse, level in the hierarchy. At each hierarchical level the polygonal contour is mirror-reflected about various candidate axes of symmetry and an optimum axis of symmetry for this hierarchical level is determined. Directional differences between the sides of the polygon, when it is mirror-reflected about a candidate axis, are used as a match criterion, and thereby as a measure of dissymmetry ***m*** (cf. Figure 4.21). At the next lower level in the hierarchical representation the result of the higher level is used as an initial estimate for the location of an axis of symmetry.

Bolles [1979] uses a tree description to represent an object. The nodes in such a tree correspond to various object features. This representation is also determined after a mirror-reflection of the object about an arbitrary axis. A measure of similarity between these two tree representations, before and after mirror-reflection, is used as a measure of symmetry ***m***.

Wechsler [1979] defines a piecewise linear axis of symmetry in a binary object-background picture. Each segment of this axis is determined as a local optimum of mirror-symmetry in terms of a mismatch criterion (see Figure 4.22). Wechsler's piecewise linear axis of symmetry is closely related to the symmetric axis of an object, as defined by Blum [1973]. The piecewise linear nature of the symmetry axis in Wechsler's method constitutes a major difference from all other methods for the measurement of symmetry ***m*** found in the literature. These methods determine a single straight line as a global axis of symmetry.

Both Freeman [1979] and Chaudhuri and Dutta Majumder [1980] use a contour representation that is closely related to the curvature function $K$. They determine this representation also for a mirror-re-

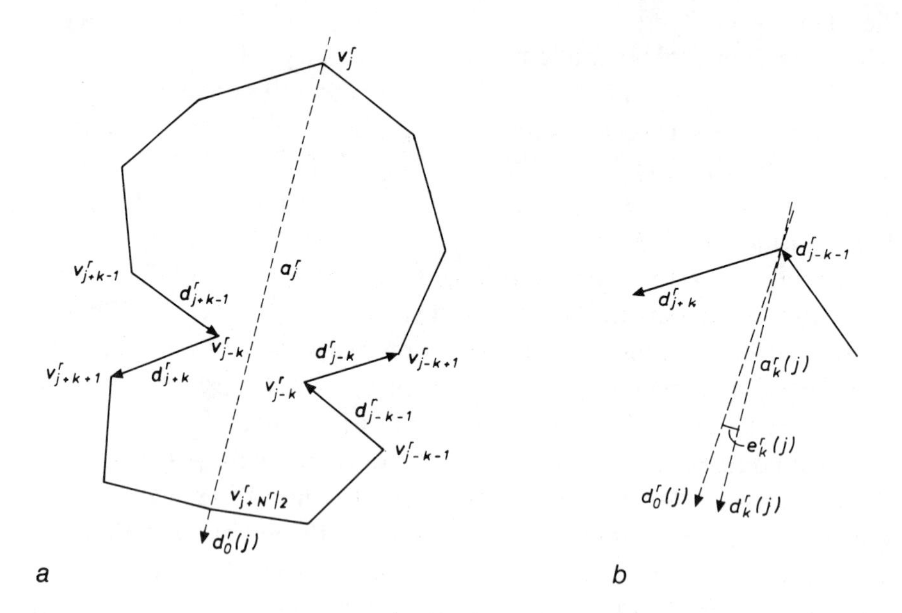


**Figure 4.21.** One stage in the determination of the axis of symmetry according to Parui and Dutta Majumder [1983]. At hierarchical stage $r$ the polygon consists of $N^r$ vertices. At lower hierarchical levels the polygon contains more vertices. The vertices at stage $r$ are equidistant, when measured along the polygon at the lowest hierarchical level (stage 1).

The polygon side formed by the vertices $v_j^r$ and $v_{j+1}^r$ has direction $d_j^r$, expressed in units of $\pi/4$. At each of $N^r/2$ consecutive polygon vertices $v_j^r$ a symmetry axis is hypothesized, which divides the polygon in two halves with an equal number of vertices, and thus of equal length ($a$). This hypothesized axis of symmetry $a_j^r$, between the vertices $v_j^r$ and $v_{j+N^r/2}^r$, has direction $d_0^r(j)$.

Between each pair of corresponding opposite sides with respect to $a_j^r$, with directions $d_{j+k}^r$ and $d_{j-k-1}^r$, a local axis of symmetry is determined in the direction of the bisecting line $a_k^r(j)$ of these two sides, as illustrated in ($b$). If the bisecting line $a_k^r(j)$ has direction $d_k^r(j)$, then the local contribution to the deviation from pure symmetry ***m*** is measured as $|d_0^r(j) - d_k^r(j)|$. The total deviation from pure symmetry ***m*** for the axis $a_j^r$ is the average over all $N^r/2$ local deviations.

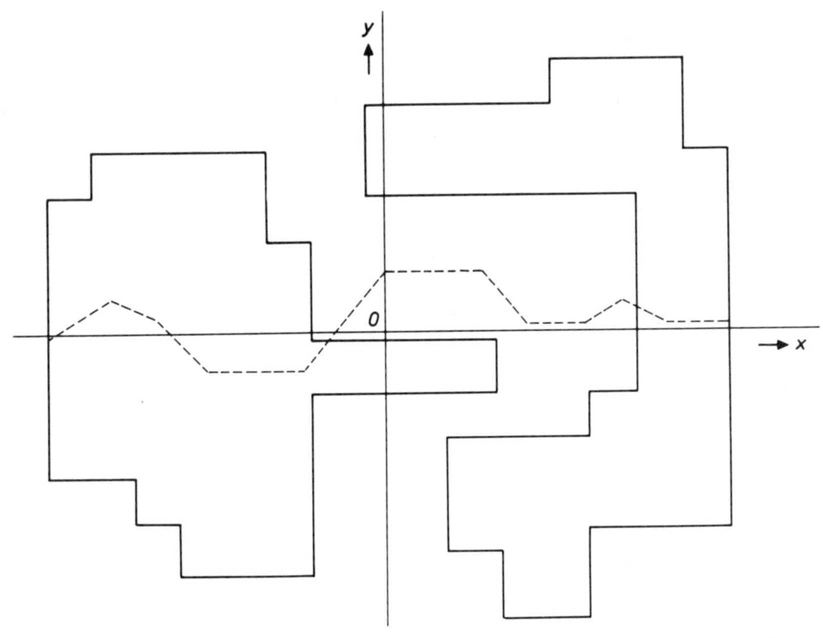


**Figure 4.22.** Piecewise linear axis of symmetry of a pair of objects according to a method proposed by Wechsler [1979]. The $x$- and $y$-axis correspond to the principal axes of the objects (determined through the method of moments as described in Section 4.3).

flected version of the same contour. A correlation-like measure between the two representations thus obtained serves as a measure of symmetry $\boldsymbol{m}$. Freeman [1979] proposes to normalize the starting point of the contour representation, whereas Chaudhuri and Dutta Majumder [1980] determine an optimal starting point along the contour, thereby varying the location of the axis of symmetry.

The measures of dissymmetry $\boldsymbol{m}$ that we propose in the following are directly based upon the definition of symmetry $\boldsymbol{m}$, which states that a contour $\gamma$ is mirror-symmetric iff there exists a line $\boldsymbol{m}$ in the plane such that when $\gamma$ is mirror-reflected about $\boldsymbol{m}$, $\gamma$ coincides with itself (cf. Definition 2.6). A natural measure of dissymmetry $\boldsymbol{m}$, in view of this definition, is obtained by performing an arbitrary mirror-reflection upon the contour and by measuring the dissimilarity between the original and the mirror-reflected contour. This approach is closely related to those of Bolles [1979], Freeman [1979], Chaudhuri and Dutta Majumder [1980] and Parui and Dutta Majumder [1983].

It follows from the foregoing that we can use the mirror-dissimilarity measures, defined in Section 4.2 (Definitions 4.4 and 4.8), for the measurement of dissymmetry $\boldsymbol{m}$.

**Definition 4.12.** *Measure of dissymmetry* $\boldsymbol{m}$ *of index p.*
Let $f$ act as a generic symbol for any of the contour representations $z$, $\dot{z}$, $\ddot{z}$, $\psi$ and $K$. Then a measure of dissymmetry $\boldsymbol{m}$ of index $p$ for a contour $\gamma$, with contour representation $f$, $f \in \mathbf{L}^p(2\pi)$, is defined as (cf. Definition 4.4)

$$d^{(p)}(f; \boldsymbol{m}) = \tilde{d}^{(p)}(f, f), \qquad 1 \leqslant p \leqslant \infty. \tag{4.5.1}$$

□

Through Eq. 4.2.29 $d^{(p)}(f; \boldsymbol{m})$ can also be written as

$$d^{(p)}(f; \boldsymbol{m}) = d^{(p)}(f, \mathcal{M}_x f). \tag{4.5.2}$$

In view of this definition it follows immediately that the range of values that $d^{(p)}(f; \boldsymbol{m})$ can assume is identical to the range of the mirror-dissimilarity measures $\tilde{d}^{(p)}(f_1, f_2)$, which in turn is equivalent to that of the dissimilarity measures $d^{(p)}(f_1, f_2)$ (cf. Eq. 4.5.2). The range of $d^{(p)}(f_1, f_2)$ has been discussed in Section 4.3.2 and depends upon the particular contour representation $f$ and the contour size normalization employed. Further theoretical properties of $d^{(p)}(f; \boldsymbol{m})$ can be found through the discussion on such properties for $d^{(p)}(f_1, f_2)$ in Section 4.4.1.

If $\hat{f}$ indicates the Fourier representation generated by $f$ then, along the same lines as we defined the mirror-dissimilarity measures based on Fourier representations $\hat{f}$ of contours in Section 4.2 (Definition 4.8), we can define a measure of dissymmetry $\boldsymbol{m}$ based on $\hat{f}$.

**Definition 4.13.** *Measure of dissymmetry* $\boldsymbol{m}$ *of index p based on Fourier representations.*
Let $\hat{f}$ be the Fourier representation of a contour representation $f$, where $f$ acts as a generic symbol for any of the contour representations $z$, $\dot{z}$, $\ddot{z}$, $\psi$ and $K$. Then a measure of dissymmetry $\boldsymbol{m}$ of index $p$ for a contour $\gamma$, with Fourier representation $\hat{f}$, $\hat{f} \in \ell^p(\mathbb{Z})$, is defined as (cf. Definition 4.8)

$$d^{(p)}(\hat{f}; \boldsymbol{m}) = \tilde{d}^{(p)}(\hat{f}, \hat{f}), \qquad 1 \leqslant p \leqslant \infty. \tag{4.5.3}$$

□

A number of properties are required for a measure of dissymmetry $\boldsymbol{m}$. The measure must be invariant for the position, size and orientation of the contour and it must be invariant for the parametric starting point on the contour. Furthermore, a measure of dissymmetry $\boldsymbol{m}$ shall assume a value zero iff the contour is mirror-symmetric, and be greater than zero otherwise. Since the measures of dissymmetry $\boldsymbol{m}$ defined above are special cases of the mirror-dissimilarity measures, all properties that apply to the mirror-dissimilarity measures are also valid for the measures of dissymmetry $\boldsymbol{m}$. In particular, all required invariance properties are valid as a result of normalization of contour position and size and of optimization over orientation and starting point in the measure. Furthermore, the measures of mirror-dissimilarity are positive semi-definite, assuming the value zero only in the case of mirror-similar contours. Thus, the validity of all properties that we required the measures of dissymmetry $\boldsymbol{m}$ to satisfy, immediately follows.

In analogy with the (mirror-)dissimilarity measures, the measures of dissymmetry $\boldsymbol{m}$ are computed in practice from $N$ equidistant samples of a contour representation. The resulting discrete measures of dissymmetry $\boldsymbol{m}$ are a special case of the discrete measures of mirror-dissimilarity in Eq. 4.2.51, i.e.

$$d^{(p)}[f; \boldsymbol{m}] = \tilde{d}^{(p)}[f, f]. \tag{4.5.4}$$

The computational complexities of the $d^{(p)}[f; \boldsymbol{m}]$, for the individual contour representations, are the same as mentioned for the corresponding $\tilde{d}^{(p)}[f_1, f_2]$, in Table 4.6. For index $p =$ the computational complexities of $d^{(2)}[f; \boldsymbol{m}]$ can be found in Table 4.8.

Similarly, the discrete measures of dissymmetry $\boldsymbol{m}$ based on Fourier representations are computed in practice from finite Fourier representations, i.e.

$$d^{(p)}[\hat{f}; \boldsymbol{m}] = \tilde{d}^{(p)}[\hat{f}, \hat{f}] \tag{4.5.5}$$

cf. Eqs. 4.2.61 and 4.5.3. For general values of $p$ and for $p = 2$ the computational complexities of $d^{(p)}[\hat{f}; \boldsymbol{m}]$, for the individual Fourier representations, are the same as mentioned in Section 4.2.4 for the corresponding $\tilde{d}^{(p)}[\hat{f}_1, \hat{f}_2]$. A reduction in computational complexity can be obtained by a normalization of the orientation and/or the parametric starting point of the contour, using the methods discussed in Section 4.3.3.

**Figure 4.23.** The set of contours $\{\gamma_1, \ldots, \gamma_{12}\}$ used in the dissymmetry measurement experiments in this section.

To obtain quantitative insight in the performance of $d^{(p)}(f;\, \boldsymbol{m})$ for various contour representations $f$, we performed an experiment. The set of contours used in this experiment, $\{\gamma_1, \ldots, \gamma_{12}\}$, is displayed in Figure 4.23. All these contours are **6**-fold rotationally symmetric and, apart from $\gamma_3$, all contours are also mirror-symmetric. Concerning the smoothness properties of these contours, there are two distinct groups: the subset $\{\gamma_1, \gamma_2, \gamma_3, \gamma_6, \gamma_9, \gamma_{12}\}$ of piecewise regular contours and the set $\{\gamma_4, \gamma_5, \gamma_7, \gamma_8, \gamma_{10}, \gamma_{11}\}$ of regular contours.

We have performed dissymmetry measurement experiments based on each of the five contour representations $z$, $\dot{z}$, $\ddot{z}$, $\psi$ and $K$. The considerations and methods used for contour representation estimation are exactly the same as in the dissimilarity measurement experiments in Section 4.4.2. We refer the reader to this section and to Appendix C. The number of contour representation samples used in the dissymmetry measurement experiments is the same as in the dissimilarity measurement experiment in Section 4.4.2: $N = 512$.

Since the effect of the index $p$ on the behavior of $d^{(p)}(f_m, f_n)$ has already been investigated in detail in Section 4.4, we have limited the dissymmetry measurement experiments to the case $p = 2$.

The values obtained for $d^{(2)}[f_n;\, \boldsymbol{m}]$, $n = 1, \ldots, 12$ for the contour representations $z$, $\dot{z}$, $\ddot{z}$, $\psi$ and $K$ are listed in Table 4.11. In Figures 4.24a and 4.24b we show the axis of symmetry and the shifted starting points found through $d^{(2)}[f_n;\, \boldsymbol{m}]$ for all five contour representations for the contours $\gamma_5$ and $\gamma_9$, respectively.

For the contour representations $\psi$ and $K$ we did not find an axis of symmetry directly since no orientation information is present in these representations. However, after finding an optimal starting point shift $\tau^*$ in $d^{(p)}(\psi;\, \boldsymbol{m})$ or $d^{(p)}(K;\, \boldsymbol{m})$, we can substitute $\tau^*$ in the translation-normalized position function $\mathcal{D}_{\zeta^*}z$ to find the angle between the axis of symmetry and the positive $x$-axis as $\frac{1}{2}\left[\arg\{\mathcal{D}_{\zeta^*}z(0)\} + \arg\{\mathcal{D}_{\zeta^*}z(\tau^*)\}\right]$. For $d^{(p)}(z;\, \boldsymbol{m})$, $d^{(p)}(\dot{z};\, \boldsymbol{m})$ and $d^{(p)}(\ddot{z};\, \boldsymbol{m})$ this angle is given by $\alpha^*/2$, where $\alpha^*$ is the optimal rotation angle found by the measure of dissymmetry $\boldsymbol{m}$.

From Table 4.11 we see that $d^{(2)}[f_n;\, \boldsymbol{m}]$ performs very well for the position function $z$. The largest mirror-dissymmetry value we find for a mirror-symmetric contour is $d^{(2)}[z_{12};\, \boldsymbol{m}] = 0.0051$ for $\gamma_{12}$, which may be a result of contour sampling errors and round-off noise. Yet this value is well over 50 times as small as $d^{(2)}[z_3;\, \boldsymbol{m}]$ for $\gamma_3$, the only contour in our test set that lacks mirror-symmetry.

In Table 4.11 we also see that for this set of contours the behavior of $d^{(2)}[\dot{z}_n; \boldsymbol{m}]$ and $d^{(2)}[\psi_n; \boldsymbol{m}]$ is similar, which corresponds to our previous experience with dissimilarity measurement on the basis of $\dot{z}$ and $\psi$ in Section 4.4.2. The ratio between the mirror-dissymmetry value found for the non-mirror-symmetric contour $\gamma_3$ and the largest value found for a mirror-symmetric contour is over 40 for measurement based on $\dot{z}$ and just over 25 for measurement based on $\psi$. The decrease of this ratio with respect to mirror-dissymmetry measurement based on $z$ is due to contour representation estimation errors in $\dot{z}$ and $\psi$. Yet the results in Table 4.11 show that mirror-dissymmetry measurement can be performed in a sufficiently reliable way on the basis of $\dot{z}$ and $\psi$.

**Table 4.11.** Values of $d^{(2)}[f_n; \boldsymbol{m}]$, $n = 1, \ldots, 12$, for the contour representations $z$, $\dot{z}$, $\ddot{z}$, $\psi$ and $K$ of the set of contours $\{\gamma_1, \ldots, \gamma_{12}\}$, displayed in Figure 4.23.

| Contour | Contour representation | | | | |
|---|---|---|---|---|---|
| | $z$ | $\dot{z}$ | $\ddot{z}$ | $\psi$ | $K$ |
| $\gamma_1$ | 0.0000 | 0.0054 | 0.0652 | 0.0062 | 0.0675 |
| $\gamma_2$ | 0.0003 | 0.0122 | 0.0652 | 0.0152 | 0.1109 |
| $\gamma_3$ | 0.2972 | 0.7244 | 0.1941 | 0.9484 | 0.4958 |
| $\gamma_4$ | 0.0000 | 0.0022 | 0.0176 | 0.0025 | 0.0171 |
| $\gamma_5$ | 0.0005 | 0.0011 | 0.0081 | 0.0013 | 0.0088 |
| $\gamma_6$ | 0.0000 | 0.0007 | 0.0034 | 0.0006 | 0.0123 |
| $\gamma_7$ | 0.0000 | 0.0051 | 0.0261 | 0.0051 | 0.0262 |
| $\gamma_8$ | 0.0000 | 0.0010 | 0.0023 | 0.0000 | 0.0018 |
| $\gamma_9$ | 0.0008 | 0.0115 | 0.0619 | 0.0277 | 0.4962 |
| $\gamma_{10}$ | 0.0000 | 0.0045 | 0.0168 | 0.0046 | 0.0166 |
| $\gamma_{11}$ | 0.0036 | 0.0165 | 0.0654 | 0.0206 | 0.1240 |
| $\gamma_{12}$ | 0.0051 | 0.0170 | 0.0641 | 0.0377 | 0.2755 |

This situation changes quite dramatically with measurements based on $\ddot{z}$ and $K$. While the ratio of dissymmetry values for non-mirror-symmetric contours and mirror-symmetric contours is still almost 3 when the measurement is based on $\ddot{z}$, the ratio decreases to just below 1 when based on $K$. This is mainly due to the unreliability of the contour representation estimates of $\ddot{z}$ and $K$. We also see from Table 4.11 that most problems occur for contours for which $\ddot{z}$ and $K$ are defined as distributions, such as $\gamma_1$, $\gamma_2$, $\gamma_9$ and $\gamma_{12}$. The results for contour $\gamma_6$, for which $\ddot{z}$ and $K$ are also defined as distributions, constitute an exception in a positive sense. On the other hand, the results based on $\ddot{z}$ and $K$ for $\gamma_{11}$, which is a regular contour but with very large negative curvatures, are remarkably bad. Furthermore, also the value $d^{(2)}[K_9; \boldsymbol{m}] = 0.4962$ (cf. Figure 4.24b) can only be explained from estimation errors in $K$ since the starting point shift found is one of the six correct starting point shifts.

In general we observe from the results in Table 4.11 that, concerning the smoothness of contours in relation to the contour representations that can usefully be applied for dissymmetry measurement, the same considerations hold as we found earlier in Section 4.4.2 for (mirror-)dissimilarity measurement.

For an evaluation of the various aspects of mirror-dissymmetry measured by $d^{(p)}(f; \boldsymbol{m})$ for different contour representations and for different values of $p$, we also refer to Section 4.4.

### *4.5.2 Measures of **n**-fold rotational dissymmetry or dissymmetry **n***

Rotational symmetry plays an essential role in many industrial parts (Perkins [1978], Bolles [1979]). It also constitutes an important feature in many biological structures, such as enzymes, viruses, etc. (Santisteban et al. [1980], Santisteban, García and Carrascosa [1981]). In Section 4.3.3 we saw that the detection of the order of rotational symmetry is necessary for a proper execution of procedures for orientation and starting point normalization (Hu [1962], Burkhardt [1979], Wallace and Wintz [1980]). If a contour has symmetry $\boldsymbol{n}$, then it can be represented compactly since we need to store only $1/\boldsymbol{n}$-th part of the contour representation of an $\boldsymbol{n}$-fold rotationally symmetric object. Finally, the detection of symmetry $\boldsymbol{n}$ may also indicate how shape completion should be performed in incomplete contours (e.g. with partially overlapping shapes). However an adaptation of the measures of dissymmetry $\boldsymbol{n}$, that are defined in this section, is necessary in that case in order to cope with incomplete contours.

Since in practice symmetry is rarely perfect, we need measures that enable us to establish the extent of $\boldsymbol{n}$-fold rotational symmetry in an object quantitatively. In the literature we find several propositions to this end.

Zahn and Roskies [1972] observed that a contour has $\boldsymbol{n}$-fold rotational symmetry iff $\hat{\psi}(k) = 0$, $\forall k \in \mathbb{Z}$ such that $k \neq 0 \bmod \boldsymbol{n}$ (see also Table 3.6 in Section 3.6). Based on this observation they proposed

$$2 \cdot \sum_{\substack{k \in \mathbb{Z} \\ k \neq 0 \bmod \boldsymbol{n}}} |\hat{\psi}(k)| \tag{4.5.6}$$

as a measure of dissymmetry $\boldsymbol{n}$.

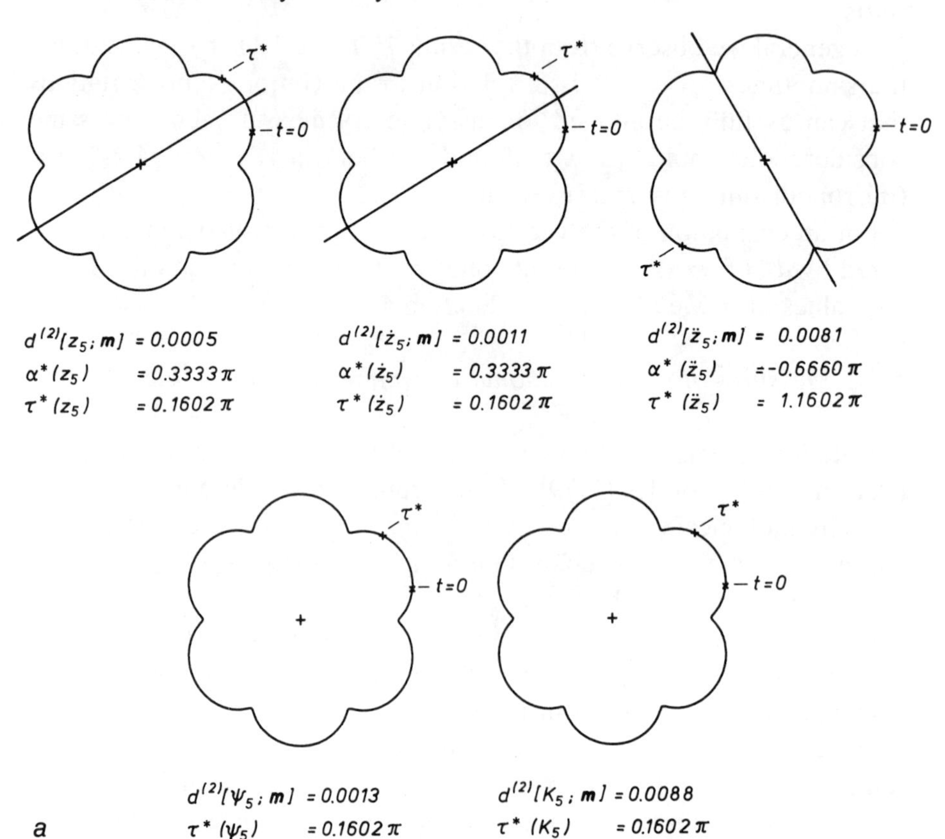


**Figure 4.24a.** Results of the measurement of dissymmetry $\boldsymbol{m}$ for the contour $\gamma_5$ from Figure 4.23 for the contour representations $z$, $\dot{z}$, $\ddot{z}$, $\psi$ and $K$.

Burkhardt [1979] has a similar proposition for a measure of dissymmetry $\boldsymbol{n}$, based on the Fourier representation $\hat{z}$. Recall from Section 3.6 that a contour has $\boldsymbol{n}$-fold rotational symmetry iff $\hat{z}(k) = 0$, $\forall k \in \mathbb{Z} - \{0\}$ such that $k \neq 1 \bmod \boldsymbol{n}$. This property constitutes the basis for Burkhardt's proposition

$$\frac{\displaystyle\sum_{\substack{k \in \mathbb{Z} - \{0\} \\ k \neq 1 \bmod \boldsymbol{n}}} |\hat{z}(k)|}{\displaystyle\sum_{k \in \mathbb{Z} - \{0\}} |\hat{z}(k)|} \tag{4.5.7}$$

as a measure of dissymmetry $\boldsymbol{n}$. The expression in the denominator of Eq. 4.5.7 ensures the scale invariance of the measure, while the exclu-

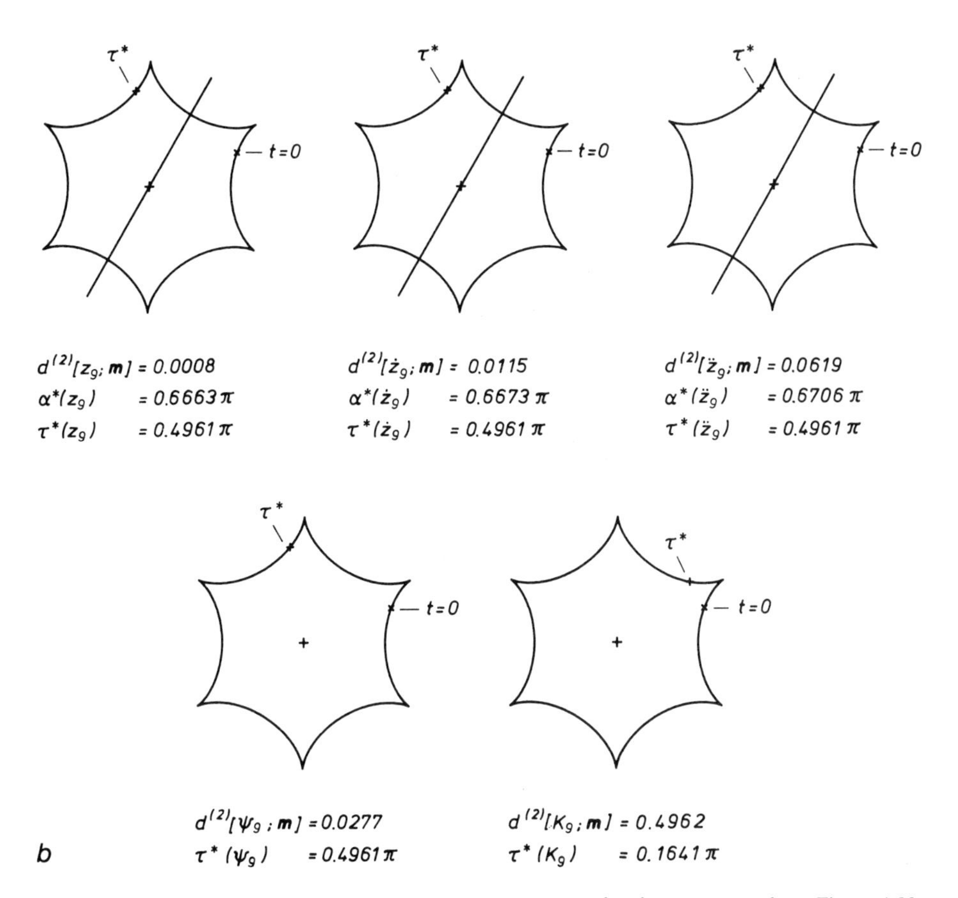


**Figure 4.24b.** Results of the measurement of dissymmetry $\boldsymbol{m}$ for the contour $\gamma_9$ from Figure 4.23 for the contour representations $z$, $\dot{z}$, $\ddot{z}$, $\psi$ and $K$.

sion of $\hat{z}(0)$ from both the numerator and the denominator takes care of the translation invariance.

By comparing Eq. 4.5.6 with Eq. 4.5.7, the conceptual similarity of these two propositions as measures of dissymmetry $\boldsymbol{n}$ is obvious. Since both measures sum over moduli of Fourier coefficients, invariance for orientation and starting point is automatically obtained (cf. Table 3.2).

Another approach to the measurement of the extent of symmetry $\boldsymbol{n}$ in an object has been given by Perkins [1978]. He represents a contour as a set of connecting line segments and circular arcs. On top of this he registers the positions of a set of equally spaced samples on the contour, $\{z[n]\}$ in our terminology, and the directions of lines perpendicular to the contour in the sample positions, $\theta[n] \pm \pi/2$ in our terminology, where $\theta$ is the tangent angle function (cf. Eq. 2.2.6). An illustration of this representation is given in Figure 4.23. Rotational symmetry $\boldsymbol{n}$ is measured by rotating the contour representation just described about the center of gravity of object over angles $m(2\pi/\boldsymbol{n})$ for $m = 1, \ldots, \boldsymbol{n} - 1$. After each rotation a measure of coincidence between the original and the rotated representation is determined. Finally, the results of the $\boldsymbol{n} - 1$ measurements are combined to obtain a global impression of the extent of symmetry $\boldsymbol{n}$ in the object.

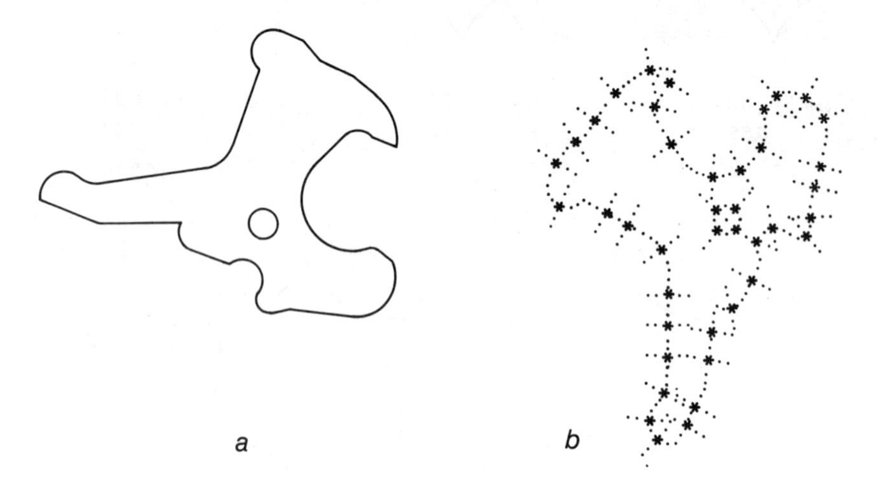


**Figure 4.25.** Illustration of contour representation according to Perkins [1978]. (*a*) displays the result of fitting straight line segments and circular arcs to input contour data, leading to a representation of 2 closed contours by 27 straight line segments and circular arcs. Each of these two approximated contours is called a concurve (Perkins [1978]). Superimposed on the concurves is a multi-sector representation (*b*). The multi-sector representation consists of a set of approximately equidistant positions along the concurves, marked by an '*' in (*b*), and by a set of perpendicular directions at these positions, marked by a short dotted line. The two concurves are represented by 37 and 4 multisectors, respectively.

Bolles [1979] introduces a tree representation for objects and determines the presence of symmetry $\boldsymbol{n}$ in an object directly from its tree representation. The nodes in this tree representation correspond to various object features. First similar subpatterns are determined within the object and subsequently it is tested whether these subpatterns are arranged in an $\boldsymbol{n}$-fold rotationally symmetric manner.

Before we actually present our propositions for measures of dissymmetry $\boldsymbol{n}$ we first mention the requirements that such measures must satisfy. Since a measure of dissymmetry $\boldsymbol{n}$ constitutes a shape property, it must be invariant for the position, size and orientation of the object and for the position of the parametric starting point on the contour (cf. Section 1.1). We also require of a measure of dissymmetry $\boldsymbol{n}$ that it assumes the value zero iff the object has $\boldsymbol{n}$-fold rotational symmetry and that its value is greater than zero otherwise.

The measures of dissymmetry $\boldsymbol{n}$ that we propose in the following are directly based upon the definition of symmetry $\boldsymbol{n}$, Definition 2.7, which states that a contour $\gamma$ is $\boldsymbol{n}$-fold rotationally symmetric iff there exists a point of order $\boldsymbol{n}$ in the plane such that, when $\gamma$ is rotated about this point, $\gamma$ coincides with itself after each rotation over an angle $2\pi/\boldsymbol{n}$. A natural measure of dissymmetry $\boldsymbol{n}$, in view of this definition, is obtained by performing a rotation of the translation-normalized contour about the origin over angles $m(2\pi/\boldsymbol{n})$ for $m = 1, \ldots, \boldsymbol{n}-1$, and by measuring the dissimilarity between the original and the rotated contour after each rotation. The results of these $\boldsymbol{n}-1$ dissimilarity measurements are subsequently averaged to obtain a global measure of dissymmetry $\boldsymbol{n}$. This approach is conceptually related to that of Perkins [1978], while further analysis will reveal that the approaches of Zahn and Roskies [1972] and Burkhardt [1979] belong to the same class.

We now formulate our proposition for a measure of dissymmetry $\boldsymbol{n}$, which is based on the property of contour representations of $\boldsymbol{n}$-fold rotationally symmetric contours, expressed in Eq. 2.4.3. We will do this in two steps. First we define the $m$-th component of dissymmetry $\boldsymbol{n}$ and subsequently we define a global measure of dissymmetry $\boldsymbol{n}$ by averaging over the components of dissymmetry $\boldsymbol{n}$.

**Definition 4.14.** *$m$-th component of dissymmetry $\boldsymbol{n}$ of index $p$.*
Let $f$ act as a generic symbol for any of the contour representations $z$, $\dot{z}$, $\ddot{z}$, $\psi$ and $K$. Then the $m$-th component of dissymmetry $\boldsymbol{n}$ for a contour

$\gamma$, with contour representation $f, f \in \mathbf{L}^p(2\pi)$, is defined as (cf. Eq. 2.4.3)

$$d^{(p)}(f; \boldsymbol{n}, m) = \|f^* - \mathcal{T}_{m(2\pi/\boldsymbol{n})}\mathcal{R}_{m(2\pi/\boldsymbol{n})}f^*\|_p$$

$$m \in \mathbb{Z}, \quad 1 \leqslant p \leqslant \infty \tag{4.5.8}$$

where $f^*$ is related to $f$ as given in Eq. 4.2.1 and Eqs. 4.2.12 and 4.2.21.

□

If $f$ stands for $z$, $\dot{z}$, $\ddot{z}$ or $K$, then $f^*$ is the translation- and scale-normalized version of $f$. If $f$ stands for $\psi$, then $f^*$ is a normalization of $f$ to a version with contour average zero. Recall from Section 2.4 that $2\pi/\boldsymbol{n}$ is the elementary angle of rotation of an $\boldsymbol{n}$-fold rotationally symmetric object. The $m$-th component of dissymmetry $\boldsymbol{n}$, $d^{(p)}(f; \boldsymbol{n}, m)$, measures the dissimilarity between a contour and a version of that contour, rotated over an angle $m(2\pi/\boldsymbol{n})$, $m \in \mathbb{Z}$. We note the following properties of $d^{(p)}(f; \boldsymbol{n}, m)$. From the properties of $f^*$ mentioned above it follows that $d^{(p)}(f; \boldsymbol{n}, m)$ is invariant for the position and size of the object. It is easily verified that $d^{(p)}(f; \boldsymbol{n}, m)$ is also invariant for the orientation of the object and for the position of the parametric starting point on its contour. If a contour possesses symmetry $\boldsymbol{n}$, then it follows from Eq. 2.4.3 that $d^{(p)}(f; \boldsymbol{n}, m) = 0$, $\forall m \in \mathbb{Z}$. We also note that $d^{(p)}(f; \boldsymbol{n}, m)$, considered as a function on $\mathbb{Z}$, is periodic, with period $\boldsymbol{n}$, and that it is an even function, i.e.

$$d^{(p)}(f; \boldsymbol{n}, m) = d^{(p)}(f; \boldsymbol{n}, -m), \qquad \forall m \in \mathbb{Z}. \tag{4.5.9}$$

Furthermore, $d^{(p)}(f; \boldsymbol{n}, m) = 0$ if $m = 0 \bmod \boldsymbol{n}$.

Since an $\boldsymbol{n}$-fold rotationally symmetric contour comes into coincidence with itself after a rotation over any multiple of $2\pi/\boldsymbol{n}$, we measure $d^{(p)}(f; \boldsymbol{n}, m)$ over all multiples of $2\pi/\boldsymbol{n}$ in a range of length $2\pi$, i.e. for $m = 1, \ldots, \boldsymbol{n} - 1$. With these measurements we obtain a global measure of dissymmetry $\boldsymbol{n}$ by averaging over the $d^{(p)}(f; \boldsymbol{n}, m)$.

**Definition 4.15.** *Measure of dissymmetry* $\boldsymbol{n}$ *of index pair* $(p, q)$.
Let $f$ act as a generic symbol for any of the contour representations $z$, $\dot{z}$, $\ddot{z}$, $\psi$ and $K$. Then a measure of dissymmetry $\boldsymbol{n}$ of index pair $(p, q)$ for a contour $\gamma$, with contour representation $f, f \in \mathbf{L}^p(2\pi)$, is defined as (cf. Eq. 2.4.3 and Definition 4.14)

$$d^{(p,q)}(f; \mathbf{1}) \equiv 0, \tag{4.5.10a}$$

$$\begin{aligned} d^{(p,q)}(f; \boldsymbol{n}) &= \left[\frac{1}{\boldsymbol{n}-1} \sum_{m=1}^{\boldsymbol{n}-1} \{d^{(p)}(f; \boldsymbol{n}, m)\}^q\right]^{1/q} \\ &= \left[\frac{1}{\boldsymbol{n}-1} \sum_{m=1}^{\boldsymbol{n}-1} \|f^* - \mathcal{T}_{m(2\pi/\boldsymbol{n})}\mathcal{R}_{m(2\pi/\boldsymbol{n})}f^*\|_p^q\right]^{1/q} \qquad \boldsymbol{n} \geqslant \mathbf{2}, \end{aligned} \tag{4.5.10b}$$

for $1 \leqslant p, q \leqslant \infty$.

□

For $\boldsymbol{n} = \mathbf{1}$ we have the trivial case of rotational symmetry. Therefore we defined $d^{(p,q)}(f; \mathbf{1}) \equiv 0$ in Eq. 4.5.10a.

In Eq. 4.5.10b we used the generalized mean for averaging over the values $d^{(p)}(f; \boldsymbol{n}, m)$ (cf. Beckenbach and Bellman [1971], Abramowitz and Stegun [1972]).

In the generalized mean the index $q$ can assume any real value. It is easily verified that the generalized mean contains the arithmetic mean ($q = 1$), the geometric mean ($\lim_{q \to 0}$) and the harmonic mean ($q = -1$) as special cases (cf. Beckenbach and Bellman [1971], pp. 3-19, Abramowitz and Stegun [1972], p. 10).

In the context of our application we have limited the range of $q$ (cf. Eq. 4.5.10b). This is necessary since for $q \leqslant 0$ we always obtain $d^{(p,q)}(f; \boldsymbol{n}) = 0$ if $d^{(p)}(f; \boldsymbol{n}, m) = 0$ for some $m \neq 0 \bmod \boldsymbol{n}$. This means for example that if we are measuring **12**-fold rotational dissymmetry for a contour that has only **6**-fold rotational symmetry, we still find $d^{(p,q)}(f; \mathbf{12}) = 0$ for $q \leqslant 0$. Clearly, for $q \leqslant 0$ the measure of dissymmetry $\boldsymbol{n}$ in Eq. 4.5.10b would not satisfy the requirement that it should only assume the value zero if the contour possesses rotational symmetry $\boldsymbol{n}$.

The interpretation of varying the value of the index $q$ in $d^{(p,q)}(f; \boldsymbol{n})$ is as follows. For $q = 1$, the arithmetic mean, we have a very global averaging over the values $d^{(p)}(f; \boldsymbol{n}, m)$. The larger the value of the index $q$, the more sensitive $d^{(p,q)}(f; \boldsymbol{n})$ becomes for the largest value among the values $d^{(p)}(f; \boldsymbol{n}, m)$. The indices $p$ and $q$ in $d^{(p,q)}(f; \boldsymbol{n})$ are independent parameters that emphasize different aspects in the measure of dissymmetry $\boldsymbol{n}$. Therefore the case $p = q$ has no special interpretation, although it leads to some computational simplifications (cf. Eq. 4.5.10b).

We now check whether $d^{(p,q)}(f; \boldsymbol{n})$ possesses the properties that we require for a measure of dissymmetry $\boldsymbol{n}$.

Since the values $d^{(p)}(f; \boldsymbol{n}, m)$ are invariant for the position, size and orientation of an object and for the position of the parametric starting point on its contour, it follows from Definition 4.15 that the same properties are valid for $d^{(p,q)}(f; \boldsymbol{n})$.

Eq. 4.5.10 shows that $d^{(p,q)}(f; \boldsymbol{n}) = 0$ iff $d^{(p)}(f; \boldsymbol{n}, m) = 0$ for $m = 1, \ldots, \boldsymbol{n} - 1$, $\boldsymbol{n} \geqslant \mathbf{2}$. It follows from Eqs. 2.4.2 and 2.4.3 and from Definition 4.14 that $d^{(p)}(f; \boldsymbol{n}, m) = 0$ for $m = 1, \ldots, \boldsymbol{n} - 1$ if the contour, represented by $f$, is $\boldsymbol{n}$-fold rotationally symmetric. If the contour is not $\boldsymbol{n}$-fold rotationally symmetric, then $d^{(p)}(f; \boldsymbol{n}, m) > 0$ at least for all $m$ that are prime to $\boldsymbol{n}$. (A number $a$ is called prime to a number $b$ if GCD $(a, b) = 1$, Shanks [1962].) Since 1 is prime to any $\boldsymbol{n} \geqslant \mathbf{2}$, we find that $d^{(p)}(f; \boldsymbol{n}, 1) > 0$ if the contour, represented by $f$, is not $\boldsymbol{n}$-fold rotationally symmetric.

From these observations we conclude that $d^{(p,q)}(f; \boldsymbol{n}) = 0$ iff the contour, represented by $f$, is $\boldsymbol{n}$-fold rotationally symmetric and that $d^{(p,q)}(f; \boldsymbol{n}) > 0$ if it is not $\boldsymbol{n}$-fold rotationally symmetric, and therefore that $d^{(p,q)}(f; \boldsymbol{n})$ possesses the properties that we require for a measure of dissymmetry $\boldsymbol{n}$.

The range of values that $d^{(p,q)}(f; \boldsymbol{n})$ can assume is analyzed as follows. Recall that $d^{(p)}(f; \boldsymbol{n}, m) = 0$ for $m = 0 \bmod \boldsymbol{n}$. Using $f^* = \mathcal{S}_{\beta^*}\mathcal{D}_{\zeta^*}f$, Eq. 4.2.1, and applying Minkowski's inequality (cf. Appendix A) to Eq. 4.5.10b yields

$$d^{(p,q)}(f; \boldsymbol{n})$$

$$= \left[\frac{1}{\boldsymbol{n} - 1} \sum_{m=1}^{\boldsymbol{n}-1} \|\mathcal{S}_{\beta^*}\mathcal{D}_{\zeta^*}f - \mathcal{T}_{m(2\pi/\boldsymbol{n})}\mathcal{R}_{m(2\pi/\boldsymbol{n})}\mathcal{S}_{\beta^*}\mathcal{D}_{\zeta^*}f\|_p^q\right]^{1/q}$$

$$\leqslant \left[\frac{2^p}{\boldsymbol{n} - 1} \sum_{m=1}^{\boldsymbol{n}-1} \|\mathcal{S}_{\beta^*}\mathcal{D}_{\zeta^*}f\|_p^q\right]^{1/p} = 2\,\mathcal{S}_{\beta^*}\|\mathcal{D}_{\zeta^*}f\|_p, \qquad \boldsymbol{n} \geqslant \mathbf{2}. \tag{4.5.11}$$

Since we proposed in Section 4.3.2 that $\mathcal{S}_{\beta^*} = \|\mathcal{D}_{\zeta^*}f\|_p^{-1}$, when $f$ stands for $z$, $\dot{z}$, $\ddot{z}$ or $K$, we find for the range of $d^{(p,q)}(f; \boldsymbol{n})$

$$0 \leqslant d^{(p,q)}(f; \boldsymbol{n}) \leqslant 2, \qquad \boldsymbol{n} \geqslant \mathbf{2}. \tag{4.5.12}$$

The measure of dissymmetry $\boldsymbol{n}$, $d^{(p,q)}(f; \boldsymbol{n})$, can be rewritten into a form, that leads to a reduction in computational complexity by a factor 2, for $\boldsymbol{n}$ odd, and a factor of almost 2, for $\boldsymbol{n}$ even. Using the $\boldsymbol{n}$-periodicity and the evenness of the values $d^{(p)}(f; \boldsymbol{n}, m)$, Eq. 4.5.9, we derive for $d^{(p,q)}(f; \boldsymbol{n})$ the expressions

$$d^{(p,q)}(f; \boldsymbol{n}) = \begin{cases} \left[\dfrac{2}{\boldsymbol{n}-1} \displaystyle\sum_{m=1}^{(\boldsymbol{n}-1)/2} \{d^{(p)}(f; \boldsymbol{n}, m)\}^q\right]^{1/q} & \text{for } \boldsymbol{n} \text{ odd}, \quad (4.5.13a) \\ \left[\dfrac{2}{\boldsymbol{n}-1} \displaystyle\sum_{m=1}^{\boldsymbol{n}/2-1} \{d^{(p)}(f; \boldsymbol{n}, m)\}^q + \dfrac{1}{\boldsymbol{n}-1}\{d^{(p)}(f; \boldsymbol{n}, \boldsymbol{n}/2)\}^q\right]^{1/q} & \text{for } \boldsymbol{n} \text{ even}. \end{cases}$$

(4.5.13b)

In practice, a measure of dissymmetry $\boldsymbol{n}$ is computed from $N$ equidistant samples of the contour representation, resulting in a discrete measure of dissymmetry $\boldsymbol{n}$, $d^{(p,q)}[f; \boldsymbol{n}]$. It follows from Eq. 4.2.48 and Definitions 4.14 and 4.15 that the computational complexity of $d^{(p,q)}[f; \boldsymbol{n}]$ is $O(\boldsymbol{n}N)$ for each of the individual contour representations indicated by $f$. Though, in practice, the expressions for $d^{(p,q)}(f; \boldsymbol{n})$ in Eqs. 4.5.13a and 4.5.13b lead to a reduction of the number of computational operations, they do not affect the order of magnitude of the number of these operations.

In the special case of $p = 2$ we can rewrite $d^{(p,q)}(f; \boldsymbol{n})$ into a form in which the computational complexity is dominated by the computation of the cyclic correlation function of $f^*$. For appropriate values of $N$, the number of contour representation samples, $d^{(2,q)}[f; \boldsymbol{n}]$ can be com-

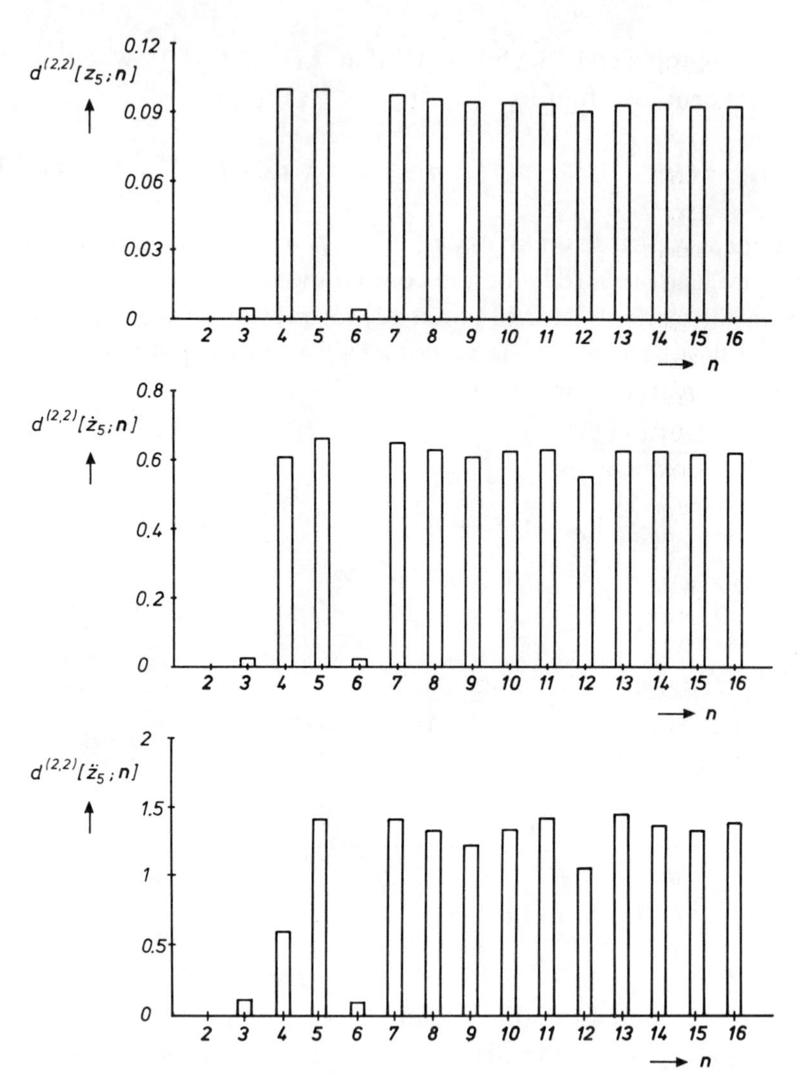


**Figure 4.26a.** Results of the measurement of dissymmetry $\boldsymbol{n}$, for $\boldsymbol{n} = \mathbf{2}, \ldots, \mathbf{16}$, for contour $\gamma_5$ from Figure 4.23. The measure of dissymmetry $\boldsymbol{n}$ used is $d^{(p,q)}[f; \boldsymbol{n}]$, with $(p, q) = (2, 2)$; $f$ stands for the contour representations $z_5$, $\dot{z}_5$, $\ddot{z}_5$, $\psi_5$ and $K_5$, respectively.

puted in $O(N \log_2 N + \boldsymbol{n})$ arithmetic operations, using the FFT algorithm to compute the cyclic correlation function (cf. Section 4.2.4). $O(N \log_2 N + \boldsymbol{n})$ is equivalent to $O(N \log_2 N)$, since for useful measurements $N$ has to be larger than $\boldsymbol{n}$. Whether this method of computation is more efficient than the direct computation of $d^{(2,q)}[f; \boldsymbol{n}]$, with a computational complexity of $O(\boldsymbol{n}N)$, merely depends upon the actual values of $\boldsymbol{n}$ and $N$.

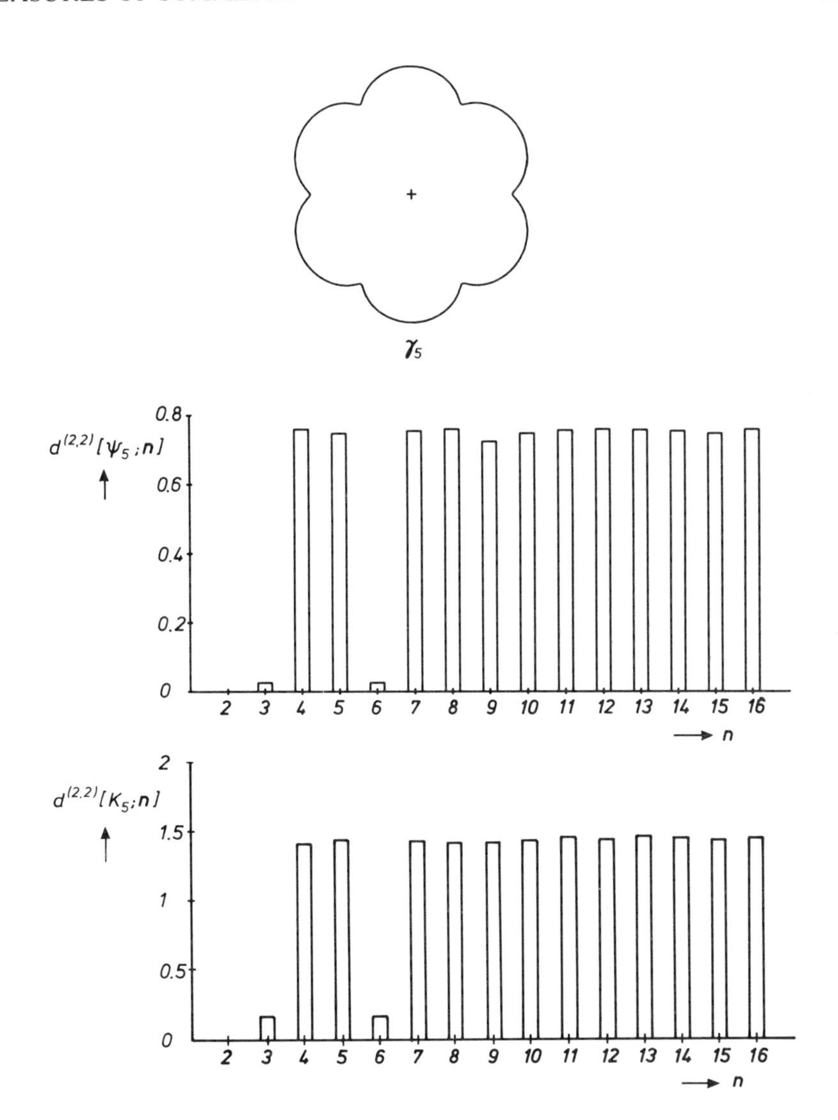


To test the performance of the measures of dissymmetry $\boldsymbol{n}$ proposed in this section, we performed an experiment with the set of contours $\{\gamma_1, \ldots, \gamma_{12}\}$, shown in Figure 4.23, which was also used in an experiment with the measures of dissymmetry $\boldsymbol{m}$. All contours in this set are **6**-fold rotationally symmetric. In Figures 4.26a and 4.26b the results of $d^{(p,q)}[f; \boldsymbol{n}]$, with index pair $(p, q) = (2, 2)$, are shown for the contours $\gamma_5$ and $\gamma_9$ respectively. We performed the measurement for the orders of rotational symmetry $\boldsymbol{n} = \mathbf{2}, \ldots, \mathbf{16}$ and for the contour representations $z$, $\dot{z}$, $\ddot{z}$, $\psi$ and $K$. We observe some remarkable phenomena from this experiment.

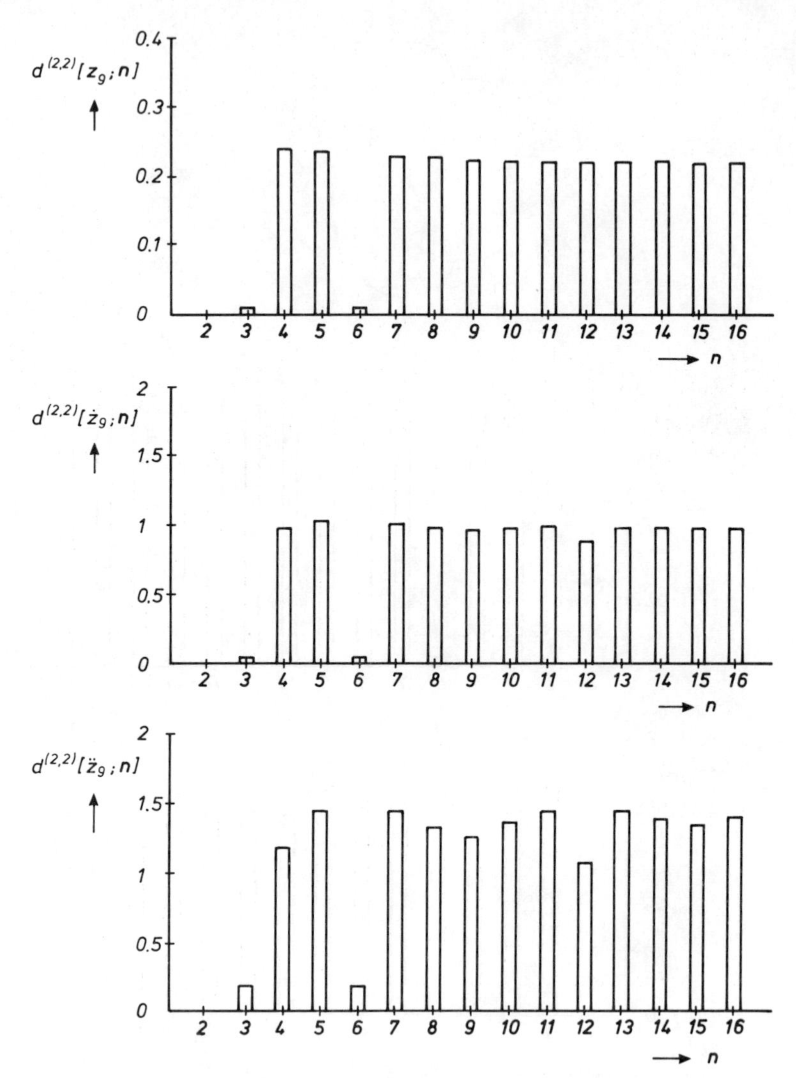


**Figure 4.26b.** Results of the measurement of dissymmetry $\boldsymbol{n}$, for $\boldsymbol{n} = \mathbf{2}, \ldots, \mathbf{16}$, for contour $\gamma_9$ from Figure 4.23. The measure of dissymmetry $\boldsymbol{n}$ used is $d^{(p,q)}[f; \boldsymbol{n}]$, with $(p, q) = (2, 2)$; $f$ stands for the contour representations $z_9$, $\dot{z}_9$, $\ddot{z}_9$, $\psi_9$ and $K_9$, respectively.

First we note that a **6**-fold rotationally symmetric contour has also symmetry **2** and symmetry **3**. In general, if a contour is ***n***-fold rotationally symmetric, then it has also rotational symmetry of all orders, greater than **1**, that are divisors of ***n***.

Next we observe from the results in Figures 4.26a and 4.26b that we have obtained relatively low values for the measurement of dissymmetry **12**. This is caused by the **6**-fold rotational symmetry of the con-

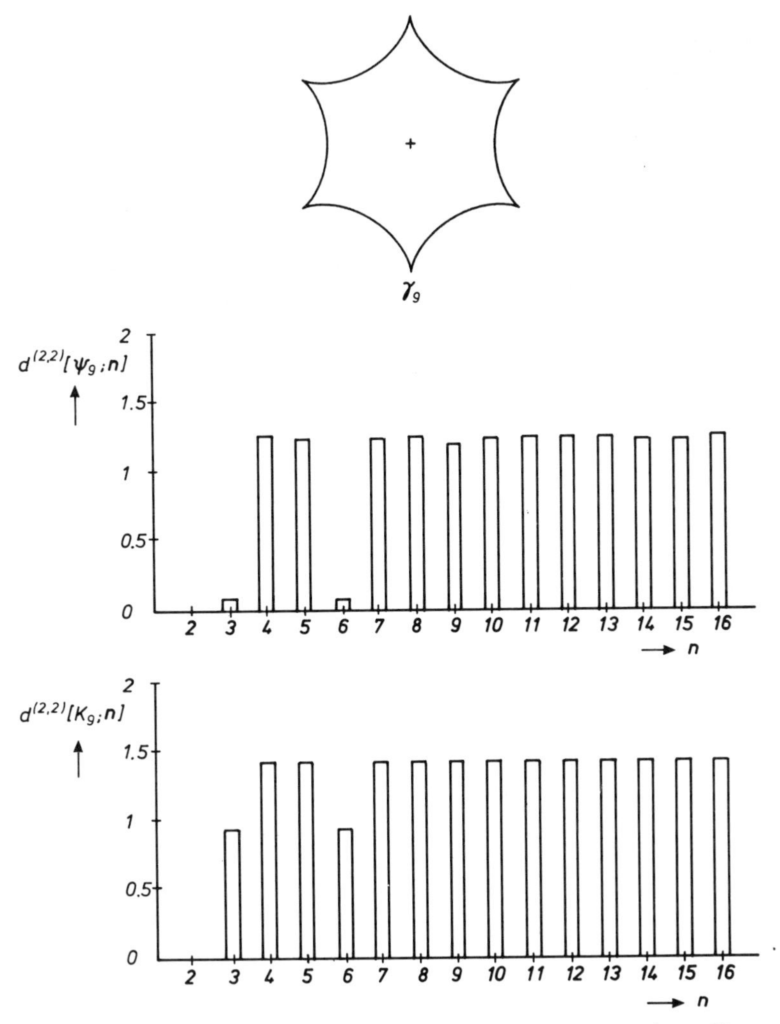


tour and the fact that 12 is a multiple of 6. To see the effects of this upon the measurement of dissymmetry **12** we analyze $d^{(p,q)}(f;\mathbf{12})$

$$d^{(p,q)}(f;\mathbf{12})$$

$$= \left[\frac{1}{11}\sum_{m=1}^{11}\{d^{(p)}(f;\mathbf{12},m)\}^q\right]^{1/q}$$

$$= \left[\frac{1}{11}\sum_{m=1}^{5}\{d^{(p)}(f;\mathbf{12},2m)\}^q + \frac{1}{11}\sum_{m=1}^{6}\{d^{(p)}(f;\mathbf{12},2m-1)\}^q\right]^{1/q}. \tag{4.5.14}$$

If $\nu \in \mathbb{N}$ is a common divisor of both $\boldsymbol{n}$ and $m$, then (cf. Eq. 4.5.8)

$$d^{(p)}(f; \boldsymbol{n}, m) = d^{(p)}\left(f; \frac{\boldsymbol{n}}{\nu}, \frac{m}{\nu}\right). \tag{4.5.15}$$

Using this property, Eq. 4.5.14 can be rewritten as

$$d^{(p,q)}(f; \mathbf{12})$$

$$= \left[\frac{1}{11} \sum_{m=1}^{5} \{d^{(p)}(f; \mathbf{6}, m)\}^q + \frac{1}{11} \sum_{m=1}^{6} \{d^{(p)}(f; \mathbf{12}, 2m-1)\}^q\right]^{1/q}$$

$$= \left[\frac{5}{11} \{d^{(p,q)}(f; \mathbf{6})\}^q + \frac{1}{11} \sum_{m=1}^{6} \{d^{(p)}(f; \mathbf{12}, 2m-1)\}^q\right]^{1/q}. \tag{4.5.16}$$

For a **6**-fold rotationally symmetric contour $d^{(p,q)}(f; \mathbf{6}) = 0$ and Eq. 4.5.16 reduces to

$$d^{(p,q)}(f; \mathbf{12}) = \left[\frac{1}{11} \sum_{m=1}^{6} \{d^{(p)}(f; \mathbf{12}, 2m-1)\}^q\right]^{1/q} \tag{4.5.17}$$

which explains the relatively low value of $d^{(p,q)}[f; \mathbf{12}]$ in Figures 4.26a and 4.26b, where we used $(p, q) = (2, 2)$.

In general, if a contour has rotational symmetry $\boldsymbol{n}$, then this influences the measurement of dissymmetries for that contour of orders that are multiples of $\boldsymbol{n}$. However, we can limit this effect by increasing the value of $q$. For example, the larger the value of $q$, the more the value of $d^{(p,q)}(f; \mathbf{12})$ will be influenced by the largest component of dissymmetry **12** among the values $d^{(p)}(f; \mathbf{12}, 2m - 1)$, $m = 1, \ldots, 6$, and thus, the smaller the influence of $d^{(p,q)}(f; \mathbf{6}) = 0$ (cf. Eq. 4.5.16).

In addition, we remark that the property mentioned in Eq. 4.5.15 can be used effectively to reduce the number of computations when $d^{(p,q)}[f; \boldsymbol{n}]$ has to be computed for several values of $\boldsymbol{n}$.

A second important observation resulting from our experiments on the measurement of rotational dissymmetry concerns the relation between the order of rotational symmetry $\boldsymbol{n}$ and the number of contour representation samples $N$. Though the contours in our experiment all are **2**-, **3**-, and **6**-fold rotationally symmetric, we found that the values for $d^{(2,2)}[f; \mathbf{3}]$ and $d^{(2,2)}[f; \mathbf{6}]$ are always substantially higher than those for $\mathbf{d}^{(2,2)}[f; \mathbf{2}]$. This is caused by the fact that neither **3** nor **6** are divisors of $N = 512$, as opposed to **2**. The values of $d^{(2,2)}[z; \mathbf{3}]$, when expressed in units of sample distance along the contour, are 0.333 for all 12 contours in Figure 4.23. If the order of rotational dissymmetry to be measured is not a divisor of the number of contour representation samples, then the starting point shifts in the discrete measures of dissymmetry $\boldsymbol{n}$, $d^{(p,q)}[f; \boldsymbol{n}]$, have to be rounded to the nearest integer number of samples. In each of the two starting point shifts in $d^{(2,2)}[z; \mathbf{3}]$this rounding is over ⅓ of a sample distance, which explains our results and at the same time confirms that the contours in this experiment may be considered as polygons at the level of a single sample distance for $N = 512$.

If we wish to measure rotational dissymmetry of contours for a number of orders $\boldsymbol{n}$, then a rounding of starting point shifts is hard to avoid. For example, for measurement of dissymmetry $\boldsymbol{n}$ of all orders up to **16** we would need as many as 720720 contour representation samples! In the special case of $p = 2$ the measure of dissymmetry $d^{(p,q)}[f; \boldsymbol{n}]$ can be computed via the Fourier domain (cf. Eqs. 4.5.10b and 4.2.42), where a starting point shift results in a phase shift in the Fourier coefficients. In floating point arithmetic these phase shifts can be executed with great precision. On the other hand, we found in our experiment that, even if we disregard the effect of rounding starting point shifts, the discriminative power of the measures of dissymmetry $\boldsymbol{n}$ for the presence or absence of rotational symmetry remains sufficient.

As a figure of merit for the discriminative power of a measure of dissymmetry $\boldsymbol{n}$ for a certain contour $\gamma$ we can use the proportion

$$\frac{\min\limits_{\{\boldsymbol{n}:\, \gamma \text{ does not have } \boldsymbol{n}\}} d^{(p,q)}[f; \boldsymbol{n}]}{\max\limits_{\{\boldsymbol{n}:\, \gamma \text{ has } \boldsymbol{n}\}} d^{(p,q)}[f; \boldsymbol{n}]}$$

i.e. the proportion of the minimum value of dissymmetry, taken over all orders of rotational symmetry that a contour does not have, and the maximum value of dissymmetry, taken over all orders of rotational symmetry that a contour does have. Since all contours in our experiment are **6**-fold rotationally symmetric and $d^{(2,2)}[f; \boldsymbol{n}]$ was measured for all orders of rotational symmetry $\boldsymbol{n}$ up to **16**, $d^{(2,2)}[f; \boldsymbol{n}]$ was minimized in the numerator of this proportion over $\boldsymbol{n} = \mathbf{4}, \mathbf{5}, \mathbf{7}, \ldots, \mathbf{16}$, and was maximized in the denominator over $\boldsymbol{n} = \mathbf{2}, \mathbf{3}$ and **6**.

In Table 4.12 we have listed the minimum and the maximum values of this figure of merit that we found in our experiment for the contour representations $z$, $\dot{z}$, $\ddot{z}$, $\psi$ and $K$. From this table we observe that the discriminative power of $d^{(2,2)}[z; \boldsymbol{n}]$, $d^{(2,2)}[\dot{z}; \boldsymbol{n}]$ and $d^{(2,2)}[\psi; \boldsymbol{n}]$, for the set of contours in Figure 4.23, is similar. The performance of $d^{(2,2)}[\ddot{z}; \boldsymbol{n}]$ and especially of $d^{(2,2)}[K; \boldsymbol{n}]$ is much worse. This is again caused by the difficulty of obtaining a reliable estimate for $\ddot{z}$ or $K$ if the contours are not sufficiently smooth. Not surprisingly, the maximum figures of merit for $d^{(2,2)}[\ddot{z}; \boldsymbol{n}]$ and $d^{(2,2)}[K; \boldsymbol{n}]$ occur for the smoothest regular contours in our test set ($\gamma_4$ and $\gamma_7$, respectively).

We can use the proportion

$$\frac{\displaystyle\max_{\{\boldsymbol{n}:\, \gamma \text{ does not have } \boldsymbol{n}\}} d^{(p,q)}[f; \boldsymbol{n}]}{\displaystyle\min_{\{\boldsymbol{n}:\, \gamma \text{ does not have } \boldsymbol{n}\}} d^{(p,q)}[f; \boldsymbol{n}]}$$

as a measure of the variability of $d^{(p,q)}[f; \boldsymbol{n}]$ for orders of rotational symmetry that a contour $\gamma$ does not possess.

In this experiment, with index pair $(p, q) = (2, 2)$, this variability is relatively low in comparison with the figures of merit in Table 4.12.

**Table 4.12.** Maximum and minimum figures of merit of measures of dissymmetry $\boldsymbol{n}$, for $\boldsymbol{n}$ up to **16**, found experimentally for the contours in Figure 4.23. The contours for which the maxima and minima occurred are given in parentheses.

| Measure of dissymmetry $\boldsymbol{n}$ | Minimum figure of merit | Maximum figure of merit |
|---|---|---|
| $d^{(2,2)}[z; \boldsymbol{n}]$ | 15.57 ($\gamma_4$) | 52.39 ($\gamma_{12}$) |
| $d^{(2,2)}[\dot{z}; \boldsymbol{n}]$ | 19.56 ($\gamma_6$) | 38.48 ($\gamma_{10}$) |
| $d^{(2,2)}[\ddot{z}; \boldsymbol{n}]$ | 5.56 ($\gamma_3$) | 13.69 ($\gamma_4$) |
| $d^{(2,2)}[\psi, \boldsymbol{n}]$ | 14.00 ($\gamma_9$) | 51.81 ($\gamma_{10}$) |
| $d^{(2,2)}[K; \boldsymbol{n}]$ | 1.19 ($\gamma_{12}$) | 21.09 ($\gamma_7$) |

It ranges from 1.001 for $d^{(2,2)}[K; \boldsymbol{n}]$ (for $\gamma_8$ and $\gamma_9$) to 1.418 for $d^{(2,2)}[\ddot{z}; \boldsymbol{n}]$ (for $\gamma_2$). The figures of merit for the measures of dissymmetry $\boldsymbol{n}$ would improve substantially in this experiment (at least by one order of magnitude) if we choose the number of contour representation samples to be a multiple of **6**.

Based on the results just mentioned we may draw the conclusion that we have defined a powerful family of measures of dissymmetry $\boldsymbol{n}$. For the contours in our test set the measures based on $z$, $\dot{z}$ and $\psi$ all perform very well. There seems to be no preference as to which of these contour representations is most favorable. If we wish to measure dissymmetry $\boldsymbol{n}$ on the basis of $\ddot{z}$ or $K$, then a contour must be sufficiently smooth. Otherwise $\ddot{z}$ and $K$ should not be used.

In practice, rotational symmetry will only be of interest up to a certain maximum order. All rotational symmetries of higher orders than this maximum are taken to be **∞**-fold, and the contour with this property is considered to be a circle. This maximum order of rotational symmetry of interest depends on the application at hand and on the two-dimensional sampling resolution used. In an inspection system for

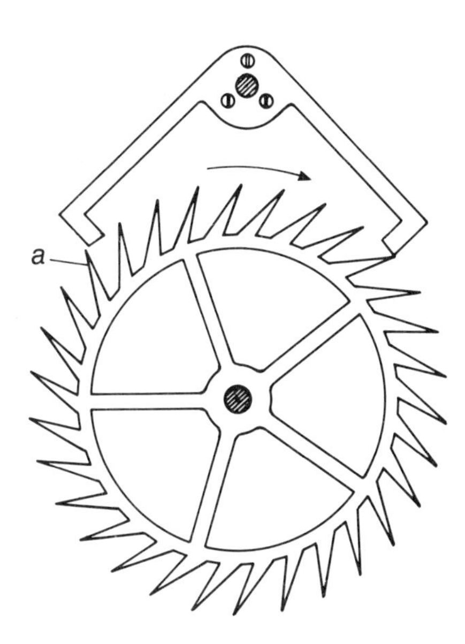


**Figure 4.27.** **30**-fold rotational symmetry in the outer contour of the escapement wheel ($a$) of a clock (dead beat escapement). The center of symmetry is formed by the center of the pivot of the escapement wheel. Note that the internal contours of the escapement wheel have a **5**-fold rotationally symmetric arrangement (from: De Carle [1959], p. 111).

industrial parts, Perkins [1978] considers rotational symmetries up to the order **16**, as we did in our experiment. However, situations exist where considerably higher orders of rotational symmetry are of interest. Consider for example the outer contour of the escapement wheel of a clock in Figure 4.27, having **30**-fold rotational symmetry on its outer boundary. Note that the crossings in the escapement wheel only lead to a **5**-fold internal rotational symmetry.

Most propositions for the measurement of dissymmetry $\boldsymbol{n}$ in a contour that can be found in the literature are on the basis of a Fourier representation of that contour. Therefore we will also define such measures and discuss the relations of these measures with the propositions in the literature. This can be done along the same lines as we defined dissimilarity measures based on the Fourier representations of contours in Section 4.2.3.

**Definition 4.16.** *m-th component of dissymmetry* $\boldsymbol{n}$ *of index p based on Fourier representations.*
Let $\hat{f}$ be the Fourier representation generated by a contour representation $f$, where $f$ acts as a generic symbol for the contour representations $z$, $\dot{z}$, $\ddot{z}$, $\psi$ and $K$. Then the $m$-th component of dissymmetry $\boldsymbol{n}$ for a contour $\gamma$, with Fourier representation $\hat{f}$, $\hat{f} \in \ell^p(\mathbb{Z})$, is defined as (cf. Eq. 3.6.3)

$$d^{(p)}(\hat{f}; \boldsymbol{n}, m) = \|\hat{f}^* - \mathcal{T}_{m(2\pi/\boldsymbol{n})}\mathcal{R}_{m(2\pi/\boldsymbol{n})}\hat{f}^*\|_p, \qquad m \in \mathbb{Z}, \quad 1 \leqslant p \leqslant \infty \tag{4.5.18}$$

where $\hat{f}^*$ is related to $\hat{f}$ as given in Eqs. 4.2.38 and 4.2.39.

□

This definition is completely analogous to that of $d^{(p)}(f; \boldsymbol{n}, m)$ in Definition 4.14. It is easily verified that all properties that we derived for $d^{(p)}(f; \boldsymbol{n}, m)$ also hold for $d^{(p)}(\hat{f}; \boldsymbol{n}, m)$. In analogy with the definition of $d^{(p,q)}(f; \boldsymbol{n})$ in Definition 4.15, we obtain a measure of dissymmetry $\boldsymbol{n}$ based on the Fourier representation $\hat{f}$ by averaging over the $d^{(p)}(\hat{f}; \boldsymbol{n}, m)$.

**Definition 4.17.** *Measure of dissymmetry* $\boldsymbol{n}$ *of index pair (p, q) based on Fourier representations.*

Let $\hat{f}$ be the Fourier representation generated by a contour representation $f$, where $f$ acts as a generic symbol for the contour representations $z, \dot{z}, \ddot{z}, \psi$ and $K$. Then a measure of dissymmetry $\boldsymbol{n}$ of index pair $(p, q)$ for a contour $\gamma$, with Fourier representation $\hat{f}, \hat{f} \in \ell^p(\mathbb{Z})$, is defined as (cf. Eq. 3.6.3 and Definition 4.16)

$$d^{(p,q)}(\hat{f}; \mathbf{1}) \equiv 0, \qquad (4.5.19a)$$

$$d^{(p,q)}(\hat{f}; \boldsymbol{n})$$

$$= \left[\frac{1}{\boldsymbol{n}-1} \sum_{m=1}^{\boldsymbol{n}-1} \{d^{(p)}(\hat{f}; \boldsymbol{n}, m)\}^q\right]^{1/q}$$

$$= \left[\frac{1}{\boldsymbol{n}-1} \sum_{m=1}^{\boldsymbol{n}-1} \|\hat{f}^* - \mathcal{T}_{m(2\pi/\boldsymbol{n})}\mathcal{R}_{m(2\pi/\boldsymbol{n})}\hat{f}^*\|_p^q\right]^{1/q} \qquad \boldsymbol{n} \geqslant \mathbf{2},$$

for $1 \leqslant p, q \leqslant \infty$. (4.5.19b)

□

The required invariance of $d^{(p,q)}(\hat{f}; \boldsymbol{n})$ for the position, size and orientation of an object and for the position of the parametric starting point on its contour can be established in the same way as we did for $d^{(p,q)}(f; \boldsymbol{n})$. Similarly it can be established that $d^{(p,q)}(\hat{f}; \boldsymbol{n}) = 0$ iff the contour, represented by $\hat{f}$, is $\boldsymbol{n}$-fold rotationally symmetric and that $d^{(p,q)}(\hat{f}; \boldsymbol{n}) > 0$ otherwise, as required.

For the range of $d^{(p,q)}(\hat{f}; \boldsymbol{n})$ we find, through Minkowski's inequality for sums (cf. Appendix A), results that are completely analogous to the corresponding results for $d^{(p,q)}(f; \boldsymbol{n})$ in Eqs. 4.5.11 and 4.5.12. Also for the properties in Eqs. 4.5.13a-4.5.17 analogues in terms of $\hat{f}$ can be derived. The effect of varying the value of the index $q$ in $d^{(p,q)}(\hat{f}; \boldsymbol{n})$ is identical to its effect in $d^{(p,q)}(f; \boldsymbol{n})$.

We will now take a closer look at $d^{(p,q)}(\hat{f}; \boldsymbol{n})$ and exhibit the resemblance of these measures to the propositions for a measure of dissymmetry $\boldsymbol{n}$ by Zahn and Roskies [1972] in Eq. 4.5.6, and by Burkhardt [1979] in Eq. 4.5.7.

First, let $\hat{f}$ stand for any of the Fourier representations $\hat{z}$, $\hat{\dot{z}}$ or $\hat{\ddot{z}}$. Then, after substitution of the representations of the similarity operators $\mathcal{T}_{m(2\pi/\boldsymbol{n})}$ and $\mathcal{R}_{m(2\pi/\boldsymbol{n})}$ (cf. Table 3.2), we obtain for $d^{(p,q)}(\hat{f};\boldsymbol{n})$

$$d^{(p,q)}(\hat{f};\boldsymbol{n})$$

$$= \left[\frac{1}{\boldsymbol{n}-1}\sum_{m=1}^{\boldsymbol{n}-1}\left\{\sum_{k\in\mathbb{Z}}\left|\hat{f}^*(k)-e^{-i(k-1)(2\pi/\boldsymbol{n})m}\hat{f}^*(k)\right|^p\right\}^{q/p}\right]^{1/q}$$

$$= \left[\frac{1}{\boldsymbol{n}-1}\sum_{m=1}^{\boldsymbol{n}-1}\left\{\sum_{k\in\mathbb{Z}}\left(\left|1-e^{-i(k-1)(2\pi/\boldsymbol{n})m}\right|^p\cdot\left|\hat{f}^*(k)\right|^p\right)\right\}^{q/p}\right]^{1/q}$$

(4.5.20)

Since

$$1-e^{-i(k-1)(2\pi/\boldsymbol{n})m}=0 \qquad \text{for } k = 1 \bmod \boldsymbol{n} \tag{4.5.21}$$

we obtain for Eq. 4.5.20

$$d^{(p,q)}(\hat{f};\boldsymbol{n})$$

$$= \left[\frac{1}{\boldsymbol{n}-1}\sum_{m=1}^{\boldsymbol{n}-1}\left\{\sum_{\substack{k\in\mathbb{Z}\\ k\neq 1 \bmod \boldsymbol{n}}}\left(\left|1-e^{-i(k-1)(2\pi/\boldsymbol{n})m}\right|^p\cdot\left|\hat{f}^*(k)\right|^p\right)\right\}^{q/p}\right]^{1/p}.$$

(4.5.22)

In this equation we see that the summation over $k$ in $d^{(p,q)}(\hat{f};\boldsymbol{n})$ involves only those Fourier coefficients that are zero-valued for an $\boldsymbol{n}$-fold rotationally symmetric contour (cf. Table 3.6). In this respect $d^{(p)}(\hat{f};\boldsymbol{n})$ is similar to the propositions by Zahn and Roskies [1972] and by Burkhardt [1979]. For the special case of index pair $(p, q) = (2, 2)$ it is straightforward to derive that $d^{(p,q)}(\hat{f};\boldsymbol{n})$ reduces to

$$d^{(2,2)}(\hat{f};\boldsymbol{n}) = \left[2\cdot\sum_{\substack{k\in\mathbb{Z}\\ k\neq 1 \bmod \boldsymbol{n}}}\left|\hat{f}^*(k)\right|^2\right]^{1/2} \tag{4.5.23}$$

which confirms the similarity with the proposition in Eq. 4.5.7 even more. Similarly, let $\hat{g}$ stand for any of the Fourier representations $\hat{\psi}$ or $\hat{K}$. Then, in analogy with the foregoing, we obtain for $d^{(p,q)}(\hat{g}; \boldsymbol{n})$

$$d^{(p,q)}(\hat{g}; \boldsymbol{n}) = \left[ \frac{1}{\boldsymbol{n}-1} \sum_{m=1}^{\boldsymbol{n}-1} \left\{ \sum_{\substack{k \in \mathbb{Z} \\ k \neq 0 \bmod \boldsymbol{n}}} \left( \left|1 - e^{-ik(2\pi/\boldsymbol{n})m}\right|^p \cdot \left|\hat{g}^*(k)\right|^p \right) \right\}^{q/p} \right]^{1/q} \tag{4.5.24}$$

which reduces, in the special case of index pair $(p, q) = (2, 2)$, to

$$d^{(2,2)}(\hat{g}; \boldsymbol{n}) = \left[ 2 \cdot \sum_{\substack{k \in \mathbb{Z} \\ k \neq 0 \bmod \boldsymbol{n}}} \left|\hat{g}^*(k)\right|^2 \right]^{1/2} . \tag{4.5.25}$$

If we compare Eq. 4.5.25 with Eq. 4.5.6 the resemblance between $d^{(p,q)}(\hat{g}; \boldsymbol{n})$ and the proposition for a measure of dissymmetry $\boldsymbol{n}$ by Zahn and Roskies [1972] is obvious.

In practice, a measure of dissymmetry $\boldsymbol{n}$, based on a Fourier representation $\hat{f}$, is computed from a finite set of $N$ Fourier coefficients, resulting in a discrete measure of dissymmetry $\boldsymbol{n}$, $d^{(p,q)}[\hat{f}; \boldsymbol{n}]$. It follows from Eqs. 4.5.22 and 4.5.24 that the computational complexity of $d^{(p,q)}[\hat{f}; \boldsymbol{n}]$ is $O(\boldsymbol{n}N)$, not taking the complexity of the computation of the $N$ Fourier coefficients into account. Thus, the order of the computational complexity of $d^{(p,q)}[\hat{f}; \boldsymbol{n}]$ is the same as that of $d^{(p,q)}[f; \boldsymbol{n}]$.

In the special case of $p = q$ for the index pair $(p, q)$, the computational complexity of $d^{(p,q)}[\hat{f}; \boldsymbol{n}]$ can be reduced to $O(\boldsymbol{n}^2 + N)$ (cf. Eqs. 4.5.22 and 4.5.24), which will usually be smaller than $O(\boldsymbol{n}N)$.

Another special case constitutes $p = 2$. It follows from Eqs. 4.5.8 and 4.5.18 and from Parseval's formula (cf. Eq. 4.2.42) that

$$d^{(2)}(f; \boldsymbol{n}, m) = d^{(2)}(\hat{f}; \boldsymbol{n}, m), \qquad \forall f \in \mathbf{L}^2(2\pi). \tag{4.5.26}$$

With this result we may conclude from Definitions 4.15 and 4.17 that

$$d^{(2,q)}(f; \boldsymbol{n}) = d^{(2,q)}(\hat{f}; \boldsymbol{n}), \qquad \boldsymbol{n} \geq \mathbf{1},$$

$$1 \leq q \leq \infty, \qquad \forall f \in \mathbf{L}^2(2\pi). \tag{4.5.27}$$

Therefore the computational complexity of $d^{(2,q)}[f; \boldsymbol{n}]$ is the same as that of $d^{(2,q)}[\hat{f}; \boldsymbol{n}]$.

In the special case of index pair $(p, q) = (2, 2)$ we find from Eqs. 4.5.23 and 4.5.25 that the computational complexity of $d^{(2,2)}[\hat{f}; \boldsymbol{n}]$ is $O(N)$. Note that the computation of the $N$ Fourier coefficients may dominate the complexity of the computation of $d^{(p,q)}[\hat{f}; \boldsymbol{n}]$.

We remark that, in view of the properties of the Fourier representations of an $\boldsymbol{n}$-fold rotationally symmetric contour

$$\left[ \sum_{\substack{k \in \mathbb{Z} \\ k \neq 1 \bmod \boldsymbol{n}}} \left| \hat{f}^*(k) \right|^p \right]^{1/p}, \qquad 1 \leqslant p \leqslant \infty \tag{4.5.28}$$

can be used as an alternative measure of dissymmetry $\boldsymbol{n}$ of index $p$ based on the Fourier representation $\hat{f}$, where $\hat{f}$ stands for any of the Fourier representations $\hat{z}$, $\hat{\dot{z}}$ or $\hat{\ddot{z}}$.

Likewise

$$\left[ \sum_{\substack{k \in \mathbb{Z} \\ k \neq 0 \bmod \boldsymbol{n}}} \left| \hat{g}^*(k) \right|^p \right]^{1/p}, \qquad 1 \leqslant p \leqslant \infty \tag{4.5.29}$$

where $\hat{g}$ stands for the Fourier representations $\hat{\psi}$ or $\hat{K}$, constitutes such an alternative measure. These alternative measures, on the other hand, cannot be given a direct interpretation in terms of dissimilarity between the original contour and the contour after a symmetry transformation, as can be done for $d^{(p,q)}(\hat{f}; \boldsymbol{n})$, the measure of dissymmetry $\boldsymbol{n}$ proposed in Definition 4.17.

### *4.5.3 Concluding remarks on symmetry measurement*

In the previous two sections, we defined measures of dissymmetry $\boldsymbol{m}$ and dissymmetry $\boldsymbol{n}$. In Definition 2.8 we described symmetry $\boldsymbol{n}\cdot\boldsymbol{m}$ or $\boldsymbol{n}$-fold compositional symmetry as the joint occurrence of symmetry $\boldsymbol{n}$ and of symmetry $\boldsymbol{m}$ in a figure. A measure of dissymmetry $\boldsymbol{n}\cdot\boldsymbol{m}$ can be obtained by combining the results of measuring dissymmetry $\boldsymbol{n}$ and dissymmetry $\boldsymbol{m}$ in an appropriate way.

In this book we have discussed only those types of symmetry that may occur in a single plane figure, and more specifically, that may occur in a plane simple closed curve. Symmetry was defined as the property that a figure comes into coincidence with itself after a symmetry transformation. We based a quantitative evaluation of the extent to which a certain type of symmetry is present in an object on the measurement of the dissimilarity between the original object and the object after the symmetry transformation. This provided us with the measures of dissymmetry.

This principle of quantitative evaluation of symmetry can also be applied to other types of plane symmetry, that involve more than one contour. For example, mirror-symmetric and/or rotationally symmetric arrangements of figures may occur in the plane (consider the arrangement of the internal contours in the escapement wheel of a clock in Figure 4.27). Also combinations of translational symmetry and/or mirror-symmetry in bands and networks can be encountered. The latter types of symmetry are frequently encountered in the creative arts and in architecture. We refer to Shubnikov and Koptsik [1974] for a theoretical account of such types of symmetries. This reference also contains numerous examples. A rich source of examples of symmetry in arrangements of figures is provided by the work of M.C. Escher (cf. MacGillavry [1965], Escher et al. [1972]).

A quantitative evaluation of the extent to which a certain type of symmetry is present in an arrangement of figures in the plane can be achieved through a straightforward generalization of the principle of symmetry measurement proposed in the previous sections. That is, we perform the symmetry transformation upon the figures in the arrangement and measure the dissimilarities between the figures in the original arrangement and in the transformed arrangements, which should have come into coincidence with each other if the given type of symmetry were present in the arrangement of figures. If necessary, this process is repeated for various realizations of the symmetry transformation. An appropriate method of averaging over the dissimilarity measurements, resulting from the individual symmetry transformations, will yield the required measure of dissymmetry.

Through this discussion we have indicated that the general principle of dissymmetry measurement, described in the previous sections, can be extended to apply to a wider class of symmetries. As we have just described the general ideas of these extensions, we will not elaborate on this topic any further.

## 4.6 Concluding remarks

In the previous sections of this chapter we have presented a detailed study of (mirror-)dissimilarity and dissymmetry measurement, based on parametric contour representations. After some introductory considerations on dissimilarity measures in Section 4.1, families of (mirror-) dissimilarity measures were introduced in Section 4.2. Our proposals generalize a number of proposals in the literature, to which we referred extensively.

An evaluation of the computational complexities of various forms of the (mirror-)dissimilarity measures in Section 4.2.4 led to a study of contour normalization methods in Section 4.3. Orientation and starting point normalization techniques mainly aim at a reduction of computational complexity. General rules for contour normalization methods were formulated: uniqueness of the normalization result and idempotency of the method for already normalized objects/contours. Though these rules constitute necessary constraints for useful normalization methods, they unfortunately do not lead to unique methods. Two major classes of normalization methods were identified: those based on moments and those based on Fourier coefficients. It was shown that, from a theoretical point of view, the methods based on Fourier coefficients are better adapted to contour representations. Despite the fact that moments do not offer the possibility to normalize the parametric starting point on a contour, they seem to perform well for the normalization of contour position and orientation. We made clear that the inverse of the norm of the contour representation, on which the (mirror-)dissimilarity measure is based, is an appropriate scale normalization parameter.

In Section 4.4 the properties of the families of (mirror-)dissimilarity measures and the relations between them were studied. Also some experimental evidence for their properties was obtained. Clustering experiments on dissimilarity measures revealed that the order of differentiation of a contour representation is a major distinguishing factor for the behavior of these measures. It also became clear that great care must be taken if one wishes to use the contour representations $\ddot{z}$ or $K$ (second order differentiation) for dissimilarity measurement because of the high noise sensitivity of these representations.

Finally in Section 4.5 we introduced methods to quantify the extent of mirror- and rotational symmetry in plane objects. By using 'coming into coincidence with itself upon a symmetry transformation' as a definition of symmetry, the concept of (mirror-)dissimilarity naturally came in. This enabled us to use the previously defined (mirror-)dissimilarity measures as elements in newly defined families of dissymmetry measures. The effectiveness of these measures was demonstrated in an experiment.

In conclusion we can state that this chapter has led to the formulation of a theoretically consistent framework for dissimilarity measurement between contours and for dissymmetry measurement in contours, which can be tailored easily to specific applications.

## References

Abramowitz, M. and I.A. Stegun, Eds. [1972]
*Handbook of Mathematical Functions with Formulas, Graphs, and Mathematical Tables*, Washington, DC: US Dept. of Commerce, National Bureau of Standards.

Aho, A.V., J.E. Hopcroft and J.D. Ullman [1974]
*The Design and Analysis of Computer Algorithms*, Reading, MA: Addison-Wesley.

Alt, F.L. [1962]
'Digital Pattern Recognition by Moments', Journ. ACM **9**: 240-258.

Anderberg, M.R. [1973]
*Cluster Analysis for Applications*, New York: Academic Press.

Asada, H. and M. Brady [1986]
'The Curvature Primal Sketch', IEEE Trans. Patt. Anal. and Mach. Intell. **PAMI-8**: 2-14.

Babaud, J., A.P. Witkin, M. Baudin and R.O. Duda [1986]
'Uniqueness of the Gaussian Kernel for Scale-Space Filtering', IEEE Trans. Patt. Anal. and Mach. Intell. **PAMI-8**: 27-33.

Bamieh, B. and R.J.P. de Figueiredo [1986]
'A General Moment-Invariants/Attributed-Graph Method for Three-Dimensional Object Recognition from a Single Image', IEEE Journ. Robot. and Automat. **RA-2:** 31-41.

Beckenbach, E.F. and R. Bellman [1971]
*Inequalities*, Third Printing, Berlin: Springer-Verlag.

Blum, H. [1973]
'Biological Shape and Visual Science (Part 1)', Journ. Theor. Biol. **38**: 205-287.

Bolles, R.C. [1979]
'Symmetry Analysis of Two-Dimensional Patterns for Computer Vision', Proc. of the Sixth Intl. Joint Conf. on Artificial Intelligence, Tokyo, Japan: 70-72.

Burkhardt, H. [1979]
*Transformationen zur Lageinvarianten Merkmalgewinnung* (Habilitationsschrift), Düsseldorf, Germany: VDI-Verlag, Fortschritt-Berichte der VDI-Zeitschriften, Reihe 10 (*Angewandte Informatik*), Nr. 7.

Casey, R.G. [1970]
'Moment Normalization of Handprinted Characters', IBM Journ. of Res. and Dev. **14**: 548-557.

Chaudhuri, B.B. and D. Dutta Majumder [1980]
'Recognition and Fuzzy Description of Sides and Symmetries of Figures by Computer', Intern. Journ. of Syst. Sci. **11**: 1435-1445.

Chen, C.-J. and Shi Q.-Y. [1980]
'Shape Features for Cancer Cell Recognition', Proc. Fifth Intl. Conf. on Patt. Recogn., Miami Beach, FL: 579-581.

Cooley, J.W. and J.W. Tukey [1965]
'An Algorithm for the Machine Calculation of Complex Fourier Series', Math. Comp. **19**: 297-301.

Crimmins, T.R. [1982]
'A Complete Set of Fourier Descriptors for Two-Dimensional Shapes', IEEE Trans. Syst., Man and Cybern. **SMC-12**: 848-855.

Cyganski, D. and J.A. Orr [1985]
'Applications of Tensor Theory to Object Recognition and Orientation Determination', IEEE Trans. Patt. Anal. and Mach. Intell. **PAMI-7**: 662-673.

Davis, L.S. [1977a]
'Understanding Shape: Angles and Sides', IEEE Trans. Comp. **C-26**: 236-242.

Davis, L.S. [1977b]
'Understanding Shape: Symmetry', IEEE Trans. Syst., Man and Cybern. **SMC-7**: 204-212.

De Boor, C. [1978]
*A Practical Guide to Splines*, New York: Springer-Verlag.

De Carle, D. [1959]
*Watch and Clock Encyclopedia*, Second Edition, London: N.A.G. Press, Ltd.

Den Hartog, J.P. [1949]
*Strength of Materials*, New York: McGraw-Hill Book Co., Inc.

Dubes, R.C. and A.K. Jain [1980]
'Clustering Methods in Exploratory Data Analysis'. In: *Advances in Computers, Volume 19*, M.C. Yovits (Ed.): 113-228, New York: Academic Press.

Dudani, S.A., K.J. Breeding and R.B. McGhee [1977]
'Aircraft Identification by Moment Invariants', IEEE Trans. Comp. **C-26**: 39-46.

Edwards, R.E. [1979]
*Fourier Series, a Modern Introduction, Volume 1*, New York: Springer-Verlag.

Edwards, R.E. [1982]
*Fourier Series, a Modern Introduction, Volume 2*, New York: Springer-Verlag.

Ehrlich, R. and B. Weinberg [1970]
'An Exact Method for Characterization of Grain Shape', Journ. of Sedim. Petrol. **40**: 205-212.

Escher, M.C., J.L. Locher, C.H.A. Broos and H.S.M. Coxeter [1972]
*The World of M.C. Escher*, New York: H.N. Abrams

Exel, K. [1978]
*Ermittlung von Fourier-Deskriptoren zur Strukturspezifischen Merkmalgewinnung* (Diplomarbeit, Institut für Mess- und Regelungstechnik in der Fakultät für Maschinenbau der Universität Karlsruhe, Karlsruhe, Germany).

Faux, I.D. and M.J. Pratt [1979]
*Computational Geometry for Design and Manufacture*, Chichester: Ellis Horwood, Ltd.

Freeman, H. [1978a]
'Shape Description via the Use of Critical Points', Patt. Recogn. **10**: 159–166.

Freeman, H. [1979]
'Use of Incremental Curvature for Describing and Analyzing Two-Dimensional Shape', Proc. IEEE Comp. Soc. Conf. on Patt. Recogn. and Image Proc., Chicago, IL: 437-444.

Freeman, H. and J.M. Glass [1969]
'On the Quantization of Line-Drawing Data', IEEE Trans. Syst., Sci. and Cybern. **SSC-5**: 70-79.

Granlund, G.H. [1972]
'Fourier Preprocessing for Hand Print Character Recognition', IEEE Trans. Comp. **C-21**: 195-201.

Hamming, R.W. [1977]
*Digital Filters*, Englewood Cliffs, NJ: Prentice-Hall, Inc.

Hardy, G.H., J.E. Littlewood and G. Pólya [1952]
*Inequalities*, Second Edition, Cambridge, England: Cambridge University Press.

Hodson, E.K., D.R. Thayer and C. Franklin [1981]
'Adaptive Gaussian Filtering and Local Frequency Estimates Using Local Curvature Analysis', IEEE Trans. Acoust., Speech and Signal Proc. **ASSP-29**: 854-859.

Horn, B.K.P. [1983]
'The Curve of Least Energy', ACM Trans. Math. Softw. **9**: 441-460.

Hu, M.K. [1962]
'Visual Pattern Recognition by Moment Invariants', IRE Trans. Inf. Th. **IT-8**: 179-187.

Kuhl, F.P. and C.R. Giardina [1982]
'Elliptic Fourier Features of a Closed Contour', Comp. Graph. and Im. Proc. **18**: 236-258.

Landau, L.D. and E.M. Lifschitz [1970]
*Theory of Elasticity*, Second Edition, Oxford: Pergamon Press.

Lawrence, J. Dennis [1972]
*A Catalog of Special Plane Curves*, New York: Dover Publications, Inc.

Luerkens, D.W., J.K. Beddow and A.F. Vetter [1982a]
'Morphological Fourier Descriptors', Powd. Technol. **31**: 209-215.

McClellan, J.H. and C.M. Rader [1979]
*Number Theory in Digital Signal Processing*, Englewood Cliffs, NJ: Prentice-Hall, Inc.

McClellan, J.H., T.W. Parks and L.R. Rabiner [1973]
'A Computer Program for Determining Optimum FIR Linear Phase Digital Filters', IEEE Trans. Audio and Electroac. **AU-21**: 506-526.

MacGillavry, C.H. [1965]
*Symmetry Aspects of M.C. Escher's Periodic Drawings*, Utrecht, The Netherlands: A. Oosthoek's Uitgeversmaatschappij N.V.

Marr, D. [1982]
*Vision*, San Francisco, CA: W.H. Freeman and Co.

Marr, D. and E. Hildreth [1980]
'Theory of Edge Detection', Trans. Roy. Soc. London **B 207**: 187-217.

Mitchell, O.R. and T.A. Grogan [1984]
'Shape Descriptors of Object Boundaries for Computer Vision'. In: *Intelligent Robots: Third International Conference on Robot Vision and Sensory Controls RoViSeC3*, D.P. Casasent and E.L. Hall (Eds.), Proc. SPIE **449**: 685-692.

Mitchell, O.R., A.P. Reeves and T.A. Grogan [1982]
'Algorithms and Architectures for Global Shape Analysis in Time-Varying Imagery'. In: *Robotics and Industrial Inspection*, D.P. Casasent (Ed.), Proc. SPIE **360**: 190-197.

Mokhtarian, F. and A. Mackworth [1986]
'Scale-Based Description and Recognition of Planar Curves and Two-Dimensional Shapes', IEEE Trans. Patt. Anal. and Mach. Intell. **PAMI-8**: 35-43.

Nagy, G. and N. Tuong [1970]
'Normalization Techniques for Handprinted Numerals', Comm. ACM **13**: 475-481.

Nguyen, N.G., R.S. Poulsen and C. Louis [1983]
'Some New Color Features and Their Application to Cervical Cell Classification', Patt. Recogn. **16**: 401-411.

Nussbaumer, H.J. [1981]
*Fast Fourier Transform and Convolution Algorithms*, Berlin: Springer-Verlag.

Oppenheim, A.V. and R.W. Schafer [1975]
*Digital Signal Processing*, Englewood Cliffs, NJ: Prentice-Hall, Inc.

Parui, S.K. and D. Dutta Majumder [1983]
'Symmetry Analysis by Computer', Patt. Recogn. **16**: 63-67.

Pavlidis, T. [1982]
*Algorithms for Graphics and Image Processing*, Berlin: Springer-Verlag.

Perkins, W.A. [1978]
'A Model-Based Vision System for Industrial Parts,' IEEE Trans. Comp. **C-27**; 126–143.

Persoon, E. and K.-S. Fu [1974]
'Shape Discrimination Using Fourier Descriptors', Proc. Second Intl. Joint Conf. on Patt. Recogn., Copenhagen, Denmark: 126-130.

Persoon, E. and K.-S. Fu [1977]
'Shape Discrimination Using Fourier Descriptors', IEEE Trans. Syst., Man and Cybern. **SMC-7**: 170-179.

Proffitt, D. [1982]
'Normalization of Discrete Planar Objects', Patt. Recogn. **15**: 137-143.

Rabiner, L.R., J.H. McClellan and T.W. Parks [1975]
'FIR Digital Filter Design Techniques Using Weighted Chebyshev Approximation', Proc. IEEE **63**: 595-610.

Reeves, A.P. and A. Rostampour [1981]
'Shape Analysis of Segmented Objects Using Moments', Proc. IEEE Comp. Soc. Conf. on Patt. Recogn. and Image Proc., Dallas, TX: 171-174.

Reeves, A.P. and B.S. Wittner [1983]
'Shape Analysis of Three Dimensional Objects Using the Method of Moments', Proc. IEEE Comp. Soc. Conf. on Comp. Vision and Patt. Recogn., Washington, DC: 20-26.

Richard, Jr., C.W. and H. Hemami [1974]
'Identification of Three-Dimensional Objects Using Fourier Descriptors of the Boundary Curve', IEEE Trans. Syst., Man and Cybern. **SMC-4**: 371-378.

Rock, I. [1973]
*Orientation and Form*, New York: Academic Press.

Santisteban, A., N. García and J.L. Carrascosa [1981]
'Digital Analysis of Axially Symmetric Images: Application to Viral Structures', Proc. of the Second Scandinavian Conference on Image Analysis, Helsinki, Finland: 450-455.

Santisteban, A., N. García, J.L. Carrascosa, J. Corral and E. Viñuela [1980]
'Digital Analysis of the Rotational Symmetries Present in Viral Particles', Proc. of Eusipco-80: First European Signal Processing Conference, Lausanne, Switzerland: 59-60.

Shanks, D. [1962]
*Solved and Unsolved Problems in Number Theory, Volume 1*, Washington, DC: Spartan Books.

Shubnikov, A.V. and V.A. Koptsik [1974]
*Symmetry in Science and Art*, New York: Plenum Press.

Singleton, R.C. [1969]
'An Algorithm for Computing the Mixed Radix Fast Fourier Transform', IEEE Trans. Audio and Electroac. **AU-17**: 93-103.

Sneath, P.H.A. and R.R. Sokal [1973]
*Numerical Taxonomy*, San Francisco, CA: W.H. Freeman and Co.

Späth, H. [1980]
*Cluster Analysis Algorithms for Data Reduction and Classification of Objects*, Chichester: Ellis Horwood, Ltd.

Spiegel, M.R. [1964]
*Complex Variables with an Introduction to Conformal Mapping and Its Applications*, Schaum's Outline Series, New York: McGraw-Hill Book Co., Inc.

Sychra, J.J., P.H. Bartels, J. Taylor, M. Bibbo and G.L. Wied [1976]
'Cytoplasmic and Nuclear Shape Analysis for Computerized Cell Recognition', Acta Cytol. **20**: 68-78.

Tai, H.T., C.C. Li and S.H. Chiang [1982]
'Application of Fourier Shape Descriptors to Classification of Fine Particles', Proc. Sixth Intl. Conf. on Patt. Recogn., Munich, Germany: 748-751.

Tang, G.Y. [1982]
'A Discrete Version of Green's Theorem', IEEE Trans. Patt. Anal. and Mach. Intell. **PAMI-4**: 242-249.

Tretter, S.A. [1976]
*Introduction to Discrete-Time Digital Signal Processing*, New York: John Wiley and Sons, Inc.

van Otterloo, P. J. [1978]
'A Feasibility Study of Automated Information Extraction from Anthropological Pictorial Data' (Thesis, Dept. of Elec. Engin., Delft University of Technology, Delft, The Netherlands).

Wallace, T.P. [1981]
'Comments on *Algorithms for Shape Analysis of Contours and Waveforms*', IEEE Trans. Patt. Anal. and Mach. Intell. **PAMI-3**: 593.

Wallace, T.P. and O.R. Mitchell [1979]
'Local and Global Shape Description of Two- and Three-Dimensional Objects', Techn. Report TR-EE 79-43, School of Electrical Engineering, Purdue University, West-Lafayette, IN.

Wallace, T.P. and P.A. Wintz [1980]
'An Efficient Three-Dimensional Aircraft Recognition Algorithm Using Normalized Fourier Descriptors', Comp. Graph. and Im. Proc. **13**: 99-126.

Wechsler, H. [1979]
'A Structural Approach to Shape Analysis Using Mirroring Axes', Comp. Graph. and Im. Proc. **9**: 246-266.

Wieleitner, H. [1908]
*Spezielle Ebene Kurven*, Leipzig: G.J. Göschen'sche Verlagshandlung.

Witkin, A.P. [1983]
'Scale Space Filtering', Proc. of the Eighth Intl. Joint Conf. on Artificial Intelligence, Karlsruhe, Germany: 1019-1022.

Wong, R.Y. and E.L. Hall [1978]
'Scene Matching with Invariant Moments', Comp. Graph. and Im. Proc. **8**: 16-24.

Young, I.T., J.E. Walker and J.E. Bowie [1974]
'An Analysis Technique for Biological Shape – I', Inform. and Contr. **25**: 357-370.

Yuille, A.L. and T.A. Poggio [1986]
'Scaling Theorems for Zero Crossings', IEEE Trans. Patt. Anal. and Mach. Intell. **PAMI-8**: 15-25.

Zahn, C.T. and R.Z. Roskies [1972]
'Fourier Descriptors for Plane Closed Curves', IEEE Trans. Comp. **C-21**: 269-281.

Zvolanek, B. [1981]
'Autonomous Ship Classification by Moment Invariants'. In: *Processing of Images and Data from Optical Sensors*, W.H. Carter (Ed.), Proc. SPIE **292**: 241-248.

# Chapter 5

# Discussion

This discussion concentrates on the general characteristics of the approach to shape analysis in 2-D imagery that was presented in the foregoing chapters. We will establish both the merits and the limitations of the approach and indicate some routes to possible extensions to overcome these limitations.

We will also mention some open problems that deserve attention, both in the context of the approach presented here and in the context of digital shape analysis in general.

## 5.1 General characteristics of the contour-oriented approach to digital shape analysis: merits and limitations

Two main topics dealt with in this book are the quantification of (mirror-)similarity between 2-D shapes and of symmetry in individual 2-D shapes. In order to do so we needed a representation for shapes. The first step was to consider shape information to be concentrated in the shape's contour(s). Next we identified five information-preserving contour representations, three of which are complex-valued, i.e. $z$, $\dot{z}$ and $\ddot{z}$, and the other two are real-valued, i.e. $\psi$ and $K$. In this context information-preserving means that the contour can be reconstructed exactly from the representation, possibly up to a translation, rotation or a scale factor. In Section 2.1 we have shown that some other contour representations that have been proposed in the literature, i.e. $r(t)$, $r'(t)$ and $R(\xi)$, have undesirable properties, for which reason we did not take them into consideration any further.

The foundation for the definition of measures of (mirror-)similarity was laid in Section 2.3 by defining shapes to be (mirror-)similar if they can be mapped into each other by means of (mirror-)similarity transformations. In Section 4.2 we used norms on differences between contour representations as a measure of the extent to which the shapes failed to come into coincidence with each other, thus defining (mirror-)dissimilarity measures. One of the merits of our way of formulating (mirror-)dissimilarity measures is that it brings together a number of earlier proposals in the literature under a general theoretical framework with well-established roots in mathematical analysis. The generality of this framework provides ample room for tuning (mirror-)dissimilarity measurement to particular circumstances and applications.

In the (mirror-)dissimilarity measures defined in Section 4.2 we optimized the orientation and the parametric starting point of one contour with respect to the other. This led to a considerable computational complexity, especially for measures based on orientation variant contour representations. By normalizing orientation and starting point the computational complexity can be reduced to the order of the number of contour representation samples. An in-depth discussion on this subject was presented in Section 4.3, where a general scheme for orientation and starting point normalization based on Fourier coefficients was described. Major dangers of normalization are that the solution is not unique or that the solution found is close to other solutions (in terms of satisfying the normalization criteria). In these cases additional constraints are needed and it may even be more desirable to optimize orientation and starting point over a limited set of normalization candidates. On the other hand, tremendous changes will occur in the years to come in terms of the computational power available. Multiprocessor architectures (cf. e.g. Uhr [1984]) will allow many operations to be executed in parallel. This will call for a reassessment of the time complexity of various computational tasks and may make an optimization of orientation and starting point feasible in practical shape analysis.

For symmetry measurement we took a route that is completely analogous to the one used for similarity measurement. Our definition of symmetry emphasizes that symmetric shapes come into coincidence with themselves upon the appropriate symmetry transformation. This definition has put symmetry measurement in the same perspective as similarity measurement. In fact, the measures of (mirror-)dissimilarity formed the basis for measures of dissymmetry, introduced in Section 4.5. For each type of symmetry, the measures of dissymmetry express the extent to which shapes fail to come into coincidence with themselves upon the associated symmetry transformations. Thus we created a general theoretical framework for the measurement of symmetry $\boldsymbol{m}$ and symmetry $\boldsymbol{n}$ that is quite similar to the theoretical framework for (mirror-)similarity measurement. A major difference is that the concept of a metric is meaningless in the context of symmetry measurement. Our approach to symmetry measurement can be extended to other types of plane symmetry in a straightforward manner.

Fourier representations pervade the literature on contour-oriented shape analysis (usually under the heading: *Fourier descriptors*). Some important observations concerning the usefulness of Fourier representations can be made on the basis of Chapters 3 and 4:

- Fourier coefficients can be useful for contour normalization (cf. Section 4.3).

- Computational efficiency of dissimilarity and dissymmetry measures can be achieved through the Fourier domain, using FFT techniques, if we choose the value of the index $p = 2$ (cf. Sections 4.2.4, 4.5.1 and 4.5.2).

- The Fourier representations of $z(t)$, $\dot{z}(t)$ and $\ddot{z}(t)$, with (normalized) arc length parameter $t$, contain an infinite number of nonzero elements (cf. Section 3.3).

- For a given level of approximation precision finite Fourier representations are often not an appropriate means of data reduction in comparison with direct representation in the contour domain (cf. Section 3.4). (Data reduction has always been one of the main motivations for using Fourier representations.)

- Each Fourier coefficient inherently contains global shape information. Therefore shape analysis based on Fourier representations cannot be adapted to local shape characteristics.

- Dissimilarity measures based on Fourier representations introduce in general different geometries in the space of equivalence classes of similar shapes in comparison to dissimilarity measures based on direct contour representations. This may provide some justification for their definition. However, the interpretation of dissimilarity measurement based on Fourier representations is far more difficult, thus limiting the usefulness of this measurement.

From these observations we draw the conclusion that in a contour-oriented approach to shape analysis dissimilarity measures are preferably defined on the basis of direct contour representations instead of Fourier representations.

Except for some examples we have assumed throughout this book that shape analysis is performed on 2-D contours, i.e. simple closed curves in the plane, represented by a contour representation with a normalized arc length parameter. This causes some limitations:

(a) no direct link to the 3-D world,

(b) shapes are assumed to be simply connected,

(c) only the shape of simple closed curves can be analyzed,

(d) shapes are compared on the basis of normalized arc length parametrizations.

In the following we discuss these limitations and indicate some routes to overcome them.

Re. (a). We have assumed that shape analysis can usefully be performed on 2-D contours. For relatively flat objects (e.g. biological cells, some industrial parts) this is certainly true. For contours that are projections of 3-D objects onto the imaging plane, our approach has implicitly assumed that we deal with perpendicular projections only. It is straightforward to incorporate into our model the possibility that a contour is the result of skewed projection (cf. e.g. Dirilten and Newman [1977], Kanade and Kender [1980], Ballard and Brown [1982], Cyganski, Orr and Pinjo [1983], Brady and Yuille [1984], Cyganski and Orr [1985], Faber and Stokely [1986], Friedberg [1986]). This additional freedom in the third dimension will of course increase the computational complexity of similarity and symmetry analysis.

Re. (b). Though our approach has assumed shapes to be topologically simple, many interesting shapes do not have this property (cf. e.g. Figures 4.25 and 4.27). To extend our approach to shapes that have holes, and possibly even subshapes inside holes, etc., we have to decompose such a topologically nonsimple shape into its constituent regions or contours and represent the shape by a structural description (cf. e.g. Buneman [1970], Pavlidis [1977a], Alexandridis and Klinger [1978], Milgram [1979], Duncan and Andriole [1986]).

An example of representing a composite shape by a tree structure is shown in Figure 5.1. At the level of individual contours our contour-oriented approach to shape analysis can be applied. The results of shape analysis for individual regions or contours can be used at higher hierarchical levels (using a structural shape analysis approach) to analyze the composite shape as a whole.

Re. (c). In many image analysis applications a shape analysis scheme based on simple closed curves may prove to be a serious limitation for a number of reasons. Objects may for example be overlapping. In this case an outer boundary will consist of segments of the boundaries of the overlapping objects. A second reason is the fact that edge detection and image segmentation procedures sometimes do not find complete shape contours, but only parts of these contours. Furthermore, shape

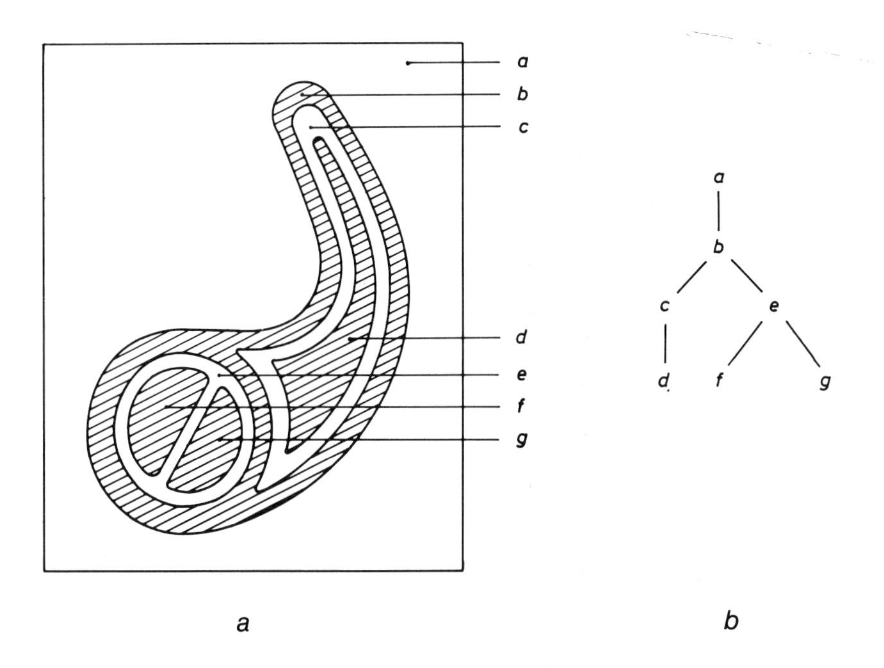


**Figure 5.1.** Example of a composite shape (*a*) represented by a tree structure (*b*). Each node in the tree may have various descriptive attributes about the region it represents. Similarly the links in the tree, that represent the adjacency relations between the regions, may have descriptive attributes.

analysis based on complete contours forces us to maintain a constant contour sampling density everywhere along the contour that is in accordance with the finest shape detail of interest.

In order to deal with these situations it is desirable to be able to analyze the shape of curve segments. We can distinguish two situations: one where one segment is completely matched with another segment and one where a segment is matched to part of another segment.

In the former situation the extension of our method is obtained by positioning the parametric starting point in endpoints of the curve segments. In dissimilarity measurement we need only to vary the parametric starting point over the two endpoints of one of the two curve segments. If the curve representations describe a curve segment from one endpoint to the other, then these representations will, in general, not be periodic. If desired (e.g. for Fourier analysis), the curve representations can be made periodic by making them describe a curve segment from one endpoint to the other *and back* (cf. Impedovo, Marangelli and Fanelli [1978], Dekking and Van Otterloo [1986] and Figure 3.6), for which we coined the term *retracing*.

The latter situation, where a segment is matched to part of another segment (cf. e.g. Turney, Mudge and Volz [1984]) is far more complex. Not only do we have to choose an appropriate starting point on one curve segment to match with one of the endpoints of an other segment, but we also have to determine the appropriate scale of arc length in one curve segment with respect to that in an other, thus increasing the computational complexity. In the process of matching a curve segment to part of an other curve segment Fourier representations cannot be of much use since Fourier coefficients contain only global shape information, as we pointed out earlier.

In both cases the extensions we propose fit in the context of the contour-oriented approach to shape analysis presented in this book.

Re. (d). The choice to use a (normalized) arc length parametrization for contour representations was made on the basis of practical considerations. In order to limit the number of degrees of freedom in contour-oriented shape matching processes this is a natural choice that sets a clear reference for such matching processes. Yet in some applications more flexibility may be desired. For example, some parts of a contour may contain more noise than other parts. In that case the contour is traversed slower in the noisier parts. Clearly this will have disturbing effects on contour-oriented similarity and symmetry measurement based on equidistant contour representation sampling, which assumes equal noise characteristics everywhere along the contour. Similar considerations apply if the amount of shape detail of interest varies along the contour. In the case of varying noise characteristics, contour smoothing/filtering, in conjunction with a contour resampling procedure, may reduce the problems. In the case of varying shape detail along the contour, different techniques are needed.

One method may be to segment the contour in parts with homogeneous characteristics in terms of some homogeneity criteria. This would enable us to maintain the linear relation between the contour segment parameter and arc length and the propositions for shape analysis of contour segments under *re.* (c) would apply. The results of shape analysis on individual contour segments may then be combined to obtain a single result, using combinations of metrics as described in Section 4.1 (e.g. Eq. 4.1.4), or using structural shape analysis techniques, similar to those for topologically nonsimple shapes (as referred to under *re.* (b)).

A second method to overcome problems in contour-oriented shape analysis, caused by the (normalized) arc length parametrization of contour representations, is to relax this parametrization convention. This can be achieved by allowing the relation between the contour parameter and arc length to become nonlinear in one of a pair of contour representations in a shape matching process. Such methods are well known in 1-D signal processing, especially in speech processing (cf. Sakoe and Chiba [1978], Ney [1981], Kuhn, Tomaschewski and Ney [1981], Ney [1982], Anderson and Gaby [1983]). In analogy with the term *time warping* in 1-D signal processing, we may call this method *arc length warping* in the context of contour-oriented shape analysis. For optical character recognition similar methods have been proposed (Fujimoto et al. [1976], Burr [1979], O'Rourke and Washington [1985]). Recently, Cheng and Fu [1987] proposed time warping for matching strings and patterns, along with a VLSI architecture to implement this method.

Dissimilarity measures need to be modified somewhat, when using arc length warping methods, in order to preserve the required properties of such measures (cf. Sections 4.1 and 4.2), which may be lost if we do not control the increased flexibility in the shape matching process. This control can be achieved for example by incorporating a parameter in the dissimilarity measure that expresses the nonlinearity of the relation between arc length and the contour parameter in the matched contour representation. Similar considerations apply to dissymmetry measures.

Another approach to cope with varying detail and noise in contours is due to Zack [1982]. He separates a contour into two components: a smooth contour with homogeneous global characteristics, that can justifiably be parametrized by (normalized) arc length, and an additional noise/detail component, that can be processed separately.

## 5.2 In search of accuracy

Important open problems still exist in the area of the practical implementation of digital image analysis techniques. For example, questions regarding the selection of a 2-D image sampling density and a contour sampling density, that will give a specified measurement accuracy, remain largely unanswered.

The model that we commonly use for image analysis, and for shape analysis in particular, is that of image regions (or objects and background) that occupy mutually exclusive regions in the (bounded) 2-D image plane and whose union comprises the entire image plane. Image functions that correspond to this image model with sharply bounded regions are not bandlimited. Therefore neither bandwidth (Shannon [1949], Jerri [1977]) nor the maximum absolute curvature $K_{max}$ of the contours (Young, Walker and Bowie [1974], van Otterloo and Gerbrands [1978]) can provide a useful criterion for a 2-D sampling theorem aimed at information preservation. As a result, the 2-D sampling process leads to an inevitable loss of information with respect to our picture model (we are temporarily disregarding the bandlimiting effects of physical imaging devices upon the image function). In signal-theoretic terms the finite sampling density causes aliasing while in geometric terms it causes quantization. Some of the effects of finite 2-D sampling are the following:

- It leads to a *tolerance region* (or *domain*, cf. Freeman and Glass [1969]) in the neighborhood of a boundary in segmented digital images. All contours that fit in this tolerance region lead to the same boundary in the segmented digital image. Thus we are confronted with a many-to-one mapping of shapes.
- Since the position and orientation of model contours with respect

to the 2-D sampling pattern is usually random, a single model contour can lead to various contours in the segmented digital image that are in general not geometrically similar. Thus we are also confronted with a one-to-many mapping of shapes.

From these observations it is clear that, given a contour in a digital image, there is uncertainty about the exact size and shape of the model contour and about the exact position and orientation of the model contour with respect to the 2-D sampling pattern.

Bribiesca and Guzman [1980] have proposed to avoid the many-to-many mapping problem by first normalizing the position and orientation of the object to be digitized, thereby evidently creating a chicken-and-egg problem in digital image analysis.

In Section 3.3 we have shown that the contour representations $z$, $\dot{z}$ and $\ddot{z}$, parametrized by (normalized) arc length, are not bandlimited. Since, in practice, dissimilarity and dissymmetry measurements are necessarily based on a finite number of contour representation samples (cf. Section 4.2.4), the contour sampling process constitutes yet another source of information loss.

As indicated above, also the bandlimiting effects of the imaging device upon the image function result in a loss of information. In fact, from a theoretical point of view, each step in the image analysis system produces a loss of information. Because of these information losses we have to deal with approximations, which must be of sufficient accuracy such that the entire measurement process can be performed with the specified accuracy. The goal of the measurement process may, for example, be dissimilarity or dissymmetry measurement, based on a certain type of contour representation.

To enable an efficient design of an image analysis system (i.e. avoiding overkill at various stages) it is extremely important to specify in the model of the scenes or objects to be analyzed what we consider to be relevant shape detail and what not. It must also be specified to which smoothness class we consider contours to belong and, if possible, what we consider to be the relevant value of $K_{max}$ of these contours, etc. On the other hand, to achieve the specified accuracy in the end result it may be clear that not only the precision of each processing step must be in accordance with the required accuracy of the end result, but that this precision must also leave room for subsequent processing steps. This observation explains why it is sensible in an image analysis system, where the emphasis is on measurement accuracy and not on picture communication, to sample images more densely than necessary according to Shannon's sampling theorem (Shannon [1949], Jerri [1977]).

The accuracy of a complete image analysis system is very difficult to analyze (and even more difficult to predict). Most reports in the literature concentrate on the accuracy of part of the image analysis system. This may still be an acceptable approach, as long as we keep in mind that the overall system will put more severe conditions on accuracy than may be apparent from an analysis of part of the system.

For an analysis of the achievable accuracy in dissimilarity and dissymmetry measurement, as proposed in Chapter 4, a good starting point might be to study the influence of the selection of:

- the 2-D image sampling density,
- the contour definition in digital pictures,
- the 1-D contour sampling density,
- processing and estimation procedures for contour representations and contour normalization parameters,

upon this accuracy, using geometrical figures of known shape.

The choice of an appropriate 2-D sampling density for shape analysis purposes has been widely studied. The quantization of the geometry of the plane by the finite 2-D sampling density has given rise to the study of digital topology and digital geometry. For these topics we refer to Rosenfeld and Kak [1982], Serra [1982] and Haas [1985], who also provide references to the relevant literature.

Pavlidis used $K_{max}$ as one of the features in a 2-D sampling theorem (cf. Pavlidis [1980b] or Pavlidis [1982], pp. 130-142). The topological structure of the image regions (i.e. objects, background, etc.) in the model picture is preserved if the conditions in this theorem are satisfied. Thus this theorem sets a minimum requirement on the 2-D sampling density.

Freeman and Glass [1969] determine the curve of minimum bending energy over all curves that lead to the same digital curve. The difference in bending energy in the original curve and this minimum bending energy curve is taken as a measure of degradation in shape detail as a result of 2-D sampling. They also use the maximum curvature, that must be recoverable to a certain accuracy after 2-D sampling, as a criterion for specifying significant detail. In Freeman [1978a] it is stated that for a well-quantized curve variations in direction over minimally 5 or maximally 13 curve segments in the digital curve should not represent significant shape detail. Wallace and Wintz [1980] also state that significant contour variations must extend over many digital curve segments, though they do not specify over how many.

The influence of the choice of a 2-D sampling density upon the accuracy with which shape parameters can be estimated has been widely studied (cf. e.g. Frolov and Maling [1969], Lloyd [1976], Kulpa [1977], Proffitt and Rosen [1979], Ellis et al. [1979], Rosen [1980], Wechsler [1981], Vossepoel and Smeulders [1982], Kulpa [1983], Ho [1983], Dorst and Smeulders [1986], Veillon [1986], Teh and Chin [1986]). Despite all these studies a sound theoretical model for the relation between the 2-D sampling density and the accuracy of parameter estimation is still lacking. The literature on geometrical probability and spatial statistics (cf. Solomon [1953], Matérn [1960], Kendall and Moran [1963], Moran [1966], Moran [1969], Miles [1972], Little [1974], Harding and Kendall [1974], Bartlett [1975], Santaló [1976], Baddeley [1977], Miles and Serra [1978], Miles [1980], Ripley [1981], Baddeley [1982]), on random set theory and integral geometry (cf. Matheron [1975], Serra [1982]) and on stereological methods (cf. Weibel [1979], Weibel [1980]) may provide valuable sources of inspiration in this respect. Rosenfeld [1984] has proposed to use 2-D fuzzy sets to take the geometrical uncertainty, resulting from a finite 2-D sampling, into account. However, for an appropriate specification of such fuzzy sets we need a probabilistic model.

For dissimilarity and dissymmetry measurement as proposed in this book, accurate methods for contour representation estimation are at least as important as those for shape parameter measurement, since this accuracy sets the limits on the discriminative power of dissimilarity and dissymmetry measures. For good performance it is essential that we abandon the geometrically discrete pixel domain (Wallace and Wintz [1980], Wallace, Mitchell and Fukunaga [1981]), i.e. the digital contour must be defined in $\mathbb{C}$ or in $\mathbb{R}^2$ to enable an effective smoothing of the geometric quantization effects of 2-D sampling.

Though there have been some reports on smoothing the digital contour to estimate the position function $z$ (cf. e.g. Dessimoz [1979], Wallace and Wintz [1980]), most reports on smoothing the digital boundary are in conjunction with the estimation of the curvature function $K$ (cf. e.g. Young, Walker and Bowie [1974], Bennett and MacDonald [1975], van Otterloo [1978], Wallace, Mitchell and Fukunaga [1981], Anderson and Bezdek [1984], Asada and Brady [1986], Mokhtarian and Mackworth [1986]). This is not surprising in view of our experience of the sensitivity of $K$ for noise, for distortion and for the finite 1-D contour sampling density. Virtually none of these reports deals with the issue of estimation accuracy.

For appropriate contour representation estimation methods it is again important that we specify the smoothness class, to which the contours, in terms of our model, belong. There should be no discrepancy between the smoothness class of the contours in our model and the type of contour representation used for dissimilarity or dissymmetry measurement (e.g. in practice do not use the curvature function to represent polygons, cf. Shirai [1973]). It may be expected that the higher the derivative involved in the contour representations the more samples will be needed for a reliable estimate.

It should be noted that there are two aspects in contour representation estimation: not only the contour representation itself, but also the relation between the contour representation and arc length is involved, since we proposed to use a (normalized) arc length parameter for contour representations. Therefore contour representation estimation procedures must take the arc length constraint into account. Dessimoz (Dessimoz [1979], Dessimoz [1980]) proposed an iterative procedure to deal with both aspects. To obtain practical guidelines for contour representation estimation, a comparative investigation of the performance of various methods from digital signal processing and numerical analysis (confer the suggestions given in Section 4.4 and the proposal in Appendix C) is desirable. Adaptive estimation techniques (cf. e.g. Hodson, Thayer and Franklin [1981]) may lead to an improved estimation accuracy. Such a comparative study should of course also involve the effects of 2-D sampling, digital boundary definition, 1-D (re)sampling, etc., and is preferably done on the basis of geometric figures with known contour representations. The dissimilarity measures proposed in this book will then constitute appropriate reference to judge the accuracy of contour representation estimation.

## References

Alexandridis, N. and A. Klinger [1978]
'Picture Decomposition, Tree Data-Structures, and Identifying Directional Symmetries as Node Combinations', Comp. Graph. and Im. Proc. **8**: 43-77.

Anderson, I.M. and J.C. Bezdek [1984]
'Curvature and Tangential Deflection of Discrete Arcs: A Theory Based on the Commutator of Scatter Matrix Pairs and Its Application to Vertex Detection in Planar Shape Data', IEEE Trans. Patt. Anal. and Mach. Intell. **PAMI-6**: 27-40.

Anderson, K.R. and J.E. Gaby [1983]
'Dynamic Waveform Matching', Inform. Sci. **31**: 221-242.

Asada, H. and M. Brady [1986]
'The Curvature Primal Sketch', IEEE Trans. Patt. Anal. and Mach. Intell. **PAMI-8**: 2-14.

Baddeley, A. [1977]
'A Fourth Note on Recent Research in Geometrical Probability', Adv. Appl. Prob. **9**: 824-860.

Baddeley, A. [1982]
'Stochastic Geometry: An Introduction and Reading-List', Intern. Statist. Rev. **50**: 179-193.

Ballard, D.H. and C.M. Brown [1982]
*Computer Vision*, Englewood Cliffs, NJ: Prentice-Hall, Inc.

Bartlett, M.S. [1975]
*The Statistical Analysis of Spatial Pattern*, London: Chapman and Hall.

Bennett, J.R. and J.S. MacDonald [1975]
'On the Measurement of Curvature in a Quantized Environment', IEEE Trans. Comp. **C-24**: 803-820.

Brady, M. and A. Yuille [1984]
'An Extremum Principle for Shape from Contour', IEEE Trans. Patt. Anal. and Mach. Intell. **PAMI-6**: 288-301.

Bribiesca, E. and A. Guzman [1980]
'How to Describe Pure Form and How to Measure Differences in Shapes Using Shape Numbers', Patt. Recogn. **12**: 101-112.

Buneman, O.P. [1970]
'A Grammar for the Topological Analysis of Plane Figures'. In: *Machine Intelligence* 5, B. Meltzer and D. Michie (Eds.): 383-393, Edinburgh: Edinburgh University Press.

Burr, D.J. [1979]
'A Technique for Comparing Curves', Proc. IEEE Comp. Soc. Conf. on Patt. Recogn. and Image Proc., Chicago, IL: 271-277.

Cheng, H.D. and K.S. Fu [1987]
'VLSI Architectures for String Matching and Pattern Matching', Patt. Recogn. **20**: 125-141.

Cyganski, D. and J.A. Orr [1985]
'Applications of Tensor Theory to Object Recognition and Orientation Determination', IEEE Trans. Patt. Anal. and Mach. Intell. **PAMI-7**: 662-673.

Cyganski, D., J.A. Orr and Z. Pinjo [1983]
'A Tensor Operator Method for Identifying the Affine Transformation Relating Image Pairs', Proc. IEEE Comp. Soc. Conf. on Comp. Vision and Patt. Recogn., Washington, DC: 361-363.

Dekking, F.M. and P.J. van Otterloo [1986]
'Fourier Coding and Reconstruction of Complicated Contours', IEEE Trans. Syst., Man and Cybern. **SMC-16**: 395-404.

Dessimoz, J.-D. [1979]
'Curve Smoothing for Improved Feature Extraction from Digitized Pictures', Sign. Proc. **1**: 205-210.

Dessimoz, J.-D. [1980]
*Traitement des Contours en Reconnaissance de Formes Visuelles: Application en Robotique* (Ph.D. Thesis, No. 387, Ecole Polytechnique Fédérale de Lausanne, Lausanne, Switzerland).

Dirilten, H. and T.G. Newman [1977]
'Pattern Matching under Affine Transformations', IEEE Trans. Comp. **C-26**: 314-317.

Dorst, L. and A.W.M. Smeulders [1986]
'Length Estimators Compared'. In: *Pattern Recognition in Practice II* (Proceedings of an International Workshop held in Amsterdam, The Netherlands, 1985), E.S. Gelsema and L.N. Kanal (Eds.): 73-78, Amsterdam, The Netherlands: Elsevier Science Publishers, B.V. (North-Holland).

Duncan, J.S. and K.P. Andriole [1986]
'The Integration of Semantic and Parametric Model Matching for Image Analysis', Proc. Eighth Intl. Conf. on Patt. Recogn., Paris, France: 887-889.

Ellis, T.J., D. Proffitt, D. Rosen and W. Rutkowski [1979]
'Measurement of the Lengths of Digitized Curved Lines', Comp. Graph. and Im. Proc. **10**: 333-347.

Faber, T.L. and E.M. Stokely [1986]
'Affine Transform Determination for 3-D Objects: A Medical Imaging Application', Proc. IEEE Comp. Soc. Conf. on Comp. Vision and Patt. Recogn., Miami Beach, FL: 440-445

Freeman, H. [1978a]
'Shape Description via the Use of Critical Points', Patt. Recogn. **10**: 159-166.

Freeman, H. and J.M. Glass [1969]
'On the Quantization of Line-Drawing Data', IEEE Trans. Syst., Sci. and Cybern. **SSC-5**: 70-79.

Friedberg, S.A. [1986]
'Finding Axes of Skewed Symmetry', Comp. Vis., Graph. and Im. Proc. **34**: 138-155.

Frolov, Y.S. and D.H. Maling [1969]
'The Accuracy of Area Measurement by Point Counting Techniques', The Cartograph. Journ. **6**: 21-35.

Fujimoto, Y., S. Kadota, S. Hayashi, M. Yamamoto, S. Yajima and M. Yasuda [1976]
'Recognition of Handprinted Characters by Nonlinear Elastic Matching', Proc. Third Intl. Joint Conf. on Patt. Recogn., Coronado, CA: 113-118.

Haas, H.P.A. [1985]
*Convexity Analysis of Hexagonally Sampled Images* (Ph.D. Thesis, Delft University of Technology, Delft, The Netherlands).

Harding, E.F. and D.G. Kendall (Eds.) [1974]
*Stochastic Geometry: A Tribute to the Memory of Rollo Davidson*, New York: John Wiley and Sons, Inc.

Ho, C.-S. [1983]
'Precision of Digital Vision Systems', IEEE Trans. Patt. Anal. and Mach. Intell. **PAMI-5**: 593-601.

Hodson, E.K., D.R. Thayer and C. Franklin [1981]
'Adaptive Gaussian Filtering and Local Frequency Estimates Using Local Curvature Analysis', IEEE Trans. Acoust., Speech and Signal Proc. **ASSP-29**: 854-859.

Impedovo, S., B. Marangelli and A.M. Fanelli [1978]
'A Fourier Descriptor Set for Recognizing Nonstylized Numerals', IEEE Trans. Syst., Man and Cybern. **SMC-8**: 640-645.

Jerri, A.J. [1977]
'The Shannon Sampling Theorem – Its Various Extensions and Applications: A Tutorial Review', Proc. IEEE **65**: 1565-1596.

Kanade, T. and J.R. Kender [1980]
'Mapping Image Properties into Shape Constraints: Affine-Transformable Patterns, and the Shape-from-Texture Paradigm', Proc. IEEE Workshop on Picture Data Description and Management, Asilomar, CA: 130–135.

Kendall, M.G. and P.A.P. Moran [1963]
*Geometrical Probability*, London: Charles Griffin & Company Limited.

Kuhn, M.H., M.H. Tomaschewski and H. Ney [1981]
'Fast Nonlinear Time Alignment for Isolated Word Recognition', Proc. IEEE Intl. Conf. on Acoust., Speech and Signal Proc. ICASSP-81, Atlanta, GA: 736-740.

Kulpa, Z. [1977]
'Area and Perimeter Measurement of Blobs in Discrete Binary Pictures', Comp. Graph. and Im. Proc. **6**: 434-451.

Kulpa, Z. [1983]
'More about Areas and Perimeters of Quantized Objects', Comp. Vis., Graph. and Im. Proc. **22**: 268-276.

Little, D.V. [1974]
'A Third Note on Recent Research in Geometrical Probability', Adv. Appl. Prob. **6**: 103-130.

Lloyd, P.R. [1976]
Quantization Error in Area Measurement', The Cartograph. Journ. **13**: 22-25.

Matérn, B. [1960]
*Spatial Variation*, Meddelanden från Statens Skogsforskningsinstitut, Band **49**, Nr. 5.

Matheron, G. [1975]
*Random Sets and Integral Geometry*, New York: John Wiley and Sons, Inc.

Miles, R.E. [1972]
'The Random Division of Space', Suppl. Adv. Appl. Prob.: 243-266.

Miles, R.E. [1980]
'A Survey of Geometrical Probability in the Plane, with Emphasis on Stochastic Image Modeling', Comp. Graph. and Im. Proc. **12**: 1-24.

Miles, R.E. and J. Serra (Eds.) [1978]
*Geometrical Probability and Biological Structures: Buffon's 200th Anniversary*, Lecture Notes in Biomathematics **23**, New York: Springer-Verlag.

Milgram, D.L. [1979]
'Constructing Trees for Region Description', Comp. Graph. and Im. Proc. **11**: 88-99.

Mokhtarian, F. and A. Mackworth [1986]
'Scale-Based Description and Recognition of Planar Curves and Two-Dimensional Shapes', IEEE Trans. Patt. Anal. and Mach. Intell. **PAMI-8**: 35-43.

Moran, P.A.P. [1966]
'A Note on Recent Research in Geometrical Probability', Journ. Appl. Prob. **3**: 453-463.

Moran, P.A.P. [1969]
'A Second Note on Recent Research in Geometrical Probability', Adv. Appl. Prob. **1**: 73-89.

Ney, H. [1981]
'Telephone-Line Speaker Recognition Using Clipped Autocorrelation Analysis', Proc. IEEE Intl. Conf. on Acoust., Speech and Signal Proc. ICASSP-81, Atlanta, GA: 188-192.

Ney, H. [1982]
'A Time Warping Approach to Fundamental Period Estimation', IEEE Trans. Syst., Man and Cybern. **SMC-12**: 383-388.

O'Rourke, J. and R. Washington [1985]
'Curve Similarity via Signatures'. In: *Computational Geometry*, G.T. Toussaint (Ed.): 295-317, Amsterdam, The Netherlands: Elsevier Science Publishers, B.V. (North-Holland).

Pavlidis, T. [1977a]
*Structural Pattern Recognition*, New York: Springer-Verlag.

Pavlidis, T. [1980b]
'Shape Analysis and Digitization'. In: *Pattern Recognition in Practice* (Proceedings of an International Workshop held in Amsterdam, The Netherlands, 1980), E.S. Gelsema and L.N. Kanal (Eds.): 123-129, Amsterdam, The Netherlands: Elsevier Science Publishers, B.V. (North-Holland).

Pavlidis, T. [1982]
*Algorithms for Graphics and Image Processing*, Berlin: Springer-Verlag.

Proffitt, D. and D. Rosen [1979]
'Metrication Errors and Coding Efficiency of Chain-Encoding Schemes for the Representation of Lines and Edges', Comp. Graph. and Im. Proc. **10**: 318-332.

Ripley, B.D. [1981]
*Spatial Statistics*, New York: John Wiley and Sons, Inc.

Rosen, D. [1980]
'On the Areas and Boundaries of Quantized Objects', Comp. Graph. and Im. Proc. **13**: 94-98.

Rosenfeld, A. [1984]
'The Fuzzy Geometry of Image Subsets', Patt. Recogn. Lett. **2**: 311-317.

Rosenfeld, A. and A.C. Kak [1982]
*Digital Picture Processing*, 2 Volumes, Second Edition, New York: Academic Press.

Sakoe, H. and S. Chiba [1978]
'Dynamic Programming Algorithm Optimization for Spoken Word Recognition', IEEE Trans. Acoust., Speech and Signal Proc. **ASSP-26**: 43-49.

Santaló, L.A. [1976]
*Integral Geometry and Geometric Probability*, Encyclopedia of Mathematics and Its Applications, Volume 1, Reading, MA: Addison-Wesley.

Serra, J. [1982]
*Image Analysis and Mathematical Morphology*, London: Academic Press.

Shannon, C.E. [1949]
'Communication in the Presence of Noise', Proc. IRE **37**: 10-21.

Shirai, Y. [1973]
'A Context Sensitive Line Finder for Recognition of Polyhedra', Artif. Intell. **4**: 95-119.

Solomon, H. [1953]
'Distribution of the Measure of a Random Two-Dimensional Set', Ann. Math. Statist. **24**: 650-656.

Teh, C.-H. and R.T. Chin [1986]
'On Digital Approximation of Moment Invariants', Comp. Vis., Graph. and Im. Proc. **33**: 318-326.

Turney, J.L., T.N. Mudge and R.A. Volz [1984]
'Experiments in Occluded Parts Recognition'. In: *Intelligent Robots: Third International Conference on Robot Vision and Sensory Controls RoViSeC3*, D.P. Casasent and E.L. Hall (Eds.), Proc. SPIE **449**: 719-725.

Uhr, L. [1984]
*Algorithm-Structured Computer Arrays and Networks*, New York: Academic Press.

van Otterloo, P.J. [1978]
'A Feasibility Study of Automated Information Extraction from Anthropological Pictorial Data' (Thesis, Dept. of Elec. Engin., Delft University of Technology, Delft, The Netherlands).

van Otterloo, P.J. and J.J. Gerbrands [1978]
'A Note on a Sampling Theorem for Simply Connected Closed Contours', Inform. and Contr. **39**: 87-91.

Veillon, F. [1986]
'Study and Comparison of Certain Shape Measures', Sign. Proc. **11**: 81-91.

Vossepoel, A.M. and A.W.M. Smeulders [1982]
'Vector Code Probability and Metrication Error in the Representation of Straight Lines of Finite Length', Comp. Graph. and Im. Proc. **20**: 347-364.

Wallace, T.P., O.R. Mitchell and K. Fukunaga [1981]
'Three-Dimensional Shape Analysis Using Local Shape Analysis', IEEE Trans. Patt. Anal. and Mach. Intell. **PAMI-3**: 310-323.

Wallace, T.P. and P.A. Wintz [1980]
'An Efficient Three-Dimensional Aircraft Recognition Algorithm Using Normalized Fourier Descriptors', Comp. Graph. and Im. Proc. **13**: 99-126.

Wechsler, H. [1981]
'How to Improve the Measurement Process Despite Digitization Effects', Proc. of the Second Scandinavian Conference on Image Analysis, Helsinki, Finland: 35-38.

Weibel, E.R. [1979]
*Stereological Methods, Volume 1: Practical Methods for Biological Morphometry*, New York: Academic Press.

Weibel, E.R. [1980]
*Stereological Methods, Volume 2: Theoretical Foundations*, New York: Academic Press.

Young, I.T., J.E. Walker and J.E. Bowie [1974]
'An Analysis Technique for Biological Shape – I', Inform. and Contr. **25**: 357-370.

Zack, G.W. [1982]
'Finding Local Boundary Characteristics by Wave Extraction', Proc. Sixth Intl. Conf. on Patt. Recogn., Munich, Germany: 458-460.

# Appendix A

# Some mathematical concepts and properties

The purpose of this appendix is to introduce some of the mathematical concepts, notations and properties that have been used at various places in this book. The treatment will be rather cursory. In many cases theorems will be stated without proof. For an in-depth treatment and more mathematical rigor we will refer to standard texts on mathematical analysis, functional analysis and Fourier series theory.

To indicate the order of magnitude of functions, we use the Landau order symbols (cf. Titchmarsh [1939], Zygmund [1959a], Apostol [1974]). Consider two functions $f$ and $g$, defined on a set $S$, with $g(x) \geqslant 0$ for all $x \in S$. By

$$f(x) = O\{g(x)\}$$

we generally mean that there is a constant $c > 0$ such that $|f(x)| < cg(x)$ for all $x \in S$. In particular

$$f(x) = O(1)$$

means that $f(x)$ is a bounded function. By

$$f(x) = o\{g(x)\}$$

as $x \to a$ we mean that $f(x)/g(x) \to 0$ as $x \to a$. In particular

$$f(x) = o\{1\}$$

as $x \to 0$ means that $f(x)$ is a function which tends to zero as $x \to 0$.

In the following we introduce some concepts and properties of function spaces and sequence spaces. We start with some basic definitions and properties, and relate these to the context of this book.

**Definition A.1.** *Metric* (Copson [1968]).
A metric on an abstract set $A$ (whose elements $a_1, a_2, a_3, \ldots$ are called *points*) is a function $d: A \times A \to \mathbb{R}$ such that for all, not necessarily distinct $a_1, a_2, a_3 \in A$:

- $d(a_1, a_2) = 0 \quad \text{iff } a_1 = a_2,$ (A.1)
- $d(a_1, a_2) \leqslant d(a_3, a_1) + d(a_3, a_2),$ (A.2)
- $d(a_1, a_2) = d(a_2, a_1),$ (A.3)
- $d(a_1, a_2) > 0 \quad \text{if } a_1 \neq a_2.$ (A.4)

The pair $(A, d)$ is called a *metric space*.

□

The properties of a metric specified in Eqs. A.1-A.4 are the ones that are usually specified to define a metric. It can be shown (Copson [1968]) that Eqs. A.1-A.2 specify the minimal conditions on a metric and that the properties in Eqs. A.3-A.4 can be derived from them.

From Eqs. A.3-A.4 we see that a metric $d$ is a symmetric and non-negative function.

In the context of this book the set $A$ consists of all $2\pi$-periodic contour representations of a certain type ($z$, $\dot{z}$, $\ddot{z}$, $\psi$ or $K$) or of all Fourier representations of a certain type ($\hat{z}$, $\hat{\dot{z}}$, $\hat{\ddot{z}}$, $\hat{\psi}$ or $\hat{K}$). In the case of $2\pi$-periodic contour representations we speak of $A$ as a *function space*. In the latter case we speak of $A$ as a *sequence space*.

By equality between a pair of elements in a function space we mean that this equality exists at least *almost everywhere*. To define the meaning of 'almost everywhere' we need the concept of a *set of measure zero*.

**Definition A.2.** *Set of measure zero* (Apostol [1974]).
A set $S$ of real numbers is said to have measure zero if, for every $\varepsilon > 0$, there is a set of intervals $(a_k, b_k)$, $k = 1, 2, \ldots$, such that

$$S \subseteq \bigcup_k (a_k, b_k) \quad \text{and} \quad \sum_k (b_k - a_k) < \varepsilon.$$

□

**Definition A.3.** *Almost everywhere (a.e.)* (Zaanen [1953], Apostol [1974]).
If a function $f$, defined on a set $S$, posseses a certain property at every point of $S$, except at most at a subset of $S$ of measure zero, then $f$ is said to satisfy this property almost everywhere on $S$.

□

For the next definition we use the concept of a *linear space*, defined over a set of scalars $\mathscr{A}$ (e.g. $\mathbb{R}$ or $\mathbb{C}$). For a definition of a linear space we refer e.g. to Banach [1955], Simmons [1963] or Wouk [1979].

**Definition A.4.** *Norm* (Wouk [1979]).
A norm on a linear space $A$ is a function $\|\cdot\| : A \to \{u \colon 0 \leqslant u < \infty\}$, such that for all $a_1$, $a_2 \in A$ and $\alpha \in \mathscr{A}$:

- $\|a_1\| = 0 \quad \text{iff } a_1 = 0,$ (A.5)
- $\|\alpha a_1\| = |\alpha| \; \|a_1\|,$ (A.6)
- $\|a_1 + a_2\| \leqslant \|a_1\| + \|a_2\|.$ (A.7)

The pair $(A, \|\cdot\|)$ is called a *normed linear space*.

□

The norm defines the notion of distance from an arbitrary element in a space to the origin, that is, the notion of 'size' of an element.

**Theorem A.1.** *Metric induced by the norm* (Simmons [1963], Lipschutz [1965]).
In a normed linear space $(A, \|\cdot\|)$ the norm of the difference between pairs $(a_1, a_2) \in A \times A$ defines a metric on $A$

$$d(a_1, a_2) = \|a_1 - a_2\|. \tag{A.8}$$

This metric is called the *induced metric* on $(A, \|\cdot\|)$. □

An important family of function spaces is characterized by the Lebesgue-integrability of its elements. Since the natural metric in these spaces is defined as a norm and the norm is based on a Lebesgue integral, these function spaces consist of equivalence classes of functions that are equal almost everywhere (a.e.) (cf. Definition A.3).

**Remark.**
A proper introduction to Lebesgue's theory of integration is beyond the scope of this appendix. Through this theory, in which Lebesgue's famous theorem on dominated convergence constitutes an important result, the notion of integrability was extended, thus yielding more integrable functions. For more details we refer e.g. to Apostol [1974], Riesz and Sz.-Nagy [1955] or to Janssen and Van der Steen [1984].

□

**Definition A.5.** *Essential supremum* (Zaanen [1953], Wouk [1979], Taylor and Lay [1980]).
The essential least upper bound or the essential supremum of a real- or complex-valued function $f$, defined on a set $S$, is the smallest number $a \geqslant 0$ such that $|f(t)| \leqslant a$ a.e. on $S$, i.e.

$$\operatorname*{ess\,sup}_S f = \min \{a\colon a \geqslant 0 \text{ and } |f(t)| \leqslant a \text{ a.e. on } S\}. \qquad \text{(A.9)}$$

If the context is clear, the subscript $S$ can be discarded.

□

**Remark.**
As mentioned above, the contour representations discussed in this thesis are $2\pi$-periodic functions (cf. Chapter 2). Therefore the function spaces, that will be defined in the following, all consist of $2\pi$-periodic functions.

□

**Definition A.6.** $\mathbf{L}^p$ *spaces* (Zaanen [1953], Wouk [1979], Taylor and Lay [1980]).
For $1 \leqslant p < \infty$ we denote by $\mathbf{L}^p = \mathbf{L}^p(2\pi)$ the set of equivalence classes of Lebesgue-integrable real- or complex-valued functions $f$ such that

$|f|^p$ is integrable, i.e.

$$\int_{2\pi} |f(t)|^p \mathrm{d}t < \infty. \tag{A.10}$$

In addition, $\mathbf{L}^\infty = \mathbf{L}^\infty(2\pi)$ denotes the set of equivalence classes of measurable real- or complex-valued functions, that are essentially bounded, i.e.

$$\text{ess sup } |f| < \infty. \tag{A.11}$$

□

The $\mathbf{L}^p$ spaces are sometimes called Lebesgue (function) spaces (Zaanen [1953], Taylor [1958], Aubin [1979]).

**Theorem A.2.** *Norm on* $\mathbf{L}^p$ (Taylor and Lay [1980]).
A norm on $\mathbf{L}^p(2\pi)$ is defined by

$$\|f\|_p = \left[\frac{1}{2\pi}\int_{2\pi} |f(t)|^p \mathrm{d}t\right]^{1/p}, \qquad 1 \leqslant p < \infty. \tag{A.12}$$

Further, it can be shown that

$$\|f\|_\infty = \lim_{p\to\infty} \|f\|_p = \text{ess sup } |f|. \tag{A.13}$$

Thus the essential supremum defines a norm on $\mathbf{L}^\infty$.

□

It follows immediately that the properties of a norm in Eqs. A.5 and A.6 are satisfied by Eqs. A.12 and A.13 (for the property in Eq. A.5 it is sufficient that equality to zero is satisfied almost everywhere). The validity of property A.7 of a norm (*triangle inequality*) for Eqs. A.12 and A.13, follows from *Minkowski's inequality*, which in turn is a consequence of *Hölder's inequality*.

**Theorem A.3.** *Hölder's inequality* (Hardy, Littlewood and Pólya [1952], Beckenbach and Bellman [1971]).

If $1 \leqslant p \leqslant \infty$, $1/p + 1/q = 1$, $f \in \mathbf{L}^p$ and $g \in \mathbf{L}^q$, then $f \cdot g \in \mathbf{L}^1$ and

$$\left| \frac{1}{2\pi} \int_{2\pi} f(t)g(t)\mathrm{d}t \right| \leqslant \frac{1}{2\pi} \int_{2\pi} |f(t)g(t)| \mathrm{d}t$$

$$\leqslant \left[ \frac{1}{2\pi} \int_{2\pi} |f(t)|^p \mathrm{d}t \right]^{1/p} \left[ \frac{1}{2\pi} \int_{2\pi} |g(t)|^q \mathrm{d}t \right]^{1/q}. \tag{A.14}$$

□

For $p = q = 2$ in Eq. A.14 we obtain the (Buniakowski-)Schwarz inequality.

**Theorem A.4.** *Minkowski's inequality* (Hardy, Littlewood and Pólya [1952], Beckenbach and Bellman [1971]).
If $1 \leqslant p \leqslant \infty$, $f \in \mathbf{L}^p$ and $g \in \mathbf{L}^p$, then

$$\left[ \frac{1}{2\pi} \int_{2\pi} |f(t) + g(t)|^p dt \right]^{1/p}$$

$$\leqslant \left[ \frac{1}{2\pi} \int_{2\pi} |f(t)|^p \mathrm{d}t \right]^{1/p} + \left[ \frac{1}{2\pi} \int_{2\pi} |g(t)|^p \mathrm{d}t \right]^{1/p}. \tag{A.15}$$

□

The norm $\|f\|_p$ constitutes the usual norm on $\mathbf{L}^p$. The Lebesgue spaces $\mathbf{L}^p$ constitute *Banach* spaces, i.e. complete normed linear spaces (Zaanen [1953]).

The $\mathbf{L}^p$ spaces are also metric spaces, with the metric induced by the usual norm (cf. Theorem A.1). These metrics are sometimes called Minkowski metrics (Anderberg [1973], Sneath and Sokal [1973]).

For $p = 2$ the norm and the metric are called Euclidean. For $p = \infty$ the norm is called the sup norm (Apostol [1974]) or the uniform norm (Simmons [1963]), while in the context of approximation theory, it is also called the Chebychev norm (Cheney [1966]). The metric induced by $\|\cdot\|_\infty$ is sometimes called the Chebychev metric (Anderberg [1973]).

The effect of varying the value of index $p$ to distance measurement in $\mathbb{R}^2$ is shown in Figure A.1. A *unit ball* is the set of all points at unit distance of a given point, e.g. the origin. For $1 < p < 2$ the unit ball for a given value of $p$ is a convex curve lying between the unit balls for the $\mathbf{L}^1$ and $\mathbf{L}^2$ metrics. For $2 < p < \infty$ the unit ball for a given value of $p$ is a convex curve lying between the unit balls for the $\mathbf{L}^2$ and the $\mathbf{L}^\infty$ metrics.

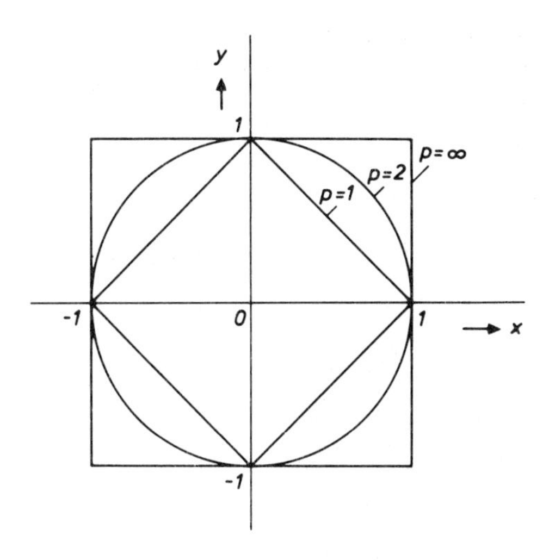


**Figure A.1.** Unit balls in $\mathbb{R}^2$, defined by the $\mathbf{L}^p$ metrics for $p = 1$, $p = 2$ and $p = \infty$.

**Remark.**

The spaces of contour representations, introduced in this book are not linear. For the spaces formed by the contour representations $z$, $\dot{z}$ and $\ddot{z}$ this has a number of reasons. The first reason is that the null element is not really part of these spaces, since it represents a degenerate contour: a single point. Secondly, these spaces of contour representations are not closed under addition, unless we discard the requirement that only simple closed contours are represented. Finally, if any of the contour representations is multiplied by a negative real, then it represents a contour with clockwise positive sense instead of counterclockwise positive sense (cf. Section 2.1). For similar reasons, the spaces of contour representations $\psi$ and $K$ are not linear.

The fact that the norm was defined in Definition A.4 as a function on a linear space does not prevent us from using the norm and its induced metric effectively to define the concepts of size/scale and distance/dissimilarity, respectively, in the spaces of contour representations. □

We continue with the definitions of some further concepts concerning function spaces.

**Definition A.7.** *Total variation* (Apostol [1974], Wouk [1979]).
Let $f$ be a scalar function, defined on $[a, b]$, and, for any $n > 0$, let $P = \{t_0, t_1, \ldots, t_n\}$, with $a = t_0 < t_1 < \cdots < t_{n-1} < t_n = b$, define a partition of $[a, b]$. We denote the set of all partitions of $[a, b]$ by $\mathcal{P}[a, b]$. Then the total variation of $f$ is defined as

$$Var\,(f) = \sup_{P \in \mathcal{P}[a, b]} \left\{ \sum_{k=1}^{n} \left| f(t_k) - f(t_{k-1}) \right| \right\}. \tag{A.16}$$

□

**Definition A.8.** *The space* **BV**$[a, b]$ *of functions of bounded variation on* $[a, b]$ (Apostol [1974], Wouk [1979]).
Let $f$ be a scalar function, defined on $[a, b]$. If there exists a positive number $M < \infty$, such that $Var\,(f) \leqslant M$, then $f$ is said to be of bounded variation on $[a, b]$.

The set of all functions of bounded variation on $[a, b]$ constitutes the space **BV**$[a, b]$.

□

**Definition A.9.** *The space* **CBV**$[a, b]$ *of continuous functions of bounded variation on* $[a, b]$ (Edwards [1979]).
The set of all functions that are both continuous and of bounded variation on $[a, b]$ constitutes the space **CBV**$[a, b]$.

□

**Definition A.10.** *The space* **AC**$[a, b]$ *of absolutely continuous functions on* $[a, b]$ (Riesz and Sz.-Nagy [1955], Apostol [1974]).
A scalar function $f$, defined on $[a, b]$ is said to be absolutely continuous on $[a, b]$ if for every $\varepsilon > 0$ there is a $\delta > 0$ such that

$$\sum_{k=1}^{n} \left| f(b_k) - f(a_k) \right| < \varepsilon \tag{A.17}$$

for every $n$ disjoint open subintervals $(a_k, b_k)$ of $[a, b]$, $n = 1, 2, \ldots,$ the sum of whose lengths $\sum_{k=1}^{n} (b_k - a_k)$ is less than $\delta$.

The set of all absolutely continuous functions on $[a, b]$ constitutes the space $\mathbf{AC}[a, b]$.

□

**Definition A.11.** *The space* $\mathbf{\Lambda}[a, b]$ *of functions that satisfy a uniform Lipschitz condition on* $[a, b]$ (Apostol [1974]).
A scalar function $f$, defined on $[a, b]$, is said to satisfy a uniform Lipschitz condition on $[a, b]$ if there exists a positive number $M < \infty$ such that

$$|f(s) - f(t)| < M|s - t| \tag{A.18}$$

for all $s$ and $t$ in $[a, b]$.

The greatest lower bound of all numbers $M$ for which Eq. A.18 is satisfied for all $s$ and $t$ in $[a, b]$ is called the *Lipschitz constant* of $f$, which we indicate by the symbol $\lambda$.

The set of all functions on $[a, b]$, that satisfy a uniform Lipschitz condition on $[a, b]$ constitutes the space $\mathbf{\Lambda}[a, b]$.

□

**Definition A.12.** *The space* $\mathbf{C}^k[a, b]$ *of k times continuously differentiable functions on* $[a, b]$ (Dunford and Schwartz [1958], Edwards [1979]).
A scalar function $f$, defined on $[a, b]$, is said to be $k$ times continuously differentiable on $[a, b]$ if the derivatives of $f$ of all orders not greater than $k$ exist and are continuous at every point in $[a, b]$.

The set of all $k$ times continuously differentiable functions on $[a, b]$ constitutes the space $\mathbf{C}^k[a, b]$.

The space $\mathbf{C}^\infty[a, b]$ of infinitely-differentiable functions on $[a, b]$ is defined as

$$\mathbf{C}^\infty[a, b] = \bigcap \{\mathbf{C}^k : k = 1, 2, \ldots\}. \tag{A.19}$$

The space $\mathbf{C}^0[a, b]$ of continuous functions on $[a, b]$, will also be denoted as $\mathbf{C}[a, b]$.

□

For the function spaces, defined in the foregoing, a number of inclusion relations exist.

**Theorem A.5.** *Inclusion relations between function spaces* (Apostol [1974], Edwards [1979]).
The inclusion relations A.20-A.22 are valid for the function spaces just defined

$$\mathbf{C}^\infty \subset \cdots \subset \mathbf{C}^{k+1} \subset \mathbf{C}^k \subset \cdots \subset \mathbf{C}^0 = \mathbf{C} \subset \mathbf{L}^\infty \subset \mathbf{L}^p \subset \mathbf{L}^q \tag{A.20}$$

where $k \in \mathbb{N} \cup \{0\}$ and where $\infty > p > q > 0$

$$\mathbf{C}^1 \subset \mathbf{\Lambda} \subset \mathbf{AC} \subset \mathbf{CBV} \subset \mathbf{BV} \subset \mathbf{L}^\infty \tag{A.21}$$

and

$$\mathbf{CBV} \subset \mathbf{C}. \tag{A.22}$$

*Proof*

The validity of the left part of Eq. A.20, i.e. $\mathbf{C}^{k+1} \subset \mathbf{C}^k$ for any integer $k \geqslant 0$ is obvious (cf. Definition A.12).

Every continuous function on $[a, b]$, $-\infty < a < b < \infty$, is bounded. Thus $\mathbf{C} \subset \mathbf{L}^\infty$. The right part of Eq. A.20, i.e. $\mathbf{L}^p \subset \mathbf{L}^q$ for $\infty \geqslant p > q > 0$, is a consequence of the inequality

$$\|f\|_q \leqslant \|f\|_p, \qquad 0 < q < p \leqslant \infty \tag{A.23}$$

which in turn is a result of Hölder's inequality, Eq. A.14 (cf. Edwards [1979], p. 28).

The validity of Eq. A.21 is verified from left to right. If $f \in \mathbf{C}^1[a, b]$, then $\dot{f}$ is bounded on $[a, b]$. Considering the definition of the derivative of a function and Eq. A.18, taking $M = \max_{t \in [a, b]} |\dot{f}(t)|$, it follows immediately that $f \in \mathbf{C}^1$ implies $f \in \mathbf{\Lambda}$. That $\mathbf{C}^1$ is properly contained in $\mathbf{\Lambda}$ follows for example from $f(t) = |t|$, defined on an interval $[a, b]$, with $0 \in [a, b]$. This establishes $\mathbf{C}^1 \subset \mathbf{\Lambda}$.

In Proposition 3.1 we established that $f \in \mathbf{\Lambda}$ implies $f \in \mathbf{AC}$. That $\mathbf{\Lambda}$ is properly contained in $\mathbf{AC}$ follows for example from $f(t) = |t|^{1/2}$, defined on an interval $[a, b]$, with $0 \in [a, b]$. Clearly, if we set $t = 0$ and let $s$ approach 0 in Eq. A.18 we see that $f(t) = |t|^{1/2}$ does not belong to $\mathbf{A}$. On the other hand, $|\dot{f}(t)| = |\text{sgn}(t)|t|^{1/2}|$ is Riemann-integrable and therefore $\dot{f}(t)$ is Lebesgue-integrable. This is a necessary and sufficient condition for $f(t)$, the indefinite integral of $\dot{f}(t)$ up to an additive constant, to belong to **AC** (cf. Riesz and Sz.-Nagy [1955], pp. 50-52). Thus we may conclude that $\mathbf{A} \subset \mathbf{AC}$.

For $\mathbf{AC} \subset \mathbf{CBV}$ we refer to Riesz and Sz.-Nagy [1955], pp. 50-52.

The validity of $\mathbf{CBV} \subset \mathbf{BV}$ is trivial (cf. Definitions A.8 and A.9). We observe from Definitions A.7 and A.8 that $f \in \mathbf{BV}[a, b]$ implies that $f$ is bounded everywhere on $[a, b]$. Thus from $f \in \mathbf{BV}[a, b]$ it follows that $f \in \mathbf{L}^\infty[a, b]$. If $f \in \mathbf{L}^\infty[a, b]$, then $f$ may be unbounded on a set of measure zero in $[a, b]$ (cf. Definitions A.3-A.6). Thus **BV** is properly contained in $\mathbf{L}^\infty$, i.e. $\mathbf{BV} \subset \mathbf{L}^\infty$.

Finally we verify the validity of Eq. A.22. It is trivial that $f \in \mathbf{CBV}$ implies $f \in \mathbf{C}$. On the other hand, a continous function need not be of bounded variation (cf. Apostol [1974], p. 129). Thus **CBV** is properly contained in **C**, i.e. $\mathbf{CBV} \subset \mathbf{C}$.

□

**Remark.**
Though Theorem A.5 is valid for $\infty > p > q > 0$ we consider only $\mathbf{L}^p$ spaces for $p \geqslant 1$ since for $0 < p < 1$ the triangle inequality (property A.7 of a norm) does not hold.

□

**Remark.**
In this book we replace the interval identifier $[a, b]$, as used in Definitions A.6-A.12 and in Theorem A.5, by $[2\pi]$ to signify the length of the fundamental parameter interval, since we deal only with $2\pi$-periodic contour representations. Often the context is clear which enables us to discard the interval identifier altogether.

□

In the following we define some concepts concerning sequence spaces over $\mathbb{Z}$. In many respects these concepts constitute duals of their counterparts in spaces of functions defined over a bounded closed interval.

**Definition A.13.** *$\ell^p$ spaces* (Simmons [1963], Wouk [1979], Taylor and Lay [1980]).

For $1 \leqslant p < \infty$ we denote by $\ell^p = \ell^p(\mathbb{Z})$ the set of real- or complex-valued sequences $\xi = \{..., \xi(-1), \xi(0), \xi(1), ...\}$, such that $|\xi(n)|^p$ is summable, i.e.

$$\sum_{n \in \mathbb{Z}} |\xi(n)|^p < \infty. \tag{A.24}$$

In addition, $\ell^\infty = \ell^\infty(\mathbb{Z})$ denotes the set of bounded real- or complex-valued sequences, i.e.

$$\sup_{n \in \mathbb{Z}} |\xi(n)| < \infty. \tag{A.25}$$

□

**Theorem A.6.** *Norm on* $\ell^p$ (Taylor and Lay [1980]).
A norm on $\ell^p(\mathbb{Z})$ is defined by

$$\|\xi\|_p = \left[\sum_{n \in \mathbb{Z}} |\xi(n)|^p\right]^{1/p}, \qquad 1 \leqslant p < \infty. \tag{A.26}$$

Further it can be shown that

$$\|\xi\|_\infty = \sup_{n \in \mathbb{Z}} |\xi(n)|. \tag{A.27}$$

Thus the supremum of a sequence defines a norm on $\ell^\infty$.

□

The validity of the triangle inequality (property A.7 of a norm) follows from Minkowski's inequality for sums, which in turn is a consequence of Hölder's inequality for sums.

**Theorem A.7.** *Hölder's inequality for sums* (Hardy, Littlewood and Pólya [1952], Beckenbach and Bellman [1971]).
If $1 \leqslant p \leqslant \infty$, $1/p + 1/q = 1$, $\xi \in \ell^p$ and $\chi \in \ell^q$, then $\xi \cdot \chi \in \ell^1$ and

$$\left|\sum_{n \in \mathbb{Z}} \xi(n)\chi(n)\right| \leqslant \sum |\xi(n)\chi(n)|$$

$$\leqslant \left[\sum_{n \in \mathbb{Z}} |\xi(n)|^p\right]^{1/p} \left[\sum_{n \in \mathbb{Z}} |\chi(n)|^q\right]^{1/q}. \tag{A.28}$$

□

For $p = q = 2$ in Eq. A.28 we obtain Cauchy's inequality.

**Theorem A.8.** *Minkowski's inequality for sums* (Hardy, Littlewood and Pólya [1952], Beckenbach and Bellman [1971]).
If $1 \leqslant p \leqslant \infty$, $\xi \in \ell^p$ and $\chi \in \ell^p$, then

$$\left[ \sum_{n \in \mathbb{Z}} |\xi(n) + \chi(n)|^p \right]^{1/p}$$

$$\leqslant \left[ \sum_{n \in \mathbb{Z}} |\xi(n)|^p \right]^{1/p} + \left[ \sum_{n \in} |\chi(n)|^p \right]^{1/p}. \qquad \text{(A.29)}$$

□

Compare Theorems A.7 and A.8 with Theorems A.3 and A.4, respectively.

The norm $\|\xi\|_p$ constitutes the usual norm on $\ell^p$. The sequence spaces $\ell^p$ constitute Banach spaces (Zaanen [1953], Taylor and Lay [1980]). The $\ell^p$ spaces are also metric spaces, with the metric induced by the usual norm (cf. Theorem A.1). The names given to the usual norm and metric on $\ell^p$, both in general and in the special cases $p = 2$ and $p = \infty$, are the same as those given to the usual norm and metric on $\mathbf{L}^p$.

**Remark.**
For the same reasons that the spaces of contour representations, introduced in this book, are not linear, the spaces of Fourier representations of contours are also not linear. Yet, the norm on $\ell^p$ and its induced metric properly define the concepts of size/scale and distance/dissimilarity, respectively, in the spaces of Fourier representations.

We introduce some additional concepts and properties concerning sequence spaces.

**Definition A.14.** $\mathbf{c}_0$ *spaces* (Edwards [1979], Wouk [1979]).
By $\mathbf{c}_0 = \mathbf{c}_0(\mathbb{Z})$ we denote the set of all bounded real- or complex-valued

sequences $\xi = \{\ldots, \xi(-1), \xi(0), \xi(1), \ldots\}$, for which

$$\lim_{|n| \to \infty} \xi(n) = 0. \tag{A.30}$$

In view of Eq. A.30, a sequence $\xi \in \mathbf{c}_0$ is called a null sequence.

□

The metric induced by $\|\xi\|_\infty$ can be used to turn $\mathbf{c}_0$ into a metric space (Wouk [1979]).

For the sequence spaces just defined the following relations exist.

**Theorem A.9.** *Inclusion relations between sequence spaces* (Edwards [1979], Wouk [1979]).
For the sequence spaces defined above we have the inclusion relations

$$\ell^q \subset \ell^p \subset \mathbf{c}_0 \subset \ell^\infty \tag{A.31}$$

where $0 < q < p < \infty$.

*Proof*

The validity of $\ell^q \subset \ell^p \subset \ell^\infty$ follows from the inequality

$$\|\xi\|_q \geqslant \|\xi\|_p \geqslant \|\xi\|_\infty \tag{A.32}$$

for $0 < q < p < \infty$. For a proof of Eq. A.32, which is sometimes called *Jensen's inequality*, we refer to Hardy, Littlewood and Pólya [1952] or Beckenbach and Bellman [1971].

To see that $\ell^p$, $0 < p < \infty$, is properly contained in $\mathbf{c}_0$ consider for example the sequence $\xi = \{\xi(n)\}_{n \in \mathbb{Z}} = \{(1 + |n|)^{-1/p}\}_{n \in \mathbb{Z}}$, which belongs to $\mathbf{c}_0$ but not to $\ell^p$, $0 < p < \infty$.

Finally, $\mathbf{c}_0$ is a subset of $\ell^\infty$ by definition. That $\mathbf{c}_0$ is a proper subset of $\ell^\infty$ is obvious from Definitions A.13 and A.14.

□

**Remark.**
For $0 < p < 1$ the triangle inequality does not hold for the $\ell^p$ spaces (in the same way as it does not hold for the $\mathbf{L}^p$ spaces for these values of $p$). Therefore we do not consider the $\ell^p$ spaces for $0 < p < 1$. □

We conclude this appendix by giving some definitions and properties of trigonometric series and Fourier series.

**Definition A.15.** *The set of trigonometric polynomials of degree at most n* (cf. Zygmund [1959a], Cheney [1966]).
This set is defined as

$$\mathbf{T}_n = \left\{ f\colon f(t) = \sum_{|k| \leqslant n} c_k \mathrm{e}^{\mathrm{i}kt}, \quad c_k \in \mathbb{C} \right\}. \tag{A.33}$$

□

Note that $\mathbf{T}_n \subset \mathbf{T}_{n+1}$ for all $n \geqslant 0$.

**Definition A.16.** *The set of trigonometric polynomials of degree at most n, free of a constant term* (cf. Cheney [1966]).
This set is defined as

$$\mathbf{t}_n = \left\{ f\colon f(t) = \sum_{0 < |k| \leqslant n} c_k \mathrm{e}^{\mathrm{i}kt}, \quad c_k \in \mathbb{C} \right\}. \tag{A.34}$$

□

It is obvious that $\mathbf{t}_n \subset \mathbf{T}_n$ for all $n \geqslant 1$.

The following theorem deals with pointwise convergence of Fourier series.

**Theorem A.10.** *Dirichlet-Jordan test* (cf. Zygmund [1959a], Edwards [1979]).

1. If $f \in \mathbf{L}^1$ is of bounded variation on some neighborhood of a point $t$, then (cf. Definitions 3.1 and 3.2)

$$\lim_{n \to \infty} (S_n f)(t) = \{f(t+0) + f(t-0)\}/2. \tag{A.35}$$

2. More in particular, at every point of continuity of $f$ we have

$$\lim_{n \to \infty} (S_n f)(t) = f(t). \tag{A.36}$$

3. If $f$ is continuous at every point of a closed interval $[a, b]$, then $(S_n f)$ converges uniformly in $[a, b]$. □

**Theorem A.11.** *Riemann-Lebesgue lemma* (cf. Zygmund [1959a], Katznelson [1968], Edwards [1979]).
For any integrable $f$ one has (cf. Definition 3.1)

$$\lim_{|n| \to \infty} \hat{f}(n) = 0. \tag{A.37}$$

□

From Theorem A.11 we see that $\hat{f} \in \mathbf{c}_0$ if $f$ is integrable.

## References

Anderberg, M.R. [1973]
*Cluster Analysis for Applications*, New York: Academic Press.

Apostol, T.M. [1974]
*Mathematical Analysis*, Second Edition, Reading, MA: Addison-Wesley.

Aubin, J.-P. [1979]
*Applied Functional Analysis*, New York: John Wiley and Sons, Inc.

Banach, S. [1955]
*Théorie des Opérations Linéaires*, New York: Chelsea Publishing Company. (Original Edition: Monografje Matematyczne, Warsaw, 1932).

Beckenbach, E.F. and R. Bellman [1971]
*Inequalities*, Third Printing, Berlin: Springer-Verlag.

Cheney, E.W. [1966]
*Introduction to Approximation Theory*, New York: McGraw-Hill Book Co., Inc.

Copson, E.T. [1968]
*Metric Spaces*, Cambridge, England: Cambridge University Press.

Dunford, N. and J.T. Schwartz [1958]
*Linear Operators, Part I: General Theory*, New York: Interscience Publishers, Inc.

Edwards, R.E. [1979]
*Fourier Series, a Modern Introduction, Volume 1*, New York: Springer-Verlag.

Hardy, G.H., J.E. Littlewood and G. Pólya [1952]
*Inequalities*, Second Edition, Cambridge, England: Cambridge University Press.

Janssen, A.J.E.M. and P. van der Steen [1984]
*Integration Theory*, Lecture Notes in Mathematics **1078**, Berlin: Springer-Verlag.

Katznelson, Y. [1968]
*An Introduction to Harmonic Analysis*, New York: John Wiley and Sons, Inc.

Lipschutz, S. [1965]
*Theory and Problems of General Topology*, Schaum's Outline Series, New York: McGraw-Hill Book Co., Inc.

Riesz, F. and B. Sz.-Nagy [1955]
*Functional Analysis*, New York: Frederick Ungar Publishing Company.

Simmons, G.F. [1963]
*Introduction to Topology and Modern Analysis*, New York: McGraw-Hill Book Co., Inc.

Sneath, P.H.A. and R.R. Sokal [1973]
*Numerical Taxonomy*, San Francisco, CA: W.H. Freeman and Co.

Taylor, A.E. [1958]
*Introduction to Functional Analysis*, New York: John Wiley and Sons, Inc.

Taylor, A.E. and D.C. Lay [1980]
*Introduction to Functional Analysis*, Second Edition, New York: John Wiley and Sons, Inc.

Titchmarsh, E.C. [1939]
*The Theory of Functions*, Second Edition, Oxford: Oxford University Press.

Wouk, A. [1979]
*A Course of Applied Functional Analysis*, New York: John Wiley and Sons, Inc.

Zaanen, A.C. [1953]
*Linear Analysis*, Amsterdam, The Netherlands: North-Holland Publishing Company.

Zygmund, A. [1959a]
*Trigonometric Series, Volume I*, Second Edition, Cambridge, England: Cambridge University Press.

# Appendix B

# A method for a fast and reliable computation of moments $m_{pq}$ of regions bounded by polygons

## B.1 Introduction

Over many years after their introduction in the context of pattern recognition and image analysis (Hu [1961], Hu [1962], Alt [1962]), moments have maintained a considerable popularity and a substantial body of literature on this subject now exists.

Usually the two-dimensional moments $m_{pq}$ of order $(p + q)$ are defined as moments of the image function $f(x, y)$ (cf. Eq. 4.3.5)

$$m_{pq} = \int_{-\infty}^{\infty} \int_{-\infty}^{\infty} f(x, y) x^p y^q \mathrm{d}x \, \mathrm{d}y. \tag{B.1}$$

The moments $m_{pq}$ of $f(x, y)$ are also called monomial moments (Boyce and Hossack [1983]), since they are defined with respect to the monomial $x^p y^q$ (Teague [1980]), or geometric moments (Vijaya Kumar and Rahenkamp [1986]).

A major application of the moments $m_{pq}$ has been the definition of moment invariants, a set of image features which are invariant under certain image transformations such as translation, scaling, rotation and contrast change (Hu [1962], Dudani, Breeding and McGhee [1977], Wong and Hall [1978], Sadjadi and Hall [1978], Maitra [1979], Reddi [1981]). In this approach the moment invariants constitute feature vectors in a multidimensional feature space.

In another approach the low order moments are used to normalize the image, in a way that is comparable to the moment-based object normalization described in Section 4.3. Subsequently either higher order moments, computed from the normalized images, are used as normalized features (Alt [1962], Smith and Wright [1970], Nill [1981], Reeves and Rostampour [1981]), or the normalized images are used in another way in image processing and analysis (Casey [1970], Abu-Mostafa and Psaltis [1985]). In the latter paper it is emphasized that, especially under noisy circumstances, the moments should be used only for normalization but not for classification purposes.

Some papers discuss the usefulness of moments for data compression (Teague [1980], Nill [1981], Boyce and Hossack [1983]).

In applications where the internal structure of objects is not important, but merely their shape, not the image function $f(x, y)$ is used in the expression for $m_{pq}$ (Eq. B.1), but the characteristic function $\chi_R(x, y)$ of an object that covers the region $R \subset \mathbb{R}^2$ (or $R \subset \mathbb{C}$) (Apostol [1974])

$$\chi_R(x, y) = \begin{cases} 1, & \text{if } (x, y) \in R \quad \text{(B.2a)} \\ 0, & \text{otherwise.} \quad \text{(B.2b)} \end{cases}$$

In that case $m_{pq}$ is sometimes called a silhouette moment of the object (Dudani, Breeding and McGhee [1977], Reeves and Rostampour [1981], Reeves and Wittner [1983]). Silhouette moments are used for shape normalization (Casey [1970], Reeves and Rostampour [1981], Gilmore and Boyd [1981], Reeves and Wittner [1983], Cyganski and Orr [1985]), for the computation of moment invariants (Reeves and Rostampour [1981], Gilmore and Boyd [1981]), or as features of the shape or inertial properties of objects (Wilson and Farrior [1976], Reeves and Rostampour [1981], Tang [1982], Miles and Tough [1983], Ho [1983]).

The computation of moments involves integration over a two-dimensional region. Especially in the case of silhouette moments it has been observed (Wilson and Farrior [1976], Tang [1982], Miles and Tough [1983], Cyganski and Orr [1985], Bamieh and de Figueiredo [1986]) that the computational complexity of moments can be reduced substantially through Green's theorem (cf. Spiegel [1964], Kreysig [1972] or Eq. 4.3.12)

$$\iint_R \left[ \frac{\partial Q(x, y)}{\partial x} - \frac{\partial P(x, y)}{\partial y} \right] \mathrm{d}x\,\mathrm{d}y = \oint_{\gamma_R} P(x, y)\mathrm{d}x + Q(x, y)\mathrm{d}y \tag{B.3}$$

where $\gamma_R$ is the contour of the region $R \subset \mathbb{R}^2$ (or equivalently, $R \subset \mathbb{C}$). In the next section we develop, through Green's theorem, formulae that enable the efficient computation of moments for a special class of objects: objects bounded by a polygon.

## B.2 Moments of objects bounded by a polygon

Usually in digital image analysis the double integral in $m_{pq}$ (Eq. B.1) is approximated by a double summation

$$m_{pq} = \sum_{i=0}^{M-1} \sum_{j=0}^{N-1} f[i, j] i^p j^q \tag{B.4}$$

where $f[i, j]$ is the sampled image function, which has dimensions $M$ and $N$, and where $i$ and $j$ indicate the discrete pixel locations. The discrete summation in Eq. B.4 is one of the sources of errors in $m_{pq}$ and causes the moment invariants to be not completely invariant under equiform transformations of the image (Abu-Mostafa and Psaltis [1985], Teh and Chin [1986]). One way to reduce these errors is to increase the sampling density.

For silhouette moments it is possible to obtain a better approximation through Green's theorem, which enables us to replace the double integral in Eq. B.1 by a contour integral. In a segmented digital picture we can define the boundary of a region for example as a polygon (cf. Figure B.1). Obviously we do not completely get rid of the geometric discreteness of the segmented image, so it still pays to increase the sampling density. However, the simple mathematical form of the region contour (i.e. a polygon) enables us to obtain an analytic, and therefore exact, expression for the silhouette moments $m_{pq}$ of the segmented region.

In the context of data reduction in image analysis, polygonal approximation is a popular technique (Montanari [1970], Ramer [1972], Pavlidis and Horowitz [1974], McClure and Vitale [1975], Ellis and Eden [1976], Pavlidis [1977b]. Sklansky and Gonzalez [1979], Williams [1981], Kurozumi and Davis [1982], Kashyap and Oommen [1983]).

The expression for polygonal silhouette moments, that will be derived in the following, constitutes a particularly efficient and reliable means for the computation of the silhouette moments of such polygonal approximations.

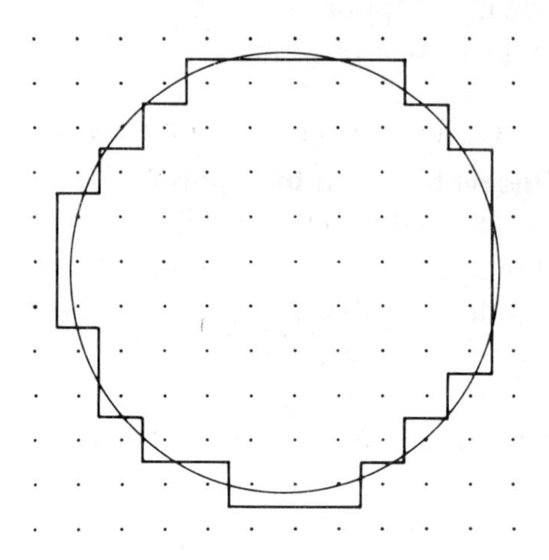

**Figure B.1.** Picture of a circle, arbitrarily positioned on a Cartesian sampling lattice. The polygon represents the boundary of the sampled circle (other polygonal boundary definitions are possible though).

We start the derivation by substituting the characteristic function $\chi_R$, Eq. B.2, for the image function $f$ in Eq. B.1

$$m_{pq} = \int_{-\infty}^{\infty}\int_{-\infty}^{\infty} \chi_R(x, y)x^p y^q \mathrm{d}x\,\mathrm{d}y = \iint_R x^p y^q \mathrm{d}x\,\mathrm{d}y. \tag{B.5}$$

We apply Green's theorem, Eq. B.3, to Eq. B.5. In Eq. B.3 we choose

$$P(x, y) = 0 \tag{B.6a}$$

$$Q(x, y) = \frac{1}{p+1} x^{p+1} y^q. \tag{B.6b}$$

Thus we obtain

$$m_{pq} = \iint_R x^p y^q \mathrm{d}x\,\mathrm{d}y = \oint_{\gamma_R} \frac{1}{p+1} x^{p+1} y^q \mathrm{d}y. \tag{B.7}$$

We now assume that $\gamma_R$ is a polygon with $N$ vertices

$$\{z_n\} = \{x_n + \mathrm{i}y_n\}, \qquad n = 0, \ldots, N - 1 \tag{B.8}$$

where the vertices have been ordered according to counterclockwise traversal of the polygon $\gamma_R$. The line segment between the vertices $z_n$ and $z_{n+1}$ can be represented parametrically as

$$z(s) = \frac{\Delta z_n}{\Delta s_n}(s - s_n) + z_n \tag{B.9}$$

where $s$ is the parameter of arc length and where

$$\Delta z_n = z_{n+1} - z_n, \tag{B.10}$$

$$s_n = \begin{cases} \sum_{m=0}^{n-1} |\Delta z_m| = \sum_{m=0}^{n-1} \Delta s_m, & n \geqslant 1 \quad \text{(B.11a)} \\ 0, & n = 0 \quad \text{(B.11b)} \\ -\sum_{m=n}^{-1} |\Delta z_m| = -\sum_{m=n}^{-1} \Delta s_m, & n \leqslant -1 \quad \text{(B.11c)} \end{cases}$$

and

$$\Delta s_n = s_{n+1} - s_n = |\Delta z_n|. \tag{B.12}$$

Note in Eq. B.9 that $z_n = z(s_n)$. Compare Eqs. B.8-B.12 with Eqs. 2.2.40-2.2.49. We also note that

$$\Delta z_n = \Delta x_n + \mathrm{i}\Delta y_n \tag{B.13}$$

where

$$\Delta x_n = x_{n+1} - x_n \tag{B.14a}$$

$$\Delta y_n = y_{n+1} - y_n. \tag{B.14b}$$

Substitution of the expressions in Eqs. B.9-B.14b into Eq. B.7 yields

$$m_{pq} = \frac{1}{p+1} \sum_{n=0}^{N-1} \int_{s_n}^{s_{n+1}} x^{p+1}(s) y^q(s) \frac{dy(s)}{ds} ds$$

$$= \frac{1}{p+1} \sum_{n=0}^{N-1} \int_0^{\Delta s_n} \left\{ \frac{\Delta x_n}{\Delta s_n} s' + x_n \right\}^{p+1} \left\{ \frac{\Delta y_n}{\Delta s_n} s' + y_n \right\}^q \frac{\Delta y_n}{\Delta s_n} ds'. \tag{B.15}$$

After application of the binomial theorem (cf. Riordan [1979], p. 1) to Eq. B.15, we obtain by integration

$$m_{pq} = \frac{1}{p+1} \sum_{n=0}^{N-1} \int_0^{\Delta s_n} \left\{ \sum_{\alpha=0}^{p+1} \binom{p+1}{\alpha} (\Delta x_n)^\alpha x_n^{p+1-\alpha} \right\}$$

$$\cdot \left\{ \sum_{\beta=0}^{q} \binom{q}{\beta} (\Delta y_n)^\beta y_n^{q-\beta} \right\} \Delta y_n \left( \frac{1}{\Delta s_n} \right)^{\alpha+\beta+1} (s')^{\alpha+\beta} ds'$$

$$= \frac{1}{p+1} \sum_{n=0}^{N-1} \sum_{\alpha=0}^{p+1} \sum_{\beta=0}^{q} \frac{1}{\alpha+\beta+1} \binom{p+1}{\alpha} \binom{q}{\beta}$$

$$\cdot (\Delta x_n)^\alpha x_n^{p+1-\alpha} (\Delta y_n)^{\beta+1} y_n^{q-\beta}. \tag{B.16}$$

Shifting the summation over $\beta$ by 1, using the fact that

$$\binom{n}{m-1} = \frac{m}{n+1} \binom{n+1}{m} \tag{B.17}$$

and inverting the order of summation gives

$$m_{pq} = \frac{1}{(p+1)(q+1)} \sum_{\alpha=0}^{p+1} \sum_{\beta=1}^{q+1} \frac{\beta}{\alpha+\beta} \binom{p+1}{\alpha} \binom{q+1}{\beta}$$

$$\cdot \sum_{n=0}^{N-1} (\Delta x_n)^\alpha x_n^{p+1-\alpha} (\Delta y_n)^\beta y_n^{q+1-\beta}. \tag{B.18}$$

Eq. B.18 constitutes the first expression for $m_{pq}$ that allows its efficient computation. The computational complexity is $O(pqN)$. In Eq. B.18 we can distinguish a data dependent part

$$T_N(\alpha, \beta; p, q) = \sum_{n=0}^{N-1} (\Delta x_n)^{\alpha} x_n^{p+1-\alpha} (\Delta y_n)^{\beta} y_n^{q+1-\beta} \tag{B.19a}$$

and a data independent part, which consists of $(p + 2)(q + 1)$ coefficients

$$d(\alpha, \beta; p, q) = \frac{1}{(p+1)(q+1)} \frac{\beta}{\alpha+\beta} \binom{p+1}{\alpha} \binom{q+1}{\beta}$$

$$\alpha = 0, \ldots, p+1, \quad \beta = 1, \ldots, q+1$$

$$p, q = 0, 1, \ldots . \tag{B.19b}$$

For any moment $m_{pq}$ the coefficients $d(\alpha, \beta; p, q)$ can be precomputed and stored in memory.

We now develop Eq. B.18 further to obtain an expression for $m_{pq}$ that involves only the coordinates of the vertices of $\gamma_R$ and not the differences $\Delta x_n$ and $\Delta y_n$. Substituting Eqs. B.14a and B.14b in Eq. B.18 and again applying the binomial theorem (cf. Riordan [1979], p. 1) yields

$$m_{pq} = \frac{1}{(p+1)(q+1)} \sum_{\alpha=0}^{p+1} \sum_{\beta=1}^{q+1} \frac{\beta}{\alpha+\beta} \binom{p+1}{\alpha} \binom{q+1}{\beta}$$

$$\cdot \sum_{n=0}^{N-1} \left\{ \sum_{\gamma=0}^{\alpha} (-1)^{\alpha-\gamma} \binom{\alpha}{\gamma} x_{n+1}^{\gamma} x_n^{\alpha-\gamma} \right\} x_n^{p+1-\alpha}$$

$$\cdot \left\{ \sum_{\delta=0}^{\beta} (-1)^{\beta-\delta} \binom{\beta}{\delta} y_{n+1}^{\delta} y_n^{\beta-\delta} \right\} y_n^{q+1-\beta}. \tag{B.20}$$

If we invert the order of summation of $\alpha$ and $\gamma$ and of $\beta$ and $\delta$, respectively and if we apply the transformations $\alpha' = \alpha - \gamma$ and $\beta' = \beta - \delta$, we obtain

$$m_{pq} = \frac{1}{(p+1)(q+1)} \sum_{\gamma=0}^{p+1} \sum_{\delta=0}^{q+1} \binom{p+1}{\gamma} \binom{q+1}{\delta}$$

$$\cdot \sum_{\alpha'=0}^{p+1-\gamma} \sum_{\beta'=0}^{q+1-\delta} (-1)^{\alpha'+\beta'} \frac{\beta'+\delta}{\alpha'+\beta'+\gamma+\delta}$$

$$\cdot \binom{p+1-\gamma}{\alpha'} \binom{q+1-\delta}{\beta'}$$

$$\cdot \sum_{n=0}^{N-1} x_{n+1}^{\gamma} x_n^{p+1-\gamma} y_{n+1}^{\delta} y_n^{q+1-\delta} \tag{B.21}$$

where, by definition

$$\frac{\beta'+\delta}{\alpha'+\beta'+\gamma+\delta} = 0, \qquad \text{if } \beta'+\delta = 0. \tag{B.22}$$

Similar to Eq. B.18, we distinguish in Eq. B.21 a data dependent part

$$S_N(\gamma, \delta; p, q) = \sum_{n=0}^{N-1} x_{n+1}^{\gamma} x_n^{p+1-\gamma} y_{n+1}^{\delta} y_n^{q+1-\delta} \tag{B.23a}$$

and a data independent part, which consists of $(p+2)(q+2)$ coefficients

$$c(\gamma, \delta; p, q) = \frac{1}{(p+1)(q+1)} \binom{p+1}{\gamma} \binom{q+1}{\delta}$$

$$\cdot \sum_{\alpha'=0}^{p+1-\gamma} \sum_{\beta'=0}^{q+1-\delta} (-1)^{\alpha'+\beta'} \frac{\beta'+\delta}{\alpha'+\beta'+\gamma+\delta}$$

$$\cdot \binom{p+1-\gamma}{\alpha'} \binom{q+1-\delta}{\beta'}. \tag{B.23b}$$

To reduce the expression for $c(\gamma, \delta; p, q)$, we define the help function

$$h(c_1, c_2; a, b) = \sum_{n=0}^{a} \sum_{m=0}^{b} (-1)^{n+m} \frac{m + c_1}{n + m + c_2} \binom{a}{n} \binom{b}{m} \tag{B.24}$$

for $a, b, c_1 \in \mathbb{N} \cup \{0\}$, $c_2 \in \mathbb{N}$. We also define the function

$$f(x) = \sum_{n=0}^{a} \sum_{m=0}^{b} (-1)^{n+m} \frac{m + c_1}{n + m + c_2} \binom{a}{n} \binom{b}{m} x^{n+m+c_2}. \tag{B.25}$$

Then we have the relation

$$h(c_1, c_2; a, b) = f(1) = \int_0^1 \dot{f}(x) \mathrm{d}x. \tag{B.26}$$

Differentiation of $f(x)$ gives

$$\begin{aligned} \dot{f}(x) &= c_1 x^{c_2 - 1} \sum_{n=0}^{a} (-1)^n \binom{a}{n} x^n \sum_{m=0}^{b} (-1)^m \binom{b}{m} x^m \\ &\quad - b x^{c_2} \sum_{n=0}^{a} (-1)^n \binom{a}{n} x^n \sum_{m=1}^{b} (-1)^{m-1} \binom{b-1}{m-1} x^{m-1} \\ &= c_1 x^{c_2 - 1} (1 - x)^{a+b} - b x^{c_2} (1 - x)^{a+b-1}. \end{aligned} \tag{B.27}$$

By substituting this result into the integral in Eq. B.26, we obtain through Gradshteyn and Ryzhik [1965], pp. 284, 938 and 950

$$\begin{aligned} h(c_1, c_2; a, b) &= c_1 \frac{(c_2 - 1)!(a + b)!}{(a + b + c_2)!} - b \frac{c_2!(a + b - 1)!}{(a + b + c_2)!} \\ &= \frac{\binom{a + b + c_2 - 2}{c_2 - 1}^{-1}}{(a + b + c_2 - 1)(a + b + c_2)} \{c_1(a + b) - c_2 b\}. \end{aligned} \tag{B.28}$$

The help function $h$ occurs in the expression for $c(\gamma, \delta; p, q)$ in Eq. B.23b, which we now can write as

$$c(\gamma, \delta; p, q) = \frac{1}{(p+1)(q+1)} \binom{p+1}{\gamma} \binom{q+1}{\delta}$$

$$\cdot\, h(\delta, \gamma + \delta; p + 1 - \gamma, q + 1 - \delta). \qquad \text{(B.29)}$$

Substitution of Eq. B.28 into Eq. B.29 yields

$$c(\gamma, \delta; p, q)$$

$$= \frac{1}{(p+1)(q+1)} \binom{p+1}{\gamma} \binom{q+1}{\delta} \frac{\binom{p+q}{\gamma+\delta-1}^{-1}}{(p+q+1)(p+q+2)}$$

$$\cdot \left\{\delta(p+q+2-\gamma-\delta) - (\gamma+\delta)(q+1-\delta)\right\}$$

$$= \frac{\binom{p+1}{\gamma}\binom{q+1}{\delta}}{\binom{p+q}{\gamma+\delta-1}} \frac{1}{(p+q+1)(p+q+2)} \left\{\frac{\delta}{q+1} - \frac{\gamma}{p+1}\right\}$$

$$= \frac{\binom{p+1}{\gamma}\binom{q+1}{\delta}}{\binom{p+q+2}{\gamma+\delta}} \frac{1}{(p+q+2-\gamma-\delta)(\gamma+\delta)} \left\{\frac{\delta}{q+1} - \frac{\gamma}{p+1}\right\}.$$

$$\text{(B.30)}$$

The derivation in Eqs. B.26-B.28 is not valid for $c_2 = 0$, which corresponds to $(\gamma, \delta) = (0, 0)$. Substituting these values into Eq. B.23b and bearing Eq. B.22 in mind we find

$$c(0, 0; p, q)$$

$$= \frac{1}{(p+1)(q+1)} \sum_{\alpha'=0}^{p+1} \sum_{\beta'=1}^{q+1} (-1)^{\alpha'+\beta'} \frac{\beta'}{\alpha'+\beta'} \binom{p+1}{\alpha'} \binom{q+1}{\beta'}$$

$$= \frac{-1}{(p+1)(q+1)} \sum_{\alpha'=0}^{p+1} (-1)^{\alpha'} \binom{p+1}{\alpha'} \sum_{\beta=0}^{q} (-1)^{\beta} \frac{\beta+1}{\alpha'+\beta+1} \binom{\alpha+1}{\beta+1}$$

$$= \frac{-1}{(p+1)} \sum_{\alpha'=0}^{p+1} (-1)^{\alpha'} \binom{p+1}{\alpha'} \sum_{\beta=0}^{q} (-1)^{\beta} \frac{1}{\alpha'+\beta+1} \binom{q}{\beta}. \qquad \text{(B.31)}$$

In analogy with Eqs. B.24-B.25 we define the help functions

$$h_1(c; b) = \sum_{n=0}^{b} (-1)^n \frac{1}{n+c} \binom{b}{n} \qquad \text{(B.32)}$$

for $b \in \mathbb{N} \cup \{0\}$, $c \in \mathbb{N}$, and

$$f_1(x) = \sum_{n=0}^{b} (-1)^n \frac{1}{n+c} \binom{b}{n} x^{n+c}. \qquad \text{(B.33)}$$

Then

$$h_1(c; b) = f_1(1) = \int_0^1 \dot{f}_1(x)\mathrm{d}x. \qquad \text{(B.34)}$$

We derive

$$\dot{f}_1(x) = \sum_{n=0}^{b} (-1)^n \binom{b}{n} x^{n+c-1} = x^{c-1}(1-x)^b. \qquad \text{(B.35)}$$

Through substitution of Eq. B.35 into Eq. B.34 and through Gradshteyn and Ryzhik [1965], pp. 284, 938 and 950, we obtain

$$h_1(c; b) = \sum_{n=0}^{b} (-1)^n \frac{1}{n+c} \binom{b}{n} = \frac{1}{c} \binom{b+c}{c}^{-1} \qquad \text{(B.36)}$$

for $b \in \mathbb{N} \cup \{0\}$, $c \in \mathbb{N}$.

With this result we find for Eq. B.31

$$c(0,0;p,q) = \frac{-1}{p+1}\sum_{\alpha'=0}^{p+1}(-1)^{\alpha'}\binom{p+1}{\alpha'}h_1(\alpha'+1;q)$$

$$= \frac{-1}{p+1}\sum_{\alpha'=0}^{p+1}(-1)^{\alpha'}\binom{p+1}{\alpha'}\frac{1}{\alpha'+1}\binom{q+\alpha'+1}{\alpha'+1}^{-1}. \tag{B.37}$$

After some rearrangements in the binomial coefficients in Eq. B.37 we obtain

$$c(0,0;p,q) = -\frac{1}{(p+1)(q+1)}\binom{p+q+2}{p+1}^{-1}\sum_{\alpha'=0}^{p+1}(-1)^{\alpha'}\binom{p+q+2}{q+1+\alpha'}$$

$$= \frac{(-1)^{-q}}{(p+1)(q+1)}\binom{p+q+2}{p+1}^{-1}\sum_{\alpha=q+1}^{p+q+2}(-1)^{\alpha}\binom{p+q+2}{\alpha}$$

$$= \frac{(-1)^{-q}}{(p+1)(q+1)}\binom{p+q+2}{p+1}^{-1}$$

$$\cdot\left\{\sum_{\alpha=0}^{p+q+2}(-1)^{\alpha}\binom{p+q+2}{\alpha} - \sum_{\alpha=0}^{q}(-1)^{\alpha}\binom{p+q+2}{\alpha}\right\}$$

$$= \frac{(-1)^{-q}}{(p+1)(q+1)}\binom{p+q+2}{p+1}^{-1}$$

$$\cdot\left\{(1-1)^{p+q+2} - (-1)^{q}\binom{p+q+1}{q}\right\} \tag{B.38}$$

where we used Gradshteyn and Ryzhik [1965], p. 3, in the last step of this derivation. From Eq. B.38 we find the result

$$c(0,0;p,q) = -\frac{1}{(p+1)(p+q+2)}. \tag{B.39}$$

The derivation of $c(\gamma, \delta; p, q)$ in Eqs. B.24-B.30 is also not valid for $(\gamma, \delta) = (p + 1, q + 1)$, i.e. $a = b = 0$ in Eq. B.24. Substituting these values in Eq. B.23b gives

$$c(p+1, q+1; p, q) = \frac{1}{(p+1)(q+1)} \frac{q+1}{p+q+2}$$

$$= \frac{1}{(p+1)(p+q+2)}. \tag{B.40}$$

So we find

$$c(p + 1, q + 1; p, q) = -c(0, 0; p, q). \tag{B.41}$$

Analyzing the data dependent part $S_N$ of $m_{pq}$ (Eq. B.23a) for $(\gamma, \delta) = (0, 0)$ gives

$$S_N(0, 0; p, q) = \sum_{n=0}^{N-1} x_n^{p+1} y_n^{q+1}$$

$$= \sum_{n=0}^{N-1} x_{n+1}^{p+1} y_{n+1}^{q+1}$$

$$= S_N(p+1, q+1; p, q) \tag{B.42}$$

where we have used the periodicity of the vertices of the polygon: $x_{n+mN} = x_n$, $y_{n+mN} = y_n$, $n \in \{0, \ldots, N-1\}$, $m \in \mathbb{Z}$. Thus we find that we can use for the data independent coefficients $c(0, 0; p, q)$ and $c(p + 1, q + 1; p, q)$ any number, as long as Eq. B.41 is satisfied, which then leads to a cancellation of the contribution of $S_N(0, 0; p, q)$ against that of $S_N(p + 1, q + 1; p, q)$. From a computational point of view it may be convenient to set both $c(0, 0; p, q) = 0$ and $c(p + 1, q + 1; p, q) = 0$.

Eqs. B.21-B.23b show that the computational complexity of $m_{pq}$, expressed in coordinates of the polygon vertices, is $O(pqN)$, the same as we found for the expression in Eq. B.18. The data independent coefficients $c(\gamma, \delta; p, q)$ need only be computed once and can be stored permanently.

Recently, Bamieh and de Figueiredo [1986] also derived a formula for the moments $m_{pq}$ of regions bounded by polygons. Though they also apply Green's theorem for the conversion of the surface integral into a contour integral, their derivation and the resulting formula are different. The resulting formula leads to a computational complexity of $O(qN)$ instead of $O(pqN)$ and can, by a minor adaptation, be further reduced to $O(\min(p, q) \cdot N)$. However, their method of computing $m_{pq}$ has some drawbacks. For each polygon side it has to be checked whether $\Delta x_n = 0$ (vertical side) or whether $\Delta y_n = 0$ (horizontal side) and, if so, the computation needs to be adapted since otherwise this would lead to a singularity. Moreover, if a polygon side is almost vertical or almost horizontal, then this may lead to overflow problems in the computation. These problems are absent in the formulas for $m_{pq}$ that we derived in this appendix.

The results obtained so far in this section can be summarized as follows. For $p, q = 0, 1, \ldots$ the moment $m_{pq}$ of a polygonal region is given by

$$m_{pq} = \sum_{\alpha=0}^{p+1} \sum_{\beta=0}^{q+1} d(\alpha, \beta; p, q)\, T_N(\alpha, \beta; p, q) \tag{B.43}$$

where $d(\alpha, \beta; p, q)$ and $T_N(\alpha, \beta; p, q)$ were defined previously, in Eqs. B.19b and B.19a respectively, as

$$d(\alpha, \beta; p, q) = \frac{1}{(p+1)(q+1)} \frac{\beta}{\alpha+\beta} \binom{p+1}{\alpha} \binom{q+1}{\beta}$$

and

$$T_N(\alpha, \beta; p, q) = \sum_{n=0}^{N-1} (\Delta x_n)^{\alpha} x_n^{p+1-\alpha} (\Delta y_n)^{\beta} y_n^{q+1-\beta}.$$

Another expression for the moment $m_{pq}$ of a polygonal region reads

$$m_{pq} = \sum_{\gamma=0}^{p+1} \sum_{\delta=0}^{q+1} c(\gamma, \delta; p, q)\, S_N(\gamma, \delta; p, q) \tag{B.44}$$

where $c(\gamma, \delta; p, q)$ was defined previously in Eq. B.30 as

$$c(\gamma, \delta; p, q) = \frac{\binom{p+1}{\gamma}\binom{q+1}{\delta}}{\binom{p+q+2}{\gamma+\delta}} \frac{1}{(p+q+2-\gamma-\delta)(\gamma+\delta)} \cdot \left\{\frac{\delta}{q+1} - \frac{\gamma}{p+1}\right\}$$

for $\gamma = 0, \ldots, p+1$ and $\delta = 0, \ldots, q+1$, with the exception of $(\gamma, \delta) = (0, 0)$ and $(\gamma, \delta) = (p+1, q+1)$ for which we found

$$c(0, 0; p, q) = -\frac{1}{(p+1)(p+q+2)}$$

and

$$c(p+1, q+1; p, q) = \frac{1}{(p+1)(p+q+2)}$$

in Eqs. B.39 and B.40 respectively, and where $S_N(\gamma, \delta; p, q)$ was defined in Eq. (B.23a) as

$$S_N(\gamma, \delta; p, q) = \sum_{n=0}^{N-1} x_{n+1}^{\gamma} x_n^{p+1-\gamma} y_{n+1}^{\delta} y_n^{q+1-\delta}.$$

In Table B.1 we list the coefficients $d(\alpha, \beta; p, q)$ in matrices $\boldsymbol{D}(p, q)$ of size $(p+2)(q+1)$ for moments of the orders $(p+q) = 0, 1, 2$ and 3. The first row in each matrix $\boldsymbol{D}$ is indexed as $\alpha = 0$, while the first column is indexed as $\beta = 1$ (cf. Eq. B.43). The column with index $\beta = 0$ is lacking since the coefficients $d(\alpha, \beta; p, q) = 0$ for $\beta = 0$ (cf. Eq. B.19b).

In Table B.2 we do the same for the coefficients $c(\gamma, \delta; p, q)$, which are listed in matrices $\boldsymbol{C}(p, q)$ of size $(p+2)(q+2)$, using similar indexing conventions. The first row in each matrix $\boldsymbol{C}$ is indexed as $\gamma = 0$, while the first column is indexed as $\delta = 0$ (cf. Eq. B.44).

The matrices $\boldsymbol{C}(p, q)$ of coefficients $c(\gamma, \delta; p, q)$ have a number of special properties, which will be derived in the following.

**Table B.1.** Matrices $\boldsymbol{D}(p, q)$ of data independent coefficients $d(\alpha, \beta; p, q)$ (Eq. B.19b) for moments of the orders $(p + q) = 0, 1, 2$ and 3.

| Moment order | Matrices $\boldsymbol{D}(p, q)$ of coefficients $d(\alpha, \beta; p, q)$ | |
|---|---|---|
| 0 | $\boldsymbol{D}(0,0) = \frac{1}{2}\begin{pmatrix} 2 \\ 1 \end{pmatrix}$ | |
| 1 | $\boldsymbol{D}(0,1) = \frac{1}{6}\begin{pmatrix} 6 & 3 \\ 3 & 2 \end{pmatrix}$; | $\boldsymbol{D}(1,0) = \frac{1}{6}\begin{pmatrix} 3 \\ 3 \\ 1 \end{pmatrix}$ |
| 2 | $\boldsymbol{D}(0,2) = \frac{1}{12}\begin{pmatrix} 12 & 12 & 4 \\ 6 & 8 & 3 \end{pmatrix}$; | $\boldsymbol{D}(1,1) = \frac{1}{24}\begin{pmatrix} 12 & 6 \\ 12 & 8 \\ 4 & 3 \end{pmatrix}$ |
| | $\boldsymbol{D}(2,0) = \frac{1}{12}\begin{pmatrix} 4 \\ 6 \\ 4 \\ 1 \end{pmatrix}$ | |
| 3 | $\boldsymbol{D}(0,3) = \frac{1}{20}\begin{pmatrix} 20 & 30 & 20 & 5 \\ 10 & 20 & 15 & 4 \end{pmatrix}$; | $\boldsymbol{D}(1,2) = \frac{1}{60}\begin{pmatrix} 30 & 30 & 10 \\ 30 & 40 & 15 \\ 10 & 15 & 6 \end{pmatrix}$ |
| | $\boldsymbol{D}(2,1) = \frac{1}{60}\begin{pmatrix} 20 & 10 \\ 30 & 20 \\ 20 & 15 \\ 5 & 4 \end{pmatrix}$; | $\boldsymbol{D}(3,0) = \frac{1}{20}\begin{pmatrix} 5 \\ 10 \\ 10 \\ 5 \\ 1 \end{pmatrix}$ |

**Theorem B.1.**
The coefficients $c(\gamma, \delta; p, q)$ of the matrices $\boldsymbol{C}(p, q)$ have the following antisymmetry property

$$c(p + 1 - \gamma, q + 1 - \delta; p, q) = -c(\gamma, \delta; p, q)$$

$$\gamma = 0, \ldots, p + 1, \quad \delta = 0, \ldots, q + 1. \tag{B.45}$$

*Proof*

By substituting $p + 1 - \gamma$ for $\gamma$ and $q + 1 - \delta$ for $\delta$ into Eq. B.30 we find

$c(p + 1 - \gamma, q + 1 - \delta; p, q)$

**Table B.2.** Matrices $\boldsymbol{C}(p, q)$ of data independent coefficients $c(\gamma, \delta; p, q)$ (Eqs. B.30, B.39 and B.40) for moments of the orders $(p + q) = 0, 1, 2$ and 3.

| Moment order | Matrices $\boldsymbol{C}(p, q)$ of coefficients $d(\gamma, \delta; p, q)$ |
|---|---|
| 0 | $\boldsymbol{C}(0,0) = \frac{1}{2}\begin{pmatrix} -1 & 1 \\ -1 & 1 \end{pmatrix}$ |
| 1 | $\boldsymbol{C}(0,1) = \frac{1}{6}\begin{pmatrix} -2 & 1 & 1 \\ -1 & -1 & 2 \end{pmatrix}$; $\boldsymbol{C}(1,0) = \frac{1}{6}\begin{pmatrix} -1 & 1 \\ -1 & 1 \\ -1 & 1 \end{pmatrix}$ |
| 2 | $\boldsymbol{C}(0,2) = \frac{1}{12}\begin{pmatrix} -3 & 1 & 1 & 1 \\ -1 & -1 & -1 & 3 \end{pmatrix}$; $\boldsymbol{C}(1,1) = \frac{1}{24}\begin{pmatrix} -3 & 2 & 1 \\ -2 & 0 & 2 \\ -1 & -2 & 3 \end{pmatrix}$ <br> $\boldsymbol{C}(2,0) = \frac{1}{12}\begin{pmatrix} -1 & 1 \\ -1 & 1 \\ -1 & 1 \\ -1 & 1 \end{pmatrix}$ |
| 3 | $\boldsymbol{C}(0,3) = \frac{1}{20}\begin{pmatrix} -4 & 1 & 1 & 1 & 1 \\ -1 & -1 & -1 & -1 & 4 \end{pmatrix}$; $\boldsymbol{C}(1,2) = \frac{1}{60}\begin{pmatrix} -6 & 3 & 2 & 1 \\ -3 & -1 & 1 & 3 \\ -1 & -2 & -3 & 6 \end{pmatrix}$ <br> $\boldsymbol{C}(2,1) = \frac{1}{60}\begin{pmatrix} -4 & 3 & 1 \\ -3 & 1 & 2 \\ -2 & -1 & 3 \\ -1 & -3 & 4 \end{pmatrix}$; $\boldsymbol{C}(3,0) = \frac{1}{20}\begin{pmatrix} -1 & 1 \\ -1 & 1 \\ -1 & 1 \\ -1 & 1 \\ -1 & 1 \end{pmatrix}$ |

$$= \frac{\binom{p+1}{p+1-\gamma}\binom{q+1}{q+1-\delta}}{\binom{p+q+2}{p+q+2-\gamma-\delta}} \frac{1}{(\gamma+\delta)(p+q+2-\gamma-\delta)}$$

$$\cdot \left\{\frac{q+1-\delta}{q+1} - \frac{p+1-\gamma}{p+1}\right\}$$

$$= -\frac{\binom{p+1}{\gamma}\binom{q+1}{\delta}}{\binom{p+q+2}{\gamma+\delta}} \frac{1}{(p+q+2-\gamma-\delta)(\gamma+\delta)} \left\{\frac{\delta}{q+1} - \frac{\gamma}{p+1}\right\}$$

$$= -c(\gamma, \delta; p, q) \qquad \text{(B.46)}$$

for $\gamma = 0, \ldots, p + 1$ and $\delta = 0, \ldots, q + 1$, with the exception of $(\gamma, \delta) = (0, 0)$ and $(\gamma, \delta) = (p + 1, q + 1)$. For the latter two pairs of $(\gamma, \delta)$ the relation in Eq. B.45 was already found in Eq. B.41.

□

Through this property the number of multiplications required to compute a moment $m_{pq}$ can be reduced by a factor of two.

**Corollary B.1.**

$$\sum_{\gamma=0}^{p+1} \sum_{\delta=0}^{q+1} c(\gamma, \delta; p, q) = 0. \tag{B.47}$$

*Proof*

This corollary is a direct consequence of Theorem B.1

$$\begin{aligned} &\sum_{\gamma=0}^{p+1} \sum_{\delta=0}^{q+1} c(\gamma, \delta; p, q) \\ &\quad = \frac{1}{2} \sum_{\gamma=0}^{p+1} \sum_{\delta=0}^{q+1} \{c(\gamma, \delta; p, q) + c(p + 1 - \gamma, q + 1 - \delta; p, q)\} \\ &\quad = 0. \end{aligned} \tag{B.48}$$

□

Through the following properties the number of arithmetic operations may be reduced even more.

**Theorem B.2.**

$$c(\gamma, \gamma; p, p) = 0, \qquad 0 < \gamma < p + 1 \tag{B.49a}$$

and

$$c\left(\frac{p+1}{2}, \frac{q+1}{2}; p, q\right) = 0, \quad \text{if both } p \text{ and } q \text{ odd.} \tag{B.49b}$$

*Proof*

In Eq. B.49a we consider the coefficients $c(\gamma, \delta; p, q)$ for which $p = q$ and $\gamma = \delta$. By applying these relations in Eq. B.30, the validity of the property in Eq. B.49a immediately becomes clear. The same is found for Eq. B.49b through a substitution of $\gamma = (p + 1)/2$ and $\delta = (q + 1)/2$ into Eq. B.30.

□

From Eqs. B.41, B.42 and B.44 we already concluded that $c(0, 0; p, q)$ and $c(p + 1, q + 1; p, q)$ can just as well both be set to zero, thus leading to a further reduction of the number of arithmetic operations.

The next two theorems deal with the properties of sums of coefficients $c(\gamma, \delta; p, q)$ over rows and columns of $\boldsymbol{C}(p, q)$.

**Theorem B.3.**
The sum of coefficients $c(\gamma, \delta; p, q)$ over a row in $\boldsymbol{C}(p, q)$ satisfies

$$\sum_{\delta=0}^{q+1} c(\gamma, \delta; p, q) = 0, \qquad 0 \leqslant \gamma \leqslant p + 1. \tag{B.50}$$

*Proof*

In order to show this property, we substitute the expression in Eq. B.23b for $c(\gamma, \delta; p, q)$ in Eq. B.50

$$\sum_{\delta=0}^{q+1} c(\gamma, \delta; p, q)$$

$$= \sum_{\delta=0}^{q+1} \frac{\binom{p+1}{\gamma}\binom{q+1}{\delta}}{(p+1)(q+1)}$$

$$\cdot \sum_{n=0}^{p+1-\gamma} \sum_{m=0}^{q+1-\delta} (-1)^{(n+m)} \frac{m+\delta}{n+m+\gamma+\delta} \binom{p+1-\gamma}{n} \binom{q+1-\delta}{m}$$

$$= \frac{\binom{p+1}{\gamma}}{(p+1)(q+1)} \sum_{n=0}^{p+1-\gamma} (-1)^n \binom{p+1-\gamma}{n}$$

$$\cdot \sum_{\delta=0}^{q+1} \binom{q+1}{\gamma} \sum_{m=0}^{q+1-\delta} (-1)^m \frac{m+\delta}{n+m+\gamma+\delta} \binom{q+1-\delta}{m}. \tag{B.51}$$

In order to solve the last summation in Eq. B.51, we introduce the help function

$$f_2(x) = \sum_{m=0}^{b} (-1)^m \frac{m+c_1}{m+c_2} \binom{b}{m} x^{m+c_2} \tag{B.52}$$

with $b,\ c_1 \in \mathbb{N} \cup \{0\},\ c_2 \in \mathbb{N}$.

Analogous to the derivations in Eqs. B.24-B.28 and Eqs. B.32-B.36 we find

$$h_2(c_1, c_2; b) = f_2(1) = \sum_{m=0}^{b} (-1)^m \frac{m+c_1}{m+c_2} \binom{b}{m}$$

$$= \begin{cases} \dfrac{c_2 - c_1}{c_2} \binom{c_2+b}{c_2}^{-1}, & b > 0, \quad \text{(B.53a)} \\[2ex] \dfrac{c_1}{c_2}, & b = 0. \quad \text{(B.53b)} \end{cases}$$

With $c_1 = \delta$, $c_2 = n + \gamma + \delta$ and $b = q + 1 - \delta$ we then obtain for the last summation in Eq. B.51

$$h_2(\delta, n+\gamma+\delta; q+1-\delta)$$

$$= \sum_{m=0}^{q+1-\delta} (-1)^m \frac{m+\delta}{n+m+\gamma+\delta} \binom{q+1-\delta}{m}$$

$$= \begin{cases} -\dfrac{n+\gamma}{n+\gamma+\delta} \dbinom{q+1+n+\gamma}{n+\gamma+\delta}^{-1}, & 0 \leqslant \delta < q+1, \quad \text{(B.54a)} \\[2ex] \dfrac{\delta}{n+\gamma+\delta} = \dfrac{q+1}{q+1+n+\gamma}, & \delta = q+1. \quad \text{(B.54b)} \end{cases}$$

Note that Eq. B.54a is not valid for $n + \gamma + \delta = 0$, since it is required that $c_2 \neq 0$. Therefore we must verify Eq. B.50 separately for the case $\gamma = 0$.

Substitution of the results in Eqs. B.54a and B.54b into Eq. B.51 yields

$$\sum_{\delta=0}^{q+1} c(\gamma, \delta; p, q)$$

$$= \frac{\binom{p+1}{\gamma}}{(p+1)(q+1)} \sum_{n=0}^{p+1-\gamma} (-1)^n \binom{p+1-\gamma}{n}$$

$$\cdot \left\{ \sum_{\delta=0}^{q} \binom{q+1}{\delta} \frac{-(n+\gamma)}{n+\gamma+\delta} \binom{q+1+n+\gamma}{n+\gamma+\delta}^{-1} + \frac{q+1}{q+1+n+\gamma} \right\}. \tag{B.55}$$

We now concentrate on the summation over $\delta$ in Eq. B.55. Rearranging factorials in this expression gives

$$\binom{q+1}{\delta} \frac{n+\gamma}{n+\gamma+\delta} \binom{q+1+n+\gamma}{n+\gamma+\delta}^{-1}$$

$$= \binom{q+1+n+\gamma}{n+\gamma}^{-1} \binom{n+\gamma-1+\delta}{n+\gamma-1}. \tag{B.56}$$

With the aid of this result and the formula

$$\sum_{k=0}^{m} \binom{n+k}{n} = \binom{n+m+1}{n+1}. \tag{B.57}$$

(cf. Gradshteyn and Ryzhik [1965], p. 3), we find for the expression in braces in Eq. B.55

$$\begin{aligned} \sum_{\delta=0}^{q} \binom{q+1}{\delta} \frac{-(n+\gamma)}{n+\gamma+\delta} \binom{q+1+n+\gamma}{n+\gamma+\delta}^{-1} + \frac{q+1}{q+1+n+\gamma} \\ = -\binom{q+1+n+\gamma}{n+\gamma}^{-1} \binom{q+n+\gamma}{n+\gamma} + \frac{q+1}{q+1+n+\gamma} \\ = -\frac{q+1}{q+1+n+\gamma} + \frac{q+1}{q+1+n+\gamma} \\ = 0. \end{aligned} \tag{B.58}$$

If we consider again Eq. B.55 it is immediately clear that with the result in Eq. B.58 we have proven this theorem for $0 < \gamma \leqslant p + 1$. We still have to verify the case $\gamma = 0$.

Since we already verified Eq. B.50 for $\gamma = p + 1$, the validity of Eq. B.50 for $\gamma = 0$ follows from Theorem B.1

$$\begin{aligned} \sum_{\delta=0}^{q+1} c(0, \delta; p, q) &= -\sum_{\delta=0}^{q+1} c(p+1, q+1-\delta; p, q) \\ &= -\sum_{\delta'=0}^{q+1} c(p+1, \delta'; p, q) \\ &= 0 \end{aligned} \tag{B.59}$$

where $\delta' = q + 1 - \delta$.

The proof of this theorem is now complete.

□

A similar theorem is valid for the sum of coefficients $c(\gamma, \delta; p, q)$ over a column in $\boldsymbol{C}(p, q)$.

**Theorem B.4.**
The sum of coefficients $c(\gamma, \delta; p, q)$ over a column in $\boldsymbol{C}(p, q)$ satisfies

$$\sum_{\gamma=0}^{p+1} c(\gamma, \delta; p, q) = 0, \qquad 0 < \delta < q + 1. \tag{B.60}$$

*Proof*

The proof of this theorem follows along similar lines as the proof of Theorem B.3. We sketch the main steps, substituting in Eq. B.60 the expression for $c(\gamma, \delta; p, q)$ in Eq. B.23b

$$\sum_{\gamma=0}^{p+1} c(\gamma, \delta; p, q)$$

$$= \sum_{\gamma=0}^{p+1} \frac{\binom{p+1}{\gamma}\binom{q+1}{\delta}}{(p+1)(q+1)}$$

$$\cdot \sum_{n=0}^{p+1-\gamma} \sum_{m=0}^{q+1-\delta} (-1)^{n+m} \frac{m+\delta}{n+m+\gamma+\delta} \binom{p+1-\gamma}{n}\binom{q+1-\delta}{m}$$

$$= \frac{\binom{q+1}{\delta}}{(p+1)(q+1)} \sum_{m=0}^{q+1-\delta} (-1)^m (m+\delta) \binom{q+1-\delta}{m}$$

$$\cdot \sum_{\gamma=0}^{p+1} \binom{p+1}{\gamma} h_1(m+\gamma+\delta; p+1-\gamma) \tag{B.61}$$

where the function $h_1$ was defined earlier in Eq. B.32. Note from Eq. B.32 that Eq. B.61 is not valid for $m + \gamma + \delta = 0$, and therefore not for $\delta = 0$.

Through Eqs. B.36 and B.57 we obtain

$$\sum_{\gamma=0}^{p+1} c(\gamma, \delta; p, q) = \frac{\binom{q+1}{\delta}}{(p+1)(q+1)} \sum_{m=0}^{q+1-\delta} (-1)^m \binom{q+1-\delta}{m}$$

$$= \frac{\binom{q+1}{\delta}}{(p+1)(q+1)} (1-1)^{q+1-\delta}$$

$$= 0, \qquad 0 < \delta < q+1 \tag{B.62}$$

which completes the proof of this theorem.

□

From Eq. B.62 we observe that the property of Theorem B.4 is not valid for $\delta = q + 1$, since

$$\sum_{\gamma=0}^{p+1} c(\gamma, q+1; p, q) = \frac{1}{(p+1)(q+1)}. \tag{B.63}$$

For $\delta = 0$ the property is not valid either (cf. also Table B.2). This follows immediately if we apply Theorem B.1 to Eq. B.63, from which we find the result

$$\sum_{\gamma=0}^{p+1} c(\gamma, 0; p, q) = -\frac{1}{(p+1)(q+1)}. \tag{B.64}$$

## References

Abu-Mostafa, Y.S. and D. Psaltis [1985]
'Image Normalization by Complex Moments', IEEE Trans. Patt. and Mach. Intell. **PAMI-7**: 46-55.

Alt, F.L. [1962]
'Digital Pattern Recognition by Moments', Journ. ACM **9**: 240-258.

Apostol, T.M. [1974]
*Mathematical Analysis*, Second Edition, Reading, MA: Addison-Wesley.

Bamieh, B. and R.J.P. de Figueiredo [1986]
'A General Moment-Invariants/Attributed-Graph Method for Three-Dimensional Object Recognition from a Single Image', IEEE Journ. of Robot. and Autom. **RA-2**: 31-41.

Boyce, J.F. and W.J. Hossack [1983]
'Moment Invariants for Pattern Recognition', Patt. Recogn. Lett. **1**: 451-456 and Patt. Recogn. Lett. **2**: 131 (Errata).

Casey, R.G. [1970]
'Moment Normalization of Handprinted Characters', IBM Journ. of Res. and Dev. **14**: 548-557.

Cyganski, D. and J.A. Orr [1985]
'Applications of Tensor Theory to Object Recognition and Orientation Determination', IEEE Trans. Patt. Anal. and Mach. Intell. **PAMI-7**: 662-673.

Dudani, S.A., K.J. Breeding and R.B. McGhee [1977]
'Aircraft Identification by Moment Invariants', IEEE Trans. Comp. **C-26**: 39-46.

Ellis, Jr., J.R. and M. Eden [1976]
'On the Number of Sides Necessary for Polygonal Approximation of Black-and-White Figures in a Plane', Inform. and Contr. **30**: 169-186.

Gilmore, J.F. and W.W. Boyd [1981]
'Building and Bridge Classification by Moment Invariants'. In: *Processing of Images and Data from Optical Sensors*, W.H. Carter (Ed.), Proc. SPIE **292**: 256-263.

Gradshteyn, I.S. and I.M. Ryzhik [1965]
*Table of Integrals, Series, and Products*, New York: Academic Press.

Ho, C.-S. [1983]
'Precision of Digital Vision Systems', IEEE Trans. Patt. Anal. and Mach. Intell. **PAMI-5**: 593-601.

Hu, M.K. [1961]
'Pattern Recognition by Moment Invariants', Proc. IRE **49**: 1428.

Hu, M.K. [1962]
'Visual Pattern Recognition by Moment Invariants', IRE Trans. Inf. Th. **IT-8**: 179-187.

Kashyap, R.L. and B.J. Oommen [1983]
'Scale Preserving Smoothing of Polygons', IEEE Trans. Patt. Anal. and Mach. Intell. **PAMI-5**: 667-671.

E. Kreysig [1972]
*Advanced Engineering Mathematics*, New York: John Wiley and Sons, Inc.

Kurozumi, Y. and W.A. Davis [1982]
'Polygonal Approximation by the Minimax Method', Comp. Graph. and Im. Proc. **19**: 248-264.

McClure, D.E. and R.A. Vitale [1975]
'Polygonal Approximation of Plane Convex Bodies', Journ. Math. Anal. and Appl. **51**: 326-358.

Maitra, S. [1979]
'Moment Invariants', Proc. IEEE **67**: 697-699.

Miles, R.G. and J.G. Tough [1983]
'A Method for the Computation of Inertial Properties for General Areas', Comp. Aid. Des. **15**: 196-200.

Montanari, U. [1970]
'A Note on Minimal Length Polygonal Approximation to a Digitized Contour', Comm. ACM **13**: 41-47.

Nill, N.B. [1981]
'Applications of Moments to Image Understanding'. In: *Techniques and Applications of Image Understanding*, J.J. Pearson (Ed.), Proc. SPIE **281**: 126-131.

Pavlidis, T. [1977b]
'Polygonal Approximations by Newton's Method', IEEE Trans. Comp. **C-26**: 800-807.

Pavlidis, T. and S.L. Horowitz [1974]
'Segmentation of Plane Curves', IEEE Trans. Comp. **C-23**: 860-870.

Ramer, U. [1972]
'An Iterative Procedure for the Polygonal Approximation of Plane Curves', Comp. Graph. and Im. Proc. **1**: 244-256.

Reddi, S.S. [1981]
'Radial and Angular Moment Invariants for Image Identification', IEEE Trans. Patt. Anal. and Mach. Intell. **PAMI-3**: 240-242.

Reeves, A.P. and A. Rostampour [1981]
'Shape Analysis of Segmented Objects Using Moments', Proc. IEEE Comp. Soc. Conf. on Patt. Recogn. and Image Proc., Dallas, TX., 1981: 171-174.

Reeves, A.P. and B.S. Wittner [1983]
'Shape Analysis of Three Dimensional Objects Using the Method of Moments', Proc. IEEE Comp. Soc. Conf. on Comp. Vision and Patt. Recogn., Washington, DC, 1983: 20-26.

Riordan, J. [1979]
*Combinatorial Identities*, Huntingdon, NY: Robert E. Krieger Publishing Company.

Sadjadi, F.A. and E.L. Hall [1978]
'Numerical Computations of Moment Invariants for Scene Analysis', Proc. IEEE Comp. Soc. Conf. on Patt. Recogn. and Image Proc., Chicago, IL: 181-187.

Sklansky, J. and V. Gonzalez [1979]
'A Parallel Mechanism for Describing Silhouettes', Proc. IEEE Comp. Soc. Conf. on Patt. Recogn. and Image Proc., Chicago, IL: 604-609.

Smith, F.W. and M.H. Wright [1970]
'Automatic Ship Photo Interpretation by the Method of Moments', IEEE Conf. Record of the Symp. on Feat. Extract. and Select. in Patt. Recogn., Argonne, IL: 145-154.

Spiegel, M.R. [1964]
*Complex Variables with an Introduction to Conformal Mapping and Its Applications*, Schaum's Outline Series, New York: McGraw-Hill Book Co., Inc.

Tang, G.Y. [1982]
'A Discrete Version of Green's Theorem', IEEE Trans. Patt. Anal. and Mach. Intell. **PAMI-4**: 242-249.

Teague, M.R. [1980]
'Image Analysis via the General Theory of Moments', Journ. Opt. Soc. Am. **70**: 920-930.

Teh, C.-H. and R.T. Chin [1986]
'On Digital Approximation of Moment Invariants', Comp. Vis., Graph. and Im. Proc. **33**: 318-326.

Vijaya Kumar, B.V.K. and C.A. Rahenkamp [1986]
'Calculation of Geometric Moments Using Fourier Plane Intensities', Appl. Opt. **25**: 997-1007.

Williams, C.M. [1981]
'Bounded Straight-Line Approximation of Digitized Planar Curves and Lines', Comp. Graph. and Im. Proc. **16**: 370-381.

Wilson, Jr., H.B. and D.S. Farrior [1976]
'Computation of Geometrical and Inertial Properties for General Areas and Volumes of Revolution', Comp. Aid. Des. **8**: 257-263.

Wong, R.Y. and E.L. Hall [1978]
'Scene Matching with Invariant Moments', Comp. Graph. and Im. Proc. **8**: 16-24.

# Appendix C

# Estimation of contour representations using polynomial filters

In Section 4.4 we described an experiment to obtain insight into the characteristics of the dissimilarity measures, defined in Section 4.2. In Section 4.5 similar experiments were described to evaluate the performance of the dissymmetry measures, introduced in that section.

In order to perform these experiments we had to estimate sets of samples of the contour representations $z$, $\dot{z}$, $\ddot{z}$, $\psi$ and $K$ (cf. Chapter 2), taken equidistantly along contours. The contour representations $\dot{z}$, $\ddot{z}$, $\psi$ and $K$ involve first and second order derivatives of the position function $z$. In the following we describe a method to obtain samples of these representations, through differentiation of piecewise polynomial approximations to position function samples $z[n]$. For the polynomial approximation we use a least squares criterion. We show that this method corresponds to applying a finite impulse response filter (FIR filter) with fixed coefficients to the position function samples $z[n]$, which leads to efficient implementations. The coefficients of the FIR filter depend on the order of the polynomials used in the fit, the order of the derivative and the number of samples $z[n]$ to which we fit the polynomial, but not on the position along the contour.

The polynomial of order $P$ that we fit to $2M + 1$ position function samples, centered at $z[n]$, with $2M + 1 > P$, is given by

$$\pi_n(t) = \sum_{p=0}^{P} c_p[n](t - t_n)^p \tag{C.1}$$

where the coefficients $c_p[n]$ are complex valued

$$c_p[n] = a_p[n] + ib_p[n]. \tag{C.2}$$

The method of least squares fit of $\pi_n(t)$ to $2M + 1$ samples of the position function $z$, centered at $z[n]$, requires that we minimize

$$D^2[n] = \sum_{m=-M}^{M} \left| z[n+m] - \sum_{p=0}^{P} c_p[n](t_{n+m} - t_n)^p \right|^2 \tag{C.3}$$

with respect to the $P + 1$ coefficients $c_p[n]$.

The gradient operator with respect to $c_p[n]$ is defined as (cf. Spiegel [1964], p. 82)

$$\nabla_{c_p[n]} = \frac{\partial}{\partial a_p[n]} + \mathrm{i}\frac{\partial}{\partial b_p[n]} = 2\frac{\partial}{\partial \bar{c}_p[n]}. \tag{C.4}$$

Applying this gradient operator to $D^2[n]$ with respect to each of the coefficients $c_p[n]$ and setting the results equal to zero yields the following system of $P + 1$ linear equations

$$\sum_{m=-M}^{M} z[n+m](t_{n+m} - t_n)^p$$

$$= \sum_{q=0}^{P} \left\{ \sum_{m=-M}^{M} (t_{n+m} - t_n)^{p+q} \right\} c_q[n], \qquad p = 0, \ldots, P. \tag{C.5}$$

We introduce the notation

$$S_q[n] = \sum_{m=-M}^{M} (t_{n+m} - t_n)^q. \tag{C.6}$$

Writing Eq. C.5 in matrix/vector form gives

$$\boldsymbol{u}[n] = \boldsymbol{T}[n]\boldsymbol{c}[n], \tag{C.7}$$

where $\boldsymbol{u}[n]$ is a vector with $P + 1$ elements

$$\boldsymbol{u}[n] = \begin{pmatrix} \sum_{m=-M}^{M} z[n+m] \\ \sum_{m=-M}^{M} z[n+m](t_{n+m} - t_n) \\ \vdots \\ \sum_{m=-M}^{M} z[n+m](t_{n+m} - t_n)^P \end{pmatrix} \tag{C.8a}$$

$\boldsymbol{T}[n]$ is a $(P + 1) \times (P + 1)$ matrix

$$\boldsymbol{T}[n] = \begin{pmatrix} S_0[n] & S_1[n] & \ldots & S_P[n] \\ S_1[n] & S_2[n] & \ldots & S_{P+1}[n] \\ \vdots & \vdots & & \vdots \\ S_P[n] & S_{P+1}[n] & \ldots & S_{2P}[n] \end{pmatrix} \tag{C.8b}$$

and $\boldsymbol{c}[n]$ is the vector of $P + 1$ coefficients $c_p[n]$ of the polynomial $\pi_n(t)$ (cf. Eq. C.1)

$$\boldsymbol{c}[n] = \begin{pmatrix} c_0[n] \\ c_1[n] \\ \vdots \\ c_P[n] \end{pmatrix}. \tag{C.8c}$$

We denote the $p$-th column in the matrix $\boldsymbol{T}[n]$ as the vector $\boldsymbol{t}_p[n]$

$$\boldsymbol{t}_p[n] = \begin{pmatrix} S_p[n] \\ S_{p+1}[n] \\ \vdots \\ S_{p+P}[n] \end{pmatrix}. \tag{C.9}$$

Furthermore, the determinant of a matrix $\boldsymbol{A}$ is denoted as $|\boldsymbol{A}|$. Then, according to Cramer's rule (cf. Apostol [1969], Cohn [1974]), the solutions of the coefficients $c_p[n]$ are given by

$$c_p[n] = \frac{|\boldsymbol{U}_p[n]|}{|\boldsymbol{T}[n]|}, \qquad p = 0, \ldots, P \tag{C.10}$$

where the matrix $\boldsymbol{U}_p[n]$ is defined as

$$\boldsymbol{U}_p[n] = \left(\boldsymbol{t}_0[n] \quad \ldots \quad \boldsymbol{t}_{p-1}[n] \quad \boldsymbol{u}[n] \quad \boldsymbol{t}_{p+1}[n] \quad \ldots \quad \boldsymbol{t}_P[n]\right). \tag{C.11}$$

We note that in the matrix $\boldsymbol{U}_p[n]$ only the $p$-th column, i.e. the vector $\boldsymbol{u}[n]$, depends upon the samples $z[n]$.

We denote the cofactor of the element $a(q, p)$ at row $q$ and column $p$ of matrix $\boldsymbol{A}$ as $A_{qp}$. For ease of notation in the context of this appendix we start counting rows and columns by index 0 instead of by index 1.

To separate the data independent elements from the data dependent elements in Eq. C.10 we expand the determinant $|\boldsymbol{U}_p[n]|$ by its $p$-th column

$$\begin{aligned} c_p[n] &= \frac{1}{|\boldsymbol{T}[n]|} \sum_{q=0}^{P} \left\{ \sum_{m=-M}^{M} z[n+m](t_{n+m} - t_n)^q \right\} (U_p[n])_{qp} \\ &= \frac{1}{|\boldsymbol{T}[n]|} \sum_{m=-M}^{M} z[n+m] \sum_{q=0}^{P} (t_{n+m} - t_n)^q (U_p[n])_{qp} \\ &= \sum_{m=-M}^{M} z[n+m] \frac{|\boldsymbol{V}_p[n, m]|}{|\boldsymbol{T}[n]|}, \qquad p = 0, \ldots, P \end{aligned} \tag{C.12}$$

where the matrix $\boldsymbol{V}_p[n, m]$ is given by

$$\boldsymbol{V}_p[n, m] = (\boldsymbol{t}_0[n] \quad \ldots \quad \boldsymbol{t}_{p-1}[n] \quad \boldsymbol{v}[n, m] \quad \boldsymbol{t}_{p+1}[n] \quad \ldots \quad \boldsymbol{t}_P[n]) \tag{C.13a}$$

with vector $\boldsymbol{v}[n, m]$

$$\boldsymbol{v}[n,m] = \begin{pmatrix} 1 \\ (t_{n+m} - t_n) \\ \vdots \\ (t_{n+m} - t_n)^P \end{pmatrix}. \tag{C.13b}$$

In our experiments, contours are parametrized according to normalized arc length and the samples $z[n]$ have been taken equidistantly along the contour. Consequentially, in this case

$$t_{n+m} - t_n = m\Delta t. \tag{C.14}$$

If $N$ is the number of samples taken along the contour, then $\Delta t = 2\pi/N$. So $t_{n+m} - t_n$ is independent of $n$. Therefore also $S_q[n]$, $\boldsymbol{T}[n]$, and $\boldsymbol{V}_p[n, m]$ are independent of $n$ and are denoted from now on as $S_q$, $\boldsymbol{T}$ and $\boldsymbol{V}_p[m]$.

As a result we can now express the coefficients $c_p[n]$ as (cf. Eq. C.12)

$$c_p[n] = \sum_{m=-M}^{M} z[n+m] \frac{|\boldsymbol{V}_p[m]|}{|\boldsymbol{T}|}$$

$$= \sum_{m'=-M}^{M} z[n-m'] \frac{|\boldsymbol{V}_p[-m']|}{|\boldsymbol{T}|}$$

$$= z[n] * h_p[n], \qquad p = 0, \dots, P \tag{C.15}$$

where

$$h_p[n] = \frac{|\boldsymbol{V}_p[-n]|}{|\boldsymbol{T}|}, \qquad n = -M, \dots, M \tag{C.16}$$

and where $*$ denotes (cyclic) convolution (cf. e.g. Oppenheim, Willsky and Young [1983]). From Eqs. C.15 and C.16 we observe that $c_p[n]$, the $p$-th coefficient of the polynomial $\pi_n(t)$ centered at $z[n]$, is the result of a (cyclic) convolution of the position function samples and an FIR filter $h_p[n]$ with $2M + 1$ coefficients, that are independent of the position along the contour.

Another consequence of the equidistant sampling along the contour (Eq. C.14) is that $S_q$ becomes (cf. Eq. C.6)

$$S_q = (\Delta t)^q \sum_{m=-M}^{M} m^q \begin{cases} = 0 & \text{for } q \text{ odd} \\ > 0 & \text{for } q \text{ even.} \end{cases} \tag{C.17}$$

As a consequence of Eq. C.17 we obtain for the elements $t(q, r)$ of the matrix $\boldsymbol{T}$ (cf. Eq. C.8b)

$$t(q, r) = S_{q+r} = 0 \qquad \text{for } q + r \text{ odd.} \tag{C.18}$$

It is easily verified that, as a result of Eq. C.18, the cofactor $T_{qr}$ of the matrix $\boldsymbol{T}$ satisfies

$$T_{qr} = 0 \qquad \text{for } q + r \text{ odd.} \tag{C.19}$$

If we expand the determinant in the numerator of the expression for $h_p[n]$ in Eq. C.16 by its $p$-th column, we obtain

$$\begin{aligned} h_p[n] &= \frac{1}{|\boldsymbol{T}|} \sum_{q=0}^{P} (-n\Delta t)^q (V_p[-n])_{qp} \\ &= \frac{1}{|\boldsymbol{T}|} \sum_{q=0}^{P} (-n\Delta t)^q T_{qp}, \qquad p = 0, \ldots, P \end{aligned} \tag{C.20}$$

since $(V_p[n])_{qp} = T_{qp}$ for $n = -M, \ldots, M$, $q = 0, \ldots, P$ (cf. Eqs. C.8b, C.13a, b).

From Eqs. C.19 and C.20 it now follows that

$$h_p[n] = h_p[-n] \qquad \text{for } p \text{ even} \tag{C.21a}$$

since for $p$ even the indices $q$ that lead to a contribution to $h_p[n]$ in Eq. C.20 are also even. Likewise we find

$$h_p[n] = -h_p[-n] \qquad \text{for } p \text{ odd.} \tag{C.21b}$$

Thus for $p$ even the coefficients $h_p[n]$ define a symmetric FIR filter, and for $p$ odd they define an antisymmetric FIR filter.

While the coefficients $h_p[n]$ need to be computed only once, the coefficients $c_p[n]$ of the polynomial $\pi_n(t)$ must be computed at each of the $N$ sample positions along the contour (cf. Eqs. C.15 and C.16). The complexity of the computation of each of the $P$ coefficients $c_p[n]$ at all $N$ sample positions along the contour is $O(N \cdot M)$, when fitting to $2M + 1$ position function samples. For sufficiently large $M$ it may be more efficient to compute the convolution sum in Eq. C.15 via multiplication in the Fourier domain using FFT methods (cf. Section 4.2). In the latter case the computational complexity is $O(N \log_2 N)$, assuming that $N > 2M + 1$ and that $N$ is a power of 2 (cf. Section 4.2).

With the result in Eq. C.17 the polynomial $\pi_n(t)$, Eq. C.1, can be computed efficiently at each position $z[n]$ along the contour. In digital image analysis, the position function samples $z[n]$ are always corrupted by distortion, caused by the discrete geometry of the two-dimensional sampling pattern. Therefore we use the polynomial $\pi_n(t)$ to smooth the position function $z$ at the $n$-th sample point, i.e. at $t = t_n$. Thus the approximated position function sample $z_a[n]$ is given by (cf. Eqs. C.1-C.2)

$$z_a[n] = \pi_n(t_n) = c_0[n] = a_0[n] + ib_0[n]. \tag{C.22}$$

In order to approximate the tangent function $\dot{z}$ at the $n$-th sample point, we take the derivative of $\pi_n(t)$ and evaluate the result at $t = t_n$. This gives (cf. Eqs. C.1-C.2)

$$\dot{z}_a[n] = \dot{\pi}_n(t_n) = c_1[n] = a_1[n] + ib_1[n]. \tag{C.23}$$

Likewise we find for the approximation of the acceleration function $\ddot{z}$:

$$\ddot{z}_a[n] = \ddot{\pi}_n(t_n) = 2c_2[n] = 2(a_2[n] + ib_2[n]). \tag{C.24}$$

Thus we observe from Eqs. C.22-C.24 that in order to approximate $z$, $\dot{z}$ or $\ddot{z}$ at the $n$-th sample position along the contour we only need to compute $c_0[n]$, $c_1[n]$ or $c_2[n]$, respectively. For the approximation of the cumulative angular function $\varphi$ we use $\dot{z}_a[n]$. Recalling that $\dot{z}_a[n] = \dot{x}_a[n] + i\dot{y}_a[n]$, the approximation for the tangent angle function $\theta$ at $t = t_n$ is (cf. Eq. C.23)

$$\theta_a[n] = \arctan\left(\frac{\dot{y}_a[n]}{\dot{x}_a[n]}\right) = \arctan\left(\frac{b_1[n]}{a_1[n]}\right), \tag{C.25}$$

where we can compute $\theta[n]$ without ambiguity in a range of length $2\pi$ from the signs of $a_1[n]$ and $b_1[n]$. From $\theta_a[n]$ we compute $\varphi_a[n]$, using the same formulas as for the polygon in Section 2.2 (Eqs. 2.2.63-2.2.65c). The periodic cumulative angular function $\psi$ is computed from $\varphi_a[n]$ by the formula (cf. Eq. 2.2.32)

$$\psi_a[n] = \varphi_a[n] - n\frac{2\pi}{N}. \tag{C.26}$$

The curvature function $K$ at the $n$-th contour sample position is approximated on the basis of Eq. 2.2.26, by using both $\dot{z}_a[n] = \dot{x}_a[n] + i\dot{y}_a[n]$ and $\ddot{z}_a[n] = \ddot{x}_a[n] + i\ddot{y}_a[n]$ (cf. Eqs. C.23-C.24)

$$\begin{aligned} K_a[n] &= \frac{\dot{x}_a[n]\ddot{y}_a[n] - \ddot{x}_a[n]\dot{y}_a[n]}{\left\{(\dot{x}_a[n])^2 + (\dot{y}_a[n])^2\right\}^{3/2}} \\ &= 2\,\frac{a_1[n]b_2[n] - a_2[n]b_1[n]}{\left\{(a_1[n])^2 + (b_1[n])^2\right\}^{3/2}}. \end{aligned} \tag{C.27}$$

In order to obtain more insight into the behavior of the polynomial filters that we derived to estimate $z_a[n]$, $\dot{z}_a[n]$ and $\ddot{z}_a[n]$ (cf. Eqs. C.15, C.16 and C.22-C.24), and into the influence of the order $P$ of the polynomial and the fitting width $2M + 1$, it may be interesting to consider their frequency responses.

The frequency response $H_p(\omega)$ of an FIR filter with impulse response $h_p[n]$, $n = -M, \ldots, M$, is defined as (cf. Hamming [1977], Oppenheim and Schafer [1975])

$$H_p(\omega) = \sum_{n=-M}^{M} h_p[n] e^{-in\omega}. \tag{C.28}$$

For symmetric FIR filters, that occur for $p$ even (cf. Eq. C.21a), $H_p(\omega)$ is real-valued

$$H_p(\omega) = h_p[0] + 2 \sum_{n=1}^{M} h_p[n] \cos(n\omega) \tag{C.29a}$$

while for antisymmetric FIR filters, that occur for $p$ odd (cf. Eq. C.21b), $H_p(\omega)$ is purely imaginary

$$H_p(\omega) = 2i \sum_{n=1}^{M} h_p[n] \sin(n\omega). \tag{C.29b}$$

Consequentially, the transfer functions $H_0(\omega)$ of the filters with impulse response $h_0[n]$ are real-valued.

In Figure C.1a we have displayed $\mathrm{Re}\{H_0(\omega)\}$ for FIR filters with impulse response $h_0[n]$ of fixed width $M = 5$ (i.e. fitting to $2M + 1 = 11$ position function samples) and of varying order of the polynomials: $P = 2$, 4 and 6.

In Figure C.1b we have done the same for FIR filters with impulse response $h_0[n]$ of fixed order of the polynomials, i.e. $P = 4$, and of varying width: $M = 3$, 5 and 7.

In Figure C.1a we observe that the higher the order of the polynomials that we fit, the higher the tangency of $\mathrm{Re}\{H_0(\omega)\}$ at $\omega = 0$ (Hamming [1977]). As a consequence, the width of the passband increases with increasing order of the polynomials.

In Figure C.1b we see that the width of the passband decreases with an increasing number of position function samples to which we fit the polynomial.

Because of the antisymmetry of $h_1[n]$ the transfer functions $H_1(\omega)$ are purely imaginary. In Figure C.2a we have displayed $\mathrm{Im}\{H_1(\omega)\}$ for

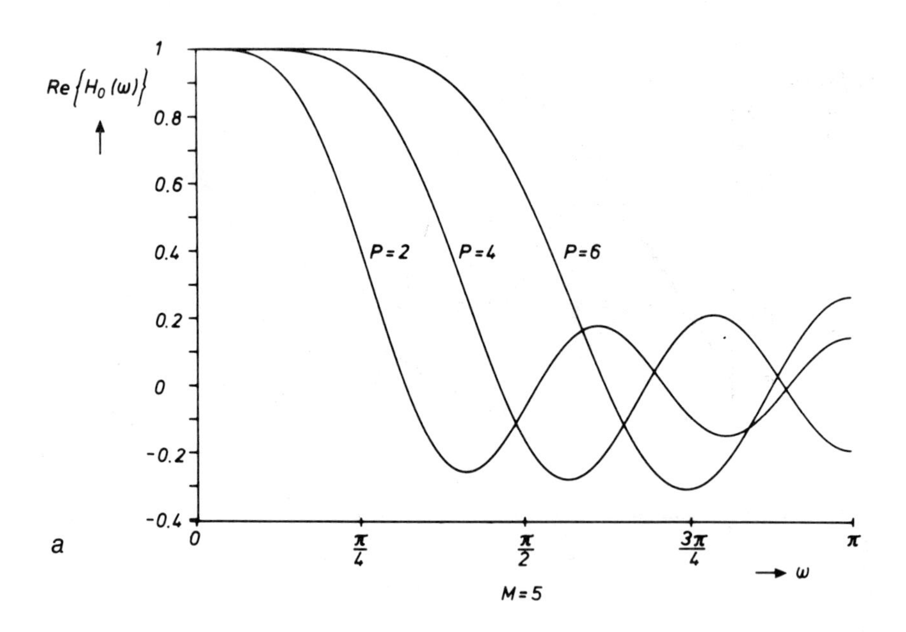


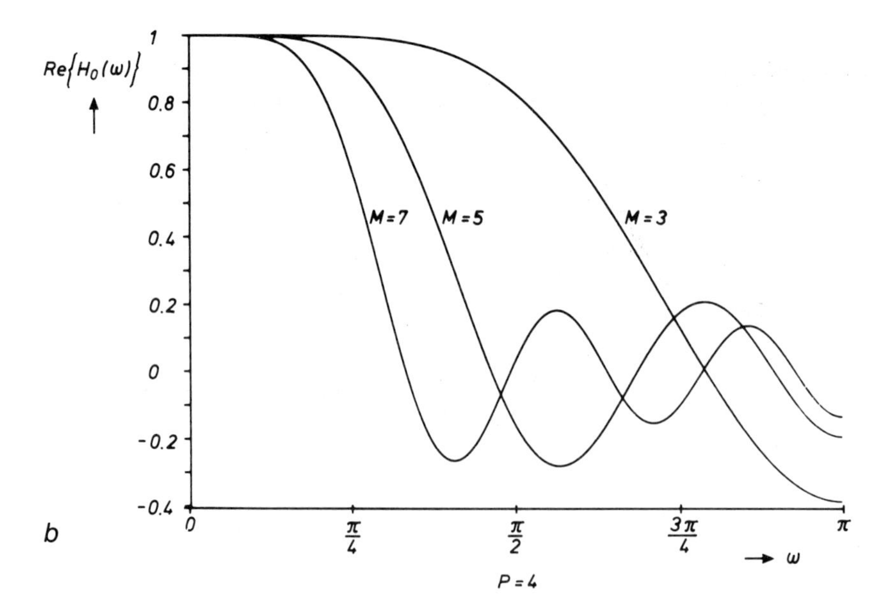


**Figure C.1.** Display of the real-valued transfer function $H_0(\omega)$ of the polynomial FIR filter with impulse response $h_0[n]$, which is used to compute the approximated position function $z_a[n]$.

In (*a*) we have varied the order of the polynomial: $P = 2$, 4 and 6, while keeping the width of the impulse response constant ($2M + 1 = 11$). In (*b*) the number of contour samples to which we fit the polynomials, and thus the width of the impulse response, has been varied ($M = 3, 5, 7$), while the order of the polynomials has been kept constant.

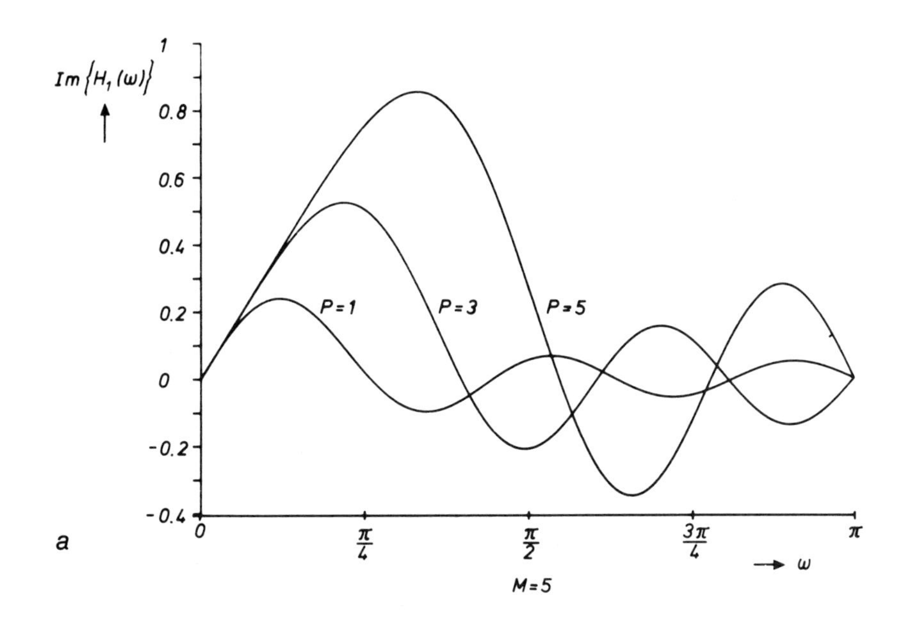


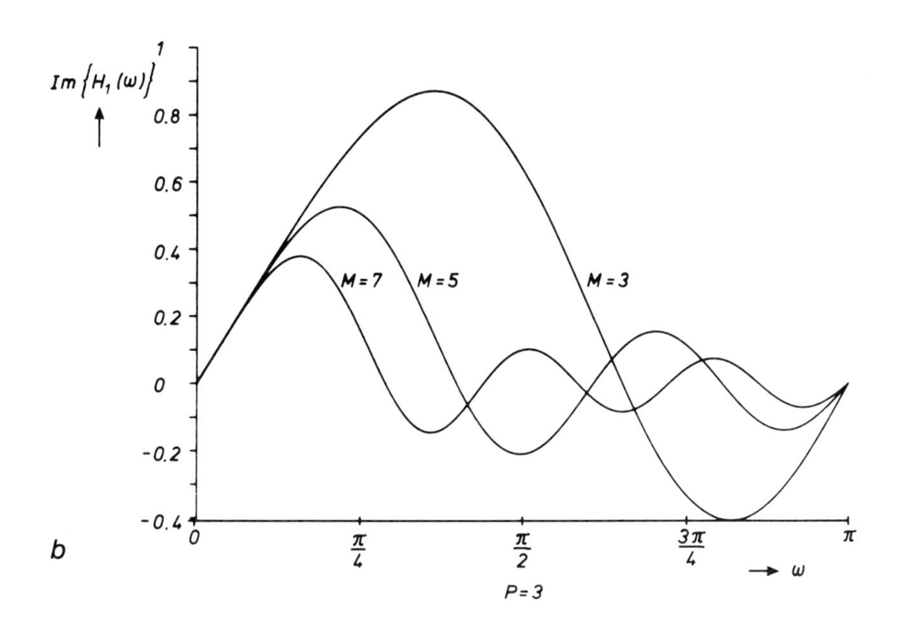


**Figure C.2.** Display of the purely imaginary transfer function $H_1(\omega)$ of the polynomial FIR filter with impulse response $h_1[n]$, which is used to compute the approximated tangent function $\dot{z}_a[n]$.

In (*a*) and (*b*) of this figure the same parameter values have been used as in (*a*) and (*b*) of Figure C.1.

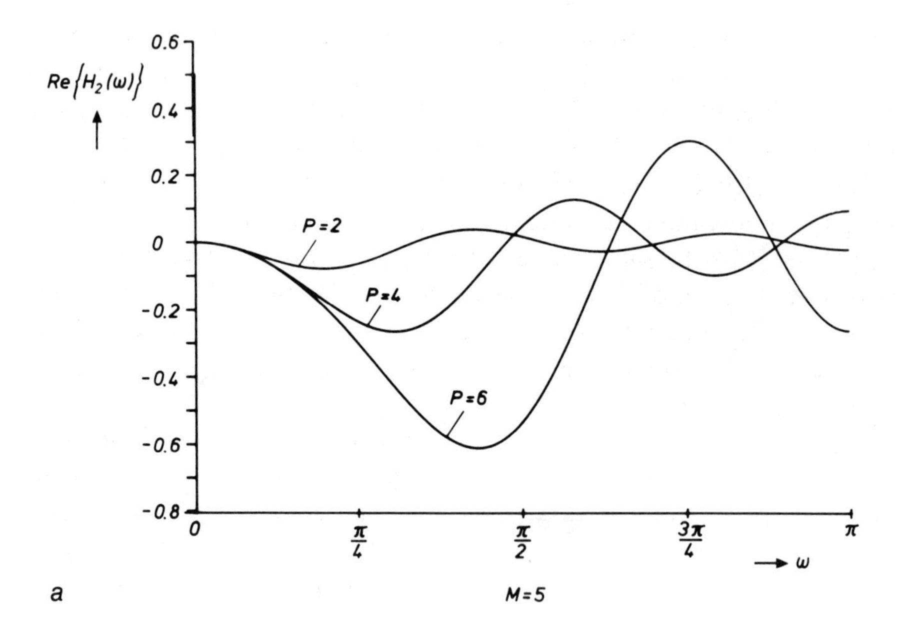


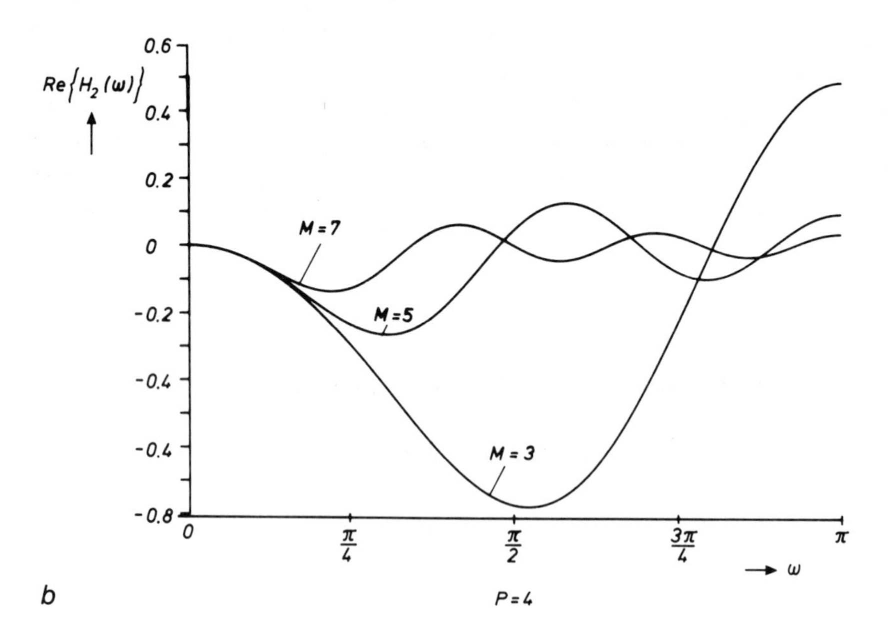


**Figure C.3.** Display of the real-valued transfer function $H_2(\omega)$ of the polynomial FIR filter with impulse response $h_2[n]$, which is used to compute the approximated acceleration function $\ddot{z}_a[n]$.
In (*a*) and (*b*) of this figure the same parameter values have been used as in (*a*) and (*b*) of Figure C.1.

filters with impulse response $h_1[n]$ of fixed width and of varying order of the polynomials: $P = 1, 3$ and 5. In Figure C.2b we have done the same for FIR filters with impulse response $h_1[n]$ of fixed order of the polynomials, i.e. $P = 3$, and of varying width: $M = 3, 5$ and 7.

The transfer functions $H_2(\omega)$ of the filters with impulse response $h_2[n]$ are real-valued. In Figures C.3a and C.3b we have displayed $\text{Re}\{H_2(\omega)\}$ for the same values of polynomial order $P$ and width $M$ as in Figures C.1a and C.1b, respectively. In Figures C.2a and C.3a we observe that, similar to $\text{Re}\{H_0(\omega)\}$ in Figure C.1a, the width of the passbands in $H_1(\omega)$ and $H_2(\omega)$ increases with increasing order $P$ of the polynomials (for fixed width $M$).

In Figures C.2b and C.3b we see that, similar to $H_0(\omega)$, an increase of the number of position function samples to which we fit the polynomial, i.e. $2M + 1$, causes the width of the passband of $H_1(\omega)$ and $H_2(\omega)$ to decrease. Note that the transfer function of the ideal differentiating filter is $H_1(\omega) = i\omega$, while that of the ideal double differentiating filter is $H_2(\omega) = -\omega^2$.

From Figures C.1a-C.3b we observe that the polynomial filters $h_0[n]$, $h_1[n]$ and $h_2[n]$, considered as low-pass (differentiating) filters, may not have the most desirable properties. However, we have to keep in mind that these filters have not been designed specifically for frequency filtering. The basic assumption that we make, when using polynomial filters, is that, at every sample position, the contour can be appropriately approximated over the fitting width by a polynomial of the order chosen.

## References

Apostol, T.M. [1969]
*Calculus, Volume 2*, Second Edition, Waltham, MA: Xerox.

Cohn, P.M. [1974]
*Algebra, Volume 1*, New York: John Wiley and Sons, Inc.

Hamming, R.W. [1977]
*Digital Filters*, Englewood Cliffs, NJ: Prentice-Hall, Inc.

Oppenheim, A.V. and R.W. Schafer [1975]
*Digital Signal Processing*, Englewood Cliffs, NJ: Prentice-Hall, Inc.

Oppenheim, A.V., A.S. Willsky and I.T. Young [1983]
*Signals and Systems*, Englewood Cliffs, NJ: Prentice-Hall, Inc.

Spiegel, M.R. [1964]
*Complex Variables with an Introduction to Conformal Mapping and Its Applications*, Schaum's Outline Series, New York: McGraw-Hill Book Co., Inc.

# Index

A000020179236